exam success

in

Physics

for Cambridge International AS & A Level

John Quill

OXFORD
UNIVERSITY PRESS

Great Clarendon Street, Oxford, OX2 6DP, United Kingdom

Oxford University Press is a department of the University of Oxford.
It furthers the University's objective of excellence in research,
scholarship, and education by publishing worldwide. Oxford is a
registered trade mark of Oxford University Press in the UK and in
certain other countries

© Oxford University Press 2018

The moral rights of the authors have been asserted

First published in 2018

British Library Cataloguing in Publication Data
Data available

978-0-19-840994-6

1 3 5 7 9 10 8 6 4 2

Paper used in the production of this book is a natural, recyclable
product made from wood grown in sustainable forests.
The manufacturing process conforms to the environmental
regulations of the country of origin.

Printed in Great Britain by Bell and Bain Ltd. Glasgow

Acknowledgements
The publishers would like to thank the following for permissions to
use copyright material:

Cover image: Istock/aryos
All photos by Shutterstock.com
Artwork by Q2A Media Pvt. Ltd. and Aptara Corp

Although we have made every effort to trace and contact all
copyright holders before publication this has not been possible in all
cases. If notified, the publisher will rectify any errors or omissions at
the earliest opportunity.

Links to third party websites are provided by Oxford in good faith
and for information only. Oxford disclaims any responsibility for
the materials contained in any third party website referenced in
this work.

Contents

Introduction

The *Exam Success* series will help you to reach your highest potential and achieve the best possible grade. Unlike traditional revision guides, these new books give advice on improving answers, helping to show you what examiners expect of candidates. All the titles are written by authors who have a great deal of experience in preparing candidates for exam success.

Exam Success in Cambridge International AS and A Level Physics covers the requirements of the AS Level and A Level Cambridge 9702 Physics syllabus. The first 26 units cover the syllabus content, while Unit 27 provides advice on experimental skills for paper 3 and help with planning, analysing and evaluating experiments for paper 5. Unit 28 consists of questions in the styles of the five exam papers, with exam tips to support your work. In addition, there is a separate Maths skills Appendix at the back of the book, providing further help and guidance with essential mathematical skills.

At the start of each unit, cross-references are given to *Physics in Context for Cambridge International AS & A Level*, should you wish to study the topic in more depth. Throughout the book, a grey vertical bar shows AS content, and an orange bar shows A Level content; key terms are also highlighted.

Each *Exam Success* book has common features to help you do your best in the exam:

Key points

☐ These summarise what you need to show that you can do in the exam. Check them off one by one when you are confident.

Worked example

These give examples of questions, and show you how best to answer them.

Remember

These include key information that you must remember if you are to achieve a high grade.

⬆ Raise your grade

Here, you can read answers by candidates who did not achieve maximum marks, as well as find out how to improve their answers.

❓ Exam-style questions

Each unit has examples of the sort of questions to expect in the exam. Answers are available on the OUP support website, together with additional exam-style questions and answers on many topics.

🔗 Link

These show where in the book you can find more information about the topic.

★ Exam tip

These provide guidance and advice to help you understand exactly what examiners are looking for.

📐 Maths skills

These remind you of the vital mathematical skills that you need in order to answer exam questions in physics.

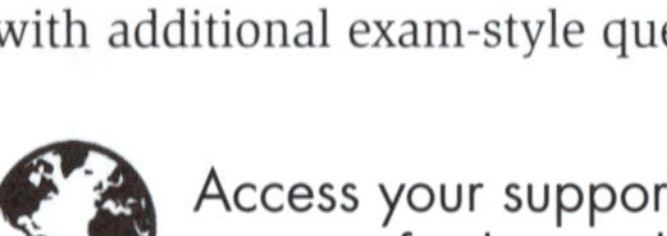

Access your support website for additional content here: www.oxfordsecondary.com/9780198409946

Key points

- ☐ Understand that physical quantities have a magnitude and a unit, and make reasonable estimates of some of them.
- ☐ Recall the SI base quantities and their units: mass (kg), length (m), time (s), current (A), and temperature (K).
- ☐ Express derived units as products or quotients of the SI base units, and use the correct units.
- ☐ Use SI base units to check the homogeneity of physical equations.
- ☐ Use prefixes and their symbols to indicate submultiples or multiples of both base and derived units: pico (p), nano (n), micro (µ), milli (m), centi (c), deci (d), kilo (k), mega (M), giga (G), and tera (T).
- ☐ Understand the conventions for labelling graph axes and table columns*.
- ☐ Distinguish between scalar and vector quantities, and give examples of each.
- ☐ Add or subtract vectors, and represent a vector as two perpendicular components.
- ☐ Recall the meaning of a mole (mol) as a specific amount of a substance.
- ☐ Understand that the Avogadro constant N_A is the number of atoms in 0.012 kg of carbon-12.
- ☐ Use molar quantities, where one mole of a substance contains Avogadro's number of particles of that substance.

Physical quantities

Physical quantities, such as kinetic energy, electric current, and temperature, are expressed by a number (the 'magnitude') and a unit.

Making estimates

Being able to estimate a quantity is an important skill that can be improved with practice. Start with quantities you know, or can estimate reasonably accurately, and combine them to estimate a value for an unknown quantity.

Worked example

The **solar constant** (the average amount of energy reaching the Earth per square metre each second) is $1.4\,\text{kW m}^{-2}$ (see Figure 1.1). The radius of the Earth $\approx 6.4 \times 10^3\,\text{km}$. Estimate the energy received per day on Earth from the Sun.

Answer

energy per day = 'area' of Earth × solar constant × time

$$= \pi \times (6.4 \times 10^6)^2 \times (1.4 \times 10^3) \times (24 \times 60 \times 60) = 10^{22}\,\text{J}$$

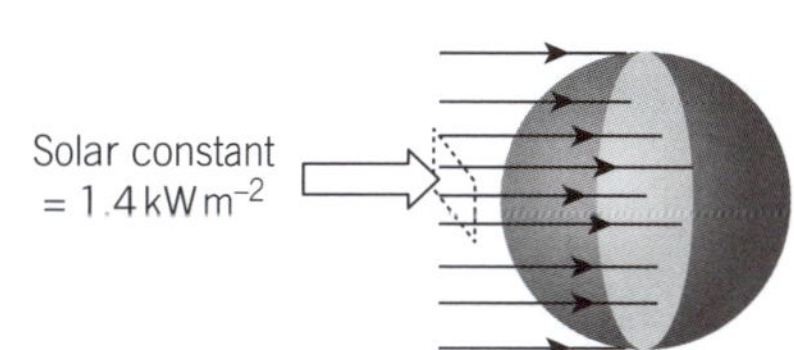

▲ **Figure 1.1** Solar constant

*covered in Appendix: Maths skills

Table 1.1 gives the value of some physical quantities that are useful to know in order to estimate other quantities.

▼ **Table 1.1** Useful quantities

Quantity	Value	Quantity	Value
Diameter of a nucleus	10^{-14} m	Atmospheric pressure	1.0×10^5 Pa
Diameter of an atom	10^{-10} m	One day	8.64×10^4 s
Wavelength of visible light	400–700 nm	One year	3.1×10^7 s
Radius of the Earth	6.4×10^6 m	Speed of a car	20–30 m s^{-1}
Mass of an electron	9.1×10^{-31} kg	Speed of sound in air	330 m s^{-1}
Mass of a proton	1.7×10^{-27} kg	Speed of light in a vacuum	3×10^8 m s^{-1}
Mass of a postage stamp	5×10^{-5} kg	Energy of an alpha particle	5 MeV
Mass of an apple	0.1 kg	Freezing point of water	0 °C
Mass of an adult	70 kg	Boiling point of water	100 °C
Density of air	1.3 kg m^{-3}	Specific heat capacity of water	4200 J kg^{-1} °C^{-1}
Density of water	10^3 kg m^{-3}	Charge on an electron e	1.6×10^{-19} C
Gravitational field strength, g	9.81 N kg^{-1} (9.81 m s^{-2})	Charge transferred by a lightning flash	1 C

SI units

The SI system of units (Système International d'Unités) is an internationally agreed system of units, built on seven base units (see Table 1.2).

The units of other quantities are derived from these seven base units.

▼ **Table 1.2** SI base units

Unit	Symbol	Base unit
metre	m	length
kilogram	kg	mass
second	s	time
ampere	A	electric current
kelvin	K	temperature
mole*	mol	amount of a substance
candela**	cd	light intensity

*A Level only ** not part of the A Level course.

Expressing derived units in terms of base units

The familiar units for quantities such as force (newtons, N), and energy (joules, J) can be 'broken down' into their base units by using an equation which links these quantities to quantities whose units are known.

Worked example

Determine the base units of force.

Answer

The base units of force can be found using $F = ma$.

units of force (N) = units of mass (kg) × units of acceleration (m s^{-2})

The SI base units of force are kg m s^{-2}.

Table 1.3 lists the derived units that you will encounter at A and AS Level.

▼ **Table 1.3** Derived units

Quantity	Symbol	Unit	SI base units
force	F	newton (N)	$kg\,m\,s^{-2}$
pressure	p	pascal (Pa)	$kg\,m^{-1}\,s^{-2}$
energy	E	joule (J)	$kg\,m^{2}\,s^{-3}$
power	P	watt (W)	$kg\,m^{2}\,s^{-1}$
frequency	f	hertz (Hz)	s^{-1}
charge	Q	coulomb (C)	$A\,s$
potential difference (p.d.)	V	volt (V)	$kg\,m^{2}\,s^{-3}\,A^{-1}$
electrical resistance	R	ohm (Ω)	$kg\,m^{2}\,s^{-3}\,A^{-2}$
capacitance	C	farad (F)	$A^{2}\,kg^{-1}\,m^{-2}\,s^{4}$
magnetic flux density*	B	tesla (T)	$kg\,s^{-2}\,A^{-1}$
magnetic flux*	Φ	weber (Wb)	$kg\,m^{2}\,s^{-2}\,A^{-1}$

*A Level only

Some quantities have no units (e.g., the refractive index of glass). These quantities are called dimensionless quantities.

Using SI base units to check equations

For any equation to be correct, a necessary condition is that the equation must be **homogeneous**; the units of the left-hand side of the equation must match those on the right-hand side.

Checking the homogeneity of an equation does not prove it is correct – it only shows whether an equation could be correct.

Worked example

A student is studying how the period of oscillation of a simple pendulum depends on its length l. She cannot recall whether the theoretical equation for the period T of the pendulum is:

$$T = 2\pi\sqrt{\frac{g}{l}} \quad \text{or} \quad T = 2\pi\sqrt{\frac{l}{g}}$$

Which equation is correct?

Answer

The units of the first equation are: $\dfrac{[ms^{-2}]^{\frac{1}{2}}}{m^{\frac{1}{2}}} = s^{-1}$

The units of the second equation are: $\dfrac{[m]^{\frac{1}{2}}}{[ms^{-2}]^{\frac{1}{2}}} = s$

and so only the second equation can be correct. (This method does not prove the equation – it cannot show whether the 2π in the equation is correct).

Using prefixes

Physics is often concerned with very small numbers, such as the diameters of atomic nuclei, and very large numbers, such as the masses of stars and planets. These can be expressed in standard form (a number between 1.0 and 10.0 multiplied by a multiple of 10); for example, the radius of the Earth is $6.37\times10^{6}\,m$.

Another way of expressing such numbers is to use prefixes (see Table 1.4).

▼ **Table 1.4** Prefixes

Prefix	Symbol	Multiple	Example
pico	p	10^{-12}	pF (picofarad)
nano	n	10^{-9}	nC (nanocoulomb)
micro	μ	10^{-6}	μA (microamp)
milli	m	10^{-3}	mV (millivolt)
centi	c	10^{-2}	cm (centimetre)
deci	d	10^{-1}	dB (decibel)
kilo	k	10^{3}	kg (kilogram)
mega	M	10^{6}	MΩ (megohm)
giga	G	10^{9}	GJ (gigajoule)
tera	T	10^{12}	TW (terawatt)

> A prefix represents a multiple of 10 (e.g., kilo means 10^{3}).

Tables and graphs

Measurements from an experiment should always be recorded in a neat table. You should make sure the table has enough columns to include any repeated readings, averages, and calculated values that you use later to analyse your results.

Each column heading should have the quantity being recorded, and the unit it is measured in, separated by a solidus (/); for example m/g or T/s. Alternatively, the units can be in brackets.

Scalar and vector quantities

Scalar quantity: A scalar quantity only has magnitude (size); for example mass, speed, distance, and temperature.

Vector quantity: A vector quantity is one which has both magnitude and a direction; for example, displacement, velocity, force, and momentum.

Adding and subtracting vectors

A vector such as velocity can be represented by an arrow, the length of the arrow indicating the *magnitude of the quantity* (the speed) and the direction of the arrow indicating the *direction of travel*. Vectors are usually indicated by bold type (**a**), or with an arrow ($\vec{a}$).

Two vectors **a** and **b** (see Figure 1.2) are added by placing one of the vectors so that it begins at the end of the other vector. The sum of the two vectors is a straight line drawn from the beginning of the first vector to the end of the second.

▲ **Figure 1.2** Adding vectors

The same procedure can be used to subtract one vector from another.

To find **a** – **b**, first draw the vector **–b** by drawing the vector **b**, but pointing in the opposite direction to **b**, and then add this to **a** (see Figure 1.3).

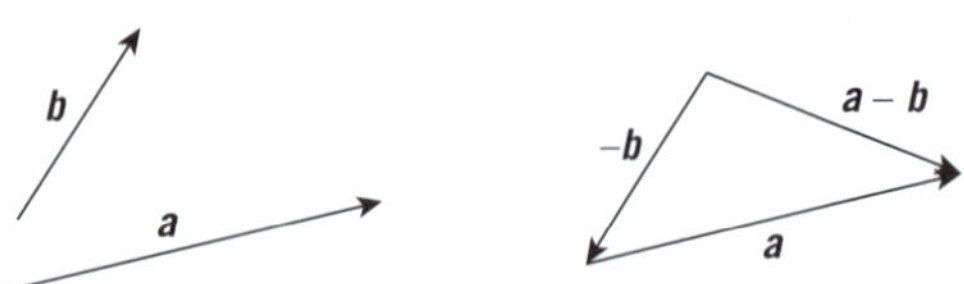

▲ **Figure 1.3** Subtracting vectors

Representing a vector as two perpendicular components (resolving vectors)

It is often useful to separate a vector, such as a force or velocity, into two components at right angles to each other.

In Figure 1.4 the horizontal component of the force F is $F\cos\theta$; the vertical component is $F\sin\theta$. The **resultant vector** (the sum of the two components) is F.

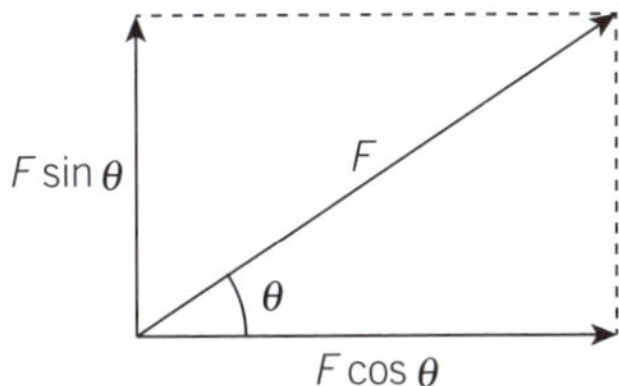

▲ **Figure 1.4** Components of a vector

> The component adjacent to the angle θ is always the cosine component.

Avogadro's constant and moles

Avogadro's constant (N_A) is defined as the number of atoms in exactly 12 g of the carbon isotope $^{12}_{6}C$.

$$N_A = 6.023 \times 10^{23}$$

The **mole (mol)** is the unit for measuring the amount of a substance. One mole of a substance consisting of identical particles is defined as the quantity of the substance containing N_A particles.

The **molar mass** of a substance is defined as the mass of the substance that contains N_A particles (1 mole of the substance). The units of molar mass are $kg\,mol^{-1}$.

> **Link**
>
> Avogadro's constant is important when studying the behaviour of gases. See Unit 10 *Ideal gases*.

Worked example

1 How many molecules of nitrogen are there in 1.0 kg of nitrogen gas?

Answer

The molar mass of nitrogen is $14.0 \times 10^{-3}\,kg\,mol^{-1}$.

$$\text{Number of atoms} = \frac{1.0}{14.0 \times 10^{-3}} \times 6.02 \times 10^{23} = 4.3 \times 10^{25}\ \text{atoms}$$

2 The molar mass of aluminium is 27 g. The density of aluminium is $2.7 \times 10^3\,kg\,m^{-3}$. Calculate the molar volume of aluminium.

Answer

$$\text{Molar volume of aluminium} = \frac{\text{molar mass}}{\text{density}} = \frac{27 \times 10^{-3}}{2.7 \times 10^3} = 1.0 \times 10^{-5}\,m^3$$

3 The density of copper is $8900\,kg\,m^{-3}$ and its molar mass is $63.5\,g\,mol^{-1}$. Calculate the number of free electrons per m^3 if each atom contributes one conduction electron.

Answer

$$\text{Number of electrons/m}^3 = \frac{8900}{63.5 \times 10^{-3}} \times 6.02 \times 10^{23}$$
$$= 8.4 \times 10^{28}\ \text{electrons/m}^3.$$

 # **Raise your grade**

1 The energy released by the fission of one uranium nucleus is 3.2×10^{-11} J.

State this energy in pJ.

3.2×10^{-11} J $= 32 \times 10^{-12}$ J;
p = pico = 10^{-12}

energy = ..32.. pJ ✔ [1]

2 The orbital period of Jupiter is 0.37 Gs. Express this time in years.

$0.37 \text{ Gs} = 0.37 \times 10^{9} \text{ s} = 3.7 \times 10^{8}$ ✔

Correct prefix of 'G' = giga = 10^9

time = ..3.7×10^{8}.. years ✘ [2]

The answer needs to be converted from seconds to years. The correct answer is 11.9 years [No. of seconds in a year = 3.1×10^{7}].

3 The speed v of sound waves in air is given by the equation:

$$v = \sqrt{\frac{\gamma p}{\rho}}$$

where p is the pressure of the air, ρ is the density of the air, and γ is a dimensionless constant.

Show that this equation is homogeneous.

units of $v = \text{m s}^{-1}$ ✔

units of $\sqrt{\dfrac{\gamma p}{\rho}} = \left(\dfrac{Nm^{-2}}{kgm^{-3}}\right)^{\frac{1}{2}} = \left(\dfrac{kgms^{-2}m^{-2}}{kgm^{-3}}\right)^{\frac{1}{2}} = (m^2 s^{-2})^{\frac{1}{2}} = ms^{-1}$ ✔

Same units, so equation is homogeneous.   Working shown clearly.   [2]

4 A ship is travelling due east at a speed of 5 km h^{-1}. A patrol vessel is 12 km due south of the ship, moving with a speed of 13 km h^{-1}, and wishes to intercept the ship.

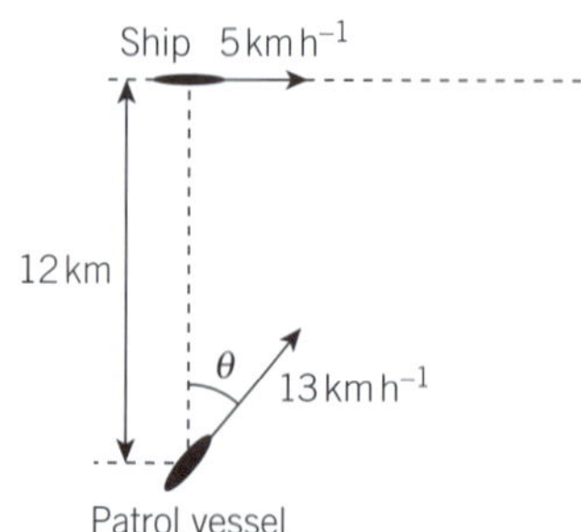

(a) At what angle θ should the patrol vessel sail in order to intercept the ship?

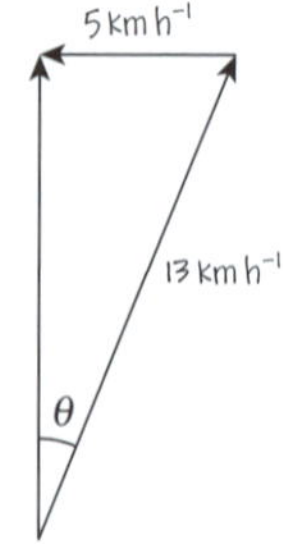

$\sin \theta = \dfrac{5}{13}$; $\theta = \sin^{-1}\left(\dfrac{5}{13}\right) = 22.6°$ ✔

Correct answer with working shown clearly.

$\theta =$ ..22.6°..

(b) How much time elapses before the patrol vessel reaches the ship?

velocity in direction of ship $= 13 \cos 22.6 = 12 \text{ km h}^{-1}$ ✔   Correct method.

time taken to travel 12 km $= \dfrac{12}{12} = 1$ hour ✔   Correct calculation.

time = ..1.0 hour.. [3]

1 The friction force F on a sphere of radius r falling with velocity v through a liquid is given by the equation:

$$F = 6\pi r \eta v$$

where η is the viscosity of the liquid.

What are the SI base units of viscosity?

A $kg\,m\,s$ **B** $kg\,m^{-1}\,s$ **C** $kg\,m\,s^{-1}$ **D** $kg\,m^{-1}\,s^{-1}$ [1]

2 The stress needed to fracture a material with a crack is given by the equation:

$$\sigma = k\sqrt{\frac{\gamma E}{d}}$$

where E is the Young modulus, d the width of the crack, and k is a dimensionless constant. For the equation to be homogeneous, what are the units of γ?

A N **B** J **C** $N\,m^{-2}$ **D** $J\,m^{-2}$ [1]

3 What is the best estimate of the kinetic energy of an Olympic 100 m runner, running at top speed?

A $0.4\,kJ$ **B** $4\,kJ$ **C** $40\,kJ$ **D** $400\,kJ$ [1]

4 Which one of the following physical quantities is a vector?

A work **B** mass **C** momentum **D** power [1]

5 Two vectors, a and b are as shown below.

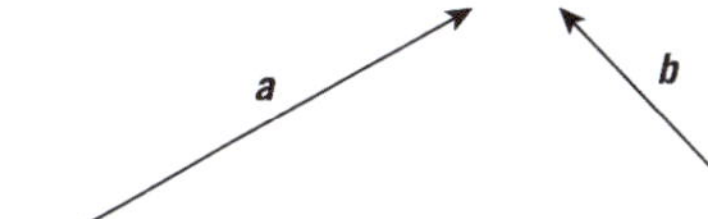

Which vector represents $a - b$?

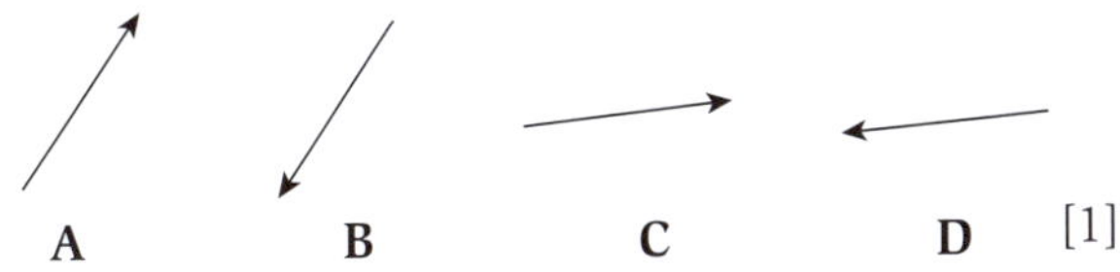

[1]

6 One light-year is the distance travelled by light in one year. The diameter of the Milky Way galaxy is approximately 100 000 light-years.

What is the best estimate of this distance in metres?

[The speed of light is $3 \times 10^8\,m\,s^{-1}$.]

A $10^{15}\,m$ **B** $10^{17}\,m$ **C** $10^{19}\,m$ **D** $10^{21}\,m$ [1]

7 **(a)** Express the following in standard form:

   (i) $470\,k\Omega$ **(ii)** $1000\,\mu F$ **(iii)** $0.05\,nm$ [3]

(b) Express the following using an appropriate prefix:

   (i) $6.4 \times 10^6\,m$

   (ii) $0.0075\,A$

   (iii) $3.0 \times 10^8\,m\,s^{-1}$ [3]

8 **(a)** Describe the difference between a *scalar* quantity and a *vector* quantity. [1]

(b) In the following list, underline all the **vector** quantities.

   weight   acceleration   stress   power   work [2]

9 The diagram shows a car travelling up a hill at constant speed.

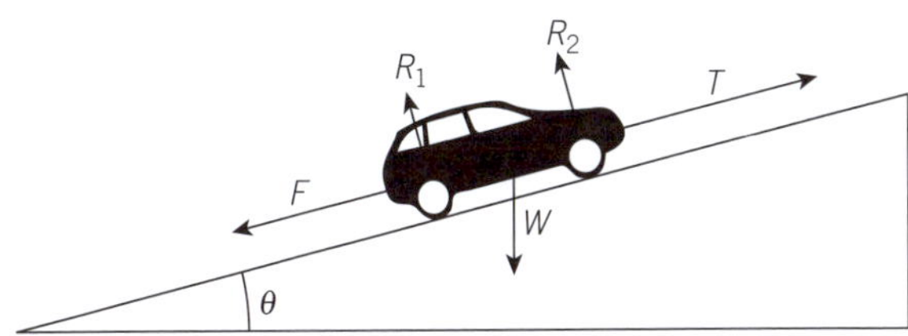

(a) State:

   (i) the horizontal component of the friction force F

   (ii) the vertical component of the engine force T

   (iii) the component of the weight force W acting down the slope. [3]

(b) By resolving forces along the slope, write down an equation relating F, T, and W. [1]

10 **(a)** State what is meant by a *mole*. [1]

(b) The density of copper is $8930\,kg\,m^{-3}$. The molar mass of copper is $63.5\,g\,mol^{-1}$. Calculate the molar volume of copper. [2]

(c) The diameter of a copper atom is $2.55 \times 10^{-10}\,m$. Assuming each copper atom occupies a cube of side $2.55 \times 10^{-10}\,m$, calculate:

   (i) the volume of one cube

   (ii) the number of cubes in one mole of copper. [3]

(d) Suggest a reason why your answer to (c)(ii) is different from the Avogadro constant. [1]

Key points

- ☐ Use techniques to measure length, volume, angle, mass, time, temperature, and electrical quantities.
- ☐ Use rulers, vernier calipers, micrometers, and protractors.
- ☐ Measure weight and, hence, mass using balances.
- ☐ Measure time using stopwatches and the time-base of a cathode-ray oscilloscope.
- ☐ Measure temperature using a thermometer.
- ☐ Use ammeters and voltmeters, selecting appropriate scales.
- ☐ Use a cathode-ray oscilloscope (c.r.o.).
- ☐ Use analogue scales and digital displays, and calibration curves.
- ☐ Understand and explain the effects of systematic errors (including zero errors) and random errors in measurements.
- ☐ Understand the difference between precision and accuracy.
- ☐ Assess the uncertainty in a derived quantity by simple addition of absolute, fractional or percentage uncertainties.
- ☐ Use a calibrated Hall probe.

Taking measurements

Measuring length

Rulers and scales

Metre rules and 30 cm rulers are usually calibrated to the nearest millimetre, and so any readings using a ruler should normally be recorded to the nearest millimetre.

If the ruler is to be used to measure a vertical distance, it should be held against a set square which is perpendicular to the ruler and the bench (see Figure 2.1). Alternatively, a plumb line (a small weight on a string) can be used to check that that the ruler is vertical. When the ruler is in the correct position it is a good idea to clamp it to a clamp stand.

A small spirit level can be used to make sure a ruler is horizontal.

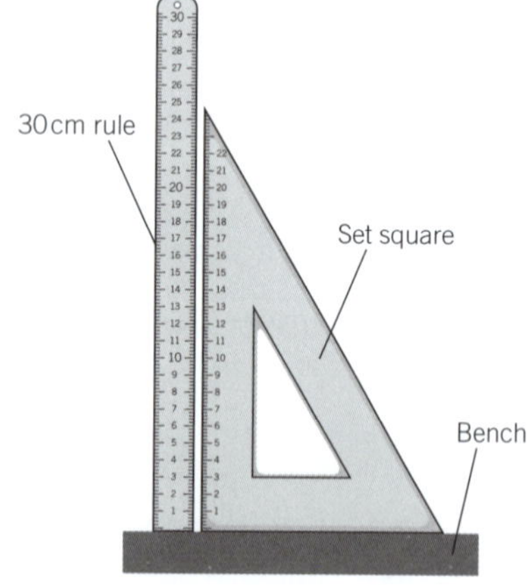

▲ **Figure 2.1** Use a set square to check a ruler is vertical

Vernier calipers

Vernier calipers (see Figure 2.2) can be used to measure distances from a few millimetres up to 10 cm or more. They can usually be read to a precision of ± 0.1 mm.

The zero mark is first used to read the main scale to the nearest millimetre. Then look for the mark on the sliding scale which is in line with a mark on the main scale – if it is 7 then 0.7 mm (0.07 cm) should be added onto the main scale reading.

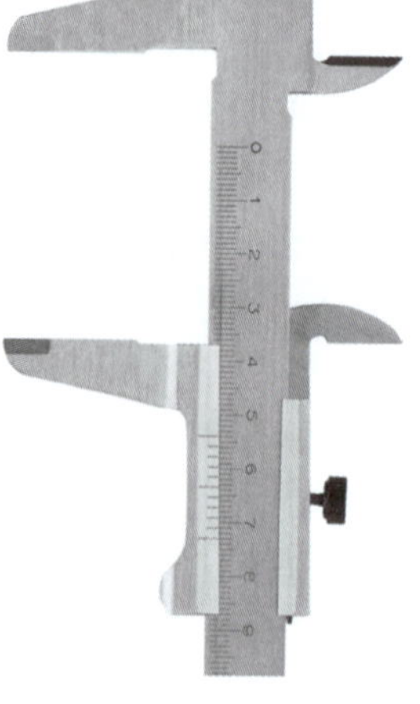

▲ **Figure 2.2** Vernier calipers

> ★ **Exam tip**
>
> Using a set square to check a ruler is vertical, or a spirit level to check it is horizontal, and then clamping the ruler in position, are often valid 'improvements' in question 2 of Paper 3 *Advanced practical skills*.

Micrometers

A micrometer (see Figure 2.3) can measure distances from 0.01 mm to a few centimetres, to a precision of ± 0.01 mm. (See Unit 27 for more details on how to use a micrometer.)

Measuring volume

The volume of regularly-shaped objects such as cubes or spheres can be found by simple measurement and calculation. For example, to find the volume V of a sphere, measure its diameter d and use the equation:

$$V = \frac{\pi d^3}{6}$$

For odd shapes, a liquid displacement method can be used. A calibrated measuring cylinder or beaker is partially filled with water. The object can then be placed into the water and the change in volume calculated. Alternatively, a displacement can may be used. Figure 2.4 shows how to measure the diameter of a tube or cylinder.

Measuring angle

Protractors usually have a precision of ± 1°, and so values of angle should be recorded to the nearest degree. If the protractor cannot be held steady (on a bench for example), it is a good idea to clamp the protractor to a clamp stand.

Measuring mass and weight

Weighing machines such as spring balances (force meters), lever balances and top-pan electronic balances all work by measuring a force (weight). Many are calibrated to display the mass of the object on the Earth's surface, using $W = mg$ where W is the weight of an object (in newtons), m the mass in kilograms, and g the acceleration of free fall ($9.81\,\mathrm{m\,s^{-2}}$).

The calibration of these instruments can be easily checked with known masses.

Measuring temperature

There are a range of instruments to measure temperature, including liquid-in-glass thermometers, thermocouples, and thermistors.

Measuring electrical quantities

Analogue and digital meters are readily available to measure electrical quantities such as current, voltage and resistance.

Analogue meters

Care needs to be taken to avoid parallax errors when reading an analogue meter (see Figure 2.5). If the meter has a mirror behind the pointer, the reflection of the pointer should always be in line with the pointer itself when taking the reading. The meter should also be checked for 'zero' error – it should read '0' when disconnected.

Digital meters

Many digital meters (see Figure 2.6) have a number of different scales to choose from; a digital ammeter may have 0–200 µA as its most sensitive range and 0–10 A as its least sensitive. For example, if you are trying to measure a current of about 10 mA, it is a good idea to start with the least sensitive range and gradually increase the sensitivity until you find a suitable range (in this example, 0–20 mA).

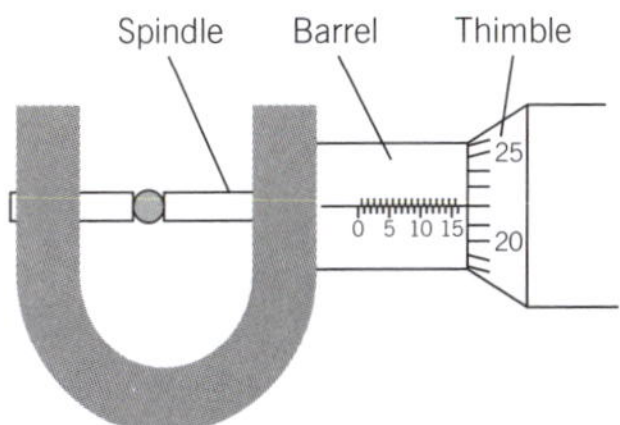

▲ **Figure 2.3** Micrometer

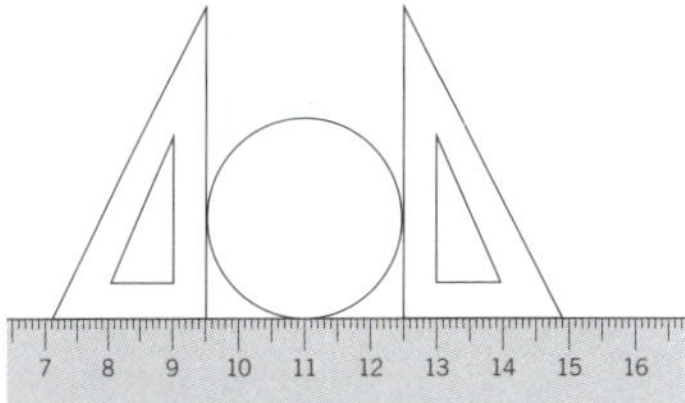

▲ **Figure 2.4** Measuring the diameter of a tube or cylinder

Link

For more details on using a protractor see Unit 27 *Practical assessment*.

Link

Instruments to measure temperature are discussed in Unit 11 *Temperature*.

Link

The use of a galvanometer in null methods is discussed in Unit 20 *Direct current circuits (d.c.)*.

▲ **Figure 2.5** Analogue meter

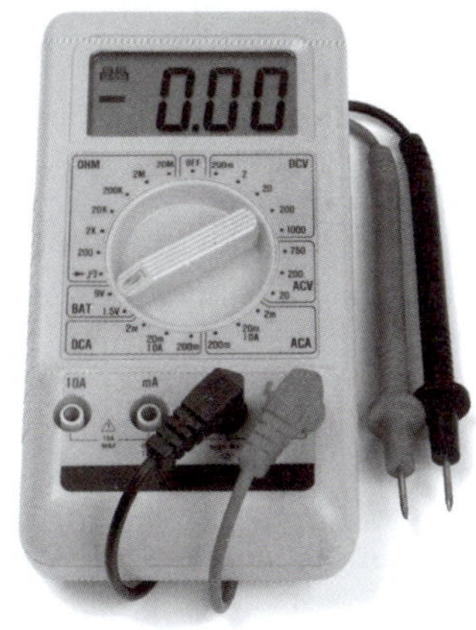

▲ **Figure 2.6** Digital meter

Measuring time

Stopwatches and stop clocks

Digital stopwatches often have a precision of ± 0.01 s, but judging the timing for an oscillation, for example, involves human error, and so the accuracy of the measurement is likely to be no better than ± 0.2 s.

When recording small time intervals, it is important to repeat the measurement several times and then find a mean value in order to reduce the random error as much as possible. When measuring the period of an oscillating system, it is good practice to measure five or ten oscillations (rather than individual oscillations), to repeat these measurements two or three times, and then to find an overall mean value. These steps reduce the percentage uncertainty in the value of the period significantly.

Cathode-ray oscilloscope

A cathode-ray oscilloscope (c.r.o.) is particularly useful for investigating voltages which are changing with time, including alternating current (a.c.) signals and the discharge of a capacitor. The two key controls on the instrument panel are:

- **y-gain:** This states the number of volts per division (volts/div.) in the y direction. A division is usually 1 cm (1 square) on the screen. The smaller the number of volts/div., the more sensitive the scale.

- **time-base:** This indicates how quickly the electron beam moves across the screen, and is usually calibrated in seconds/division (s/div.). The larger the value of the time-base, the slower the dot moves across the screen.

Figure 2.7 shows an example of an oscilloscope display:

- The y-gain is set to 20 mV/div. so the amplitude of the voltage signal is 60 mV. (The peak-to-peak voltage is 120 mV.)

- The time-base is set to 10 ms/div. Three complete cycles occur in ten divisions.

- Period T of the signal $= 10 \times \dfrac{10}{3} = 33.3\,\text{ms}$

- Frequency $f = \dfrac{1}{T} = \dfrac{1}{33.3 \times 10^{-3}} = 30\,\text{Hz}$

Using a calibrated Hall probe

A Hall probe is used to measure and detect magnetic fields. It is a thin slice of semiconductor material that is usually mounted on the end of a plastic rod (see Figure 2.8). When it is placed perpendicular to a constant magnetic field, a steady current passes through the slice and a potential difference (known as the Hall voltage) is produced across the slice. The p.d. is proportional to the size of the magnetic flux density.

In order to measure a magnetic flux density, the Hall probe must first be calibrated using a known magnetic field. See Unit 22 for more details on how to calibrate a Hall probe.

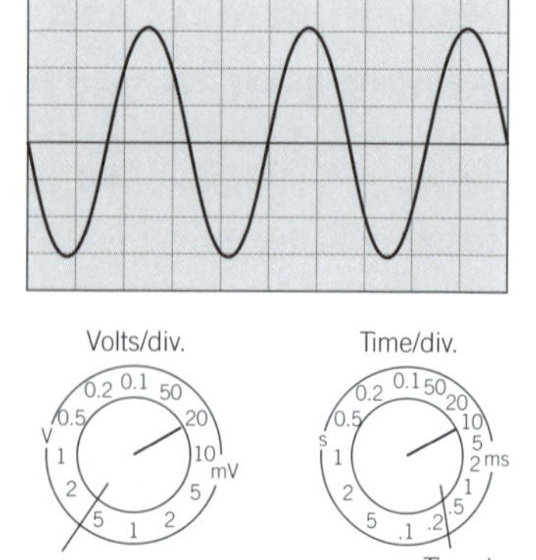

▲ **Figure 2.7** Oscilloscope controls

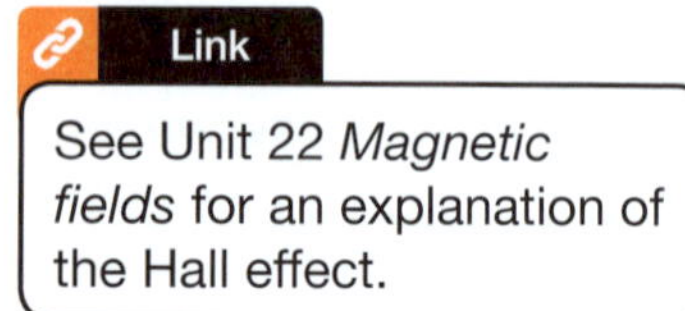

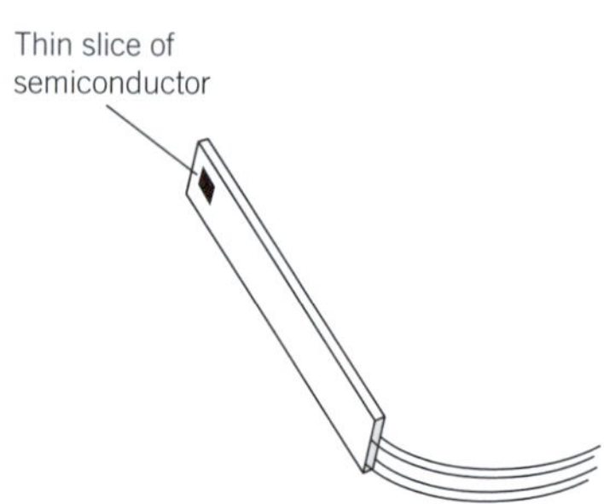

▲ **Figure 2.8** Hall probe

Calibration curves

Many instruments give outputs which are proportional to the quantity being measured, and so can easily be calibrated. Others, such as a thermocouple or thermistor used to measure temperature, give non-linear outputs, and so need calibration curves to convert the output of the instrument to the quantity being measured.

Worked example

A thermistor has the calibration curve shown in Figure 2.9.

a) What is the temperature T when the resistance of the thermistor is $6.0\,\text{k}\Omega$?

b) What is the resistance R when the temperature is $70\,°\text{C}$?

Answer

a) From the graph, when $R = 6.0\,\text{k}\Omega$ the temperature is $39\,°\text{C}$.

b) From the graph, when $T = 70\,°\text{C}$, $R = 2000\,\Omega$.

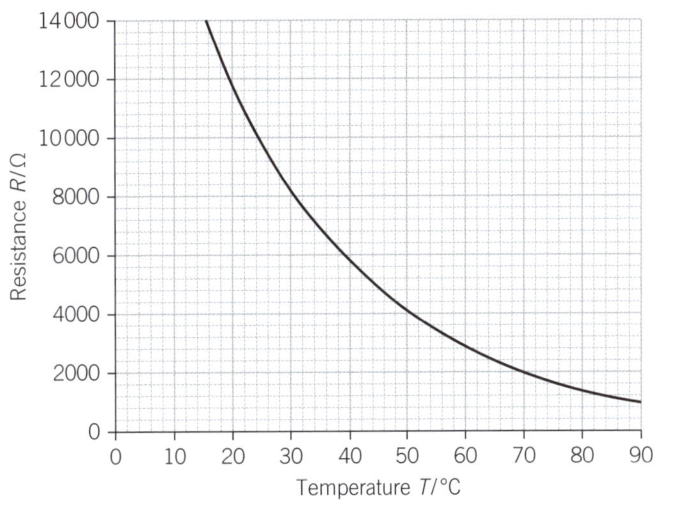

▲ **Figure 2.9** A calibration curve

Errors and uncertainties

Systematic and random errors

Systematic errors

Systematic errors cause all the recordings of a measurement to be displaced one way or the other from the true or accurate value (see Figure 2.10). Causes include zero errors in instruments, incorrectly calibrated scales, or changes in environmental conditions such as temperature. Using a micrometer with a zero error to measure the diameter of a resistance wire, for example, will give a very precise value (to the nearest 0.01 mm) but not a very accurate value if no allowance is made for the zero error.

Systematic errors can be reduced or eliminated by, for example, checking for any zero error in a micrometer or vernier calipers, or using two ammeters in series to check they read the same value.

Random errors

Random errors occur principally because of the limitations of the experimenter; for example, in judging the start and finish of an oscillation. A random error means that the values of a measurement are scattered in a random fashion (see Figure 2.11). The error can be reduced by taking several values and calculating a mean. If a range of values are recorded, a reasonable estimate of the absolute uncertainty of the measurement is half the range.

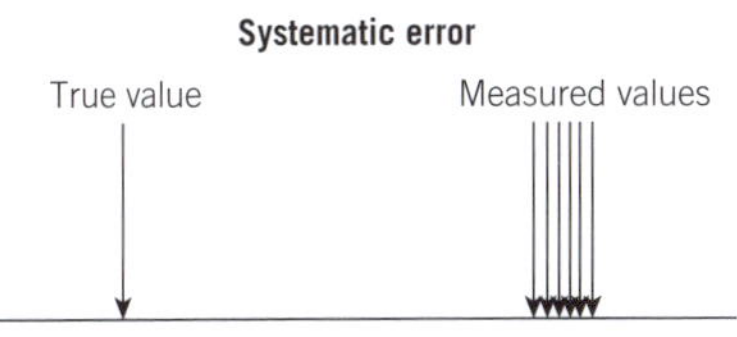

▲ **Figure 2.10** Precise but inaccurate

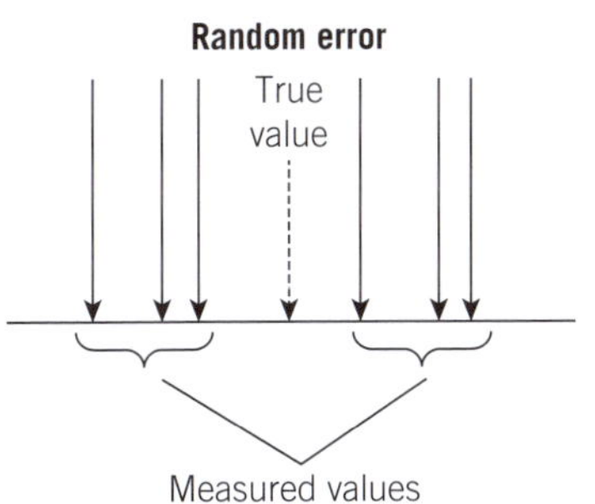

▲ **Figure 2.11** Accurate but imprecise

Precision and accuracy

Measurements should ideally be both precise and accurate, with all the readings grouped closely around the 'true' value. If several values are close together and one value is significantly different from all the others (an **outlier**), it can usually be rejected as an **anomalous result**.

Precision

A precise measurement is the degree to which the measurement is **repeatable**. The **precision** of an instrument is the smallest non-zero reading (the smallest division) that can be measured by the instrument or the size of the smallest division on a measuring instrument. A micrometer is a more precise instrument than a metre rule because it can measure to a precision of $\pm\,0.01$ mm, whereas it is only possible to record values to the nearest millimetre using a metre rule.

Accuracy

Accuracy is how close the value(s) are to the true value. It is a measure of the confidence an experimenter has in a measurement. An accurate measurement can be obtained using a correctly calibrated instrument skilfully (e.g., by avoiding parallax errors). Accuracy is expressed by the absolute or percentage uncertainty in a measurement.

Uncertainty

Suppose five measurements of the diameter d of a glass marble are recorded:

2.52 cm   2.49 cm   2.48 cm   2.51 cm   2.49 cm

The mean value is 2.498 cm and the range of values (largest to smallest) is 0.04 cm. The diameter of the marble should be recorded as:

$$2.50 \pm 0.02 \text{ cm}$$

The mean value is only given to the nearest 0.01 cm because the uncertainty in the value is 0.02 cm. The value is stating that:

- d lies between 2.48 cm and 2.52 cm

- the uncertainty in the value is 0.02 cm (half the range).

The **percentage uncertainty** in the value of d can be found from the equation:

$$\text{percentage uncertainty} = \frac{\text{absolute uncertainty}}{\text{mean value}} \times 100\%$$

In this example, the percentage uncertainty in d is:

$$\frac{0.02}{2.50} \times 100 = 0.8\%$$

To calculate the total percentage uncertainty in a quantity that depends on several measurements, add the individual percentage uncertainties, but first you need to multiply each value by the measurement's index (power). For example, if a quantity p is given by:

$$p = \frac{x^3 y^{\frac{1}{2}}}{z^4}$$

the percentage uncertainty in p is equal to:

$$3 \times \%\ \text{uncertainty in } x \quad + \quad \tfrac{1}{2} \times \%\ \text{uncertainty in } y \quad + \quad 4 \times \%\ \text{uncertainty in } z$$

↑ Raise your grade

An experiment is performed to measure the Young modulus E of copper using a long, thin copper wire. The diameter d and length ℓ_0 of the wire are measured, and then a tensile force F is applied to the wire. The extension x of the wire is then recorded. The measurements made are shown in the first two columns of the table.

Measurement	Value	Percentage uncertainty
Length of wire ℓ_0	0.953 ± 0.002 m	$\dfrac{0.002}{0.953} \times 100 = 0.2\%$
Diameter of wire d	0.21 ± 0.01 mm	$\dfrac{0.01}{0.21} \times 100 = 4.8\%$
Tensile force F	5.15 ± 0.05 N	$\dfrac{0.05}{5.15} \times 100 = 1.0\%$
Extension of wire x	1.2 ± 0.1 mm	$\dfrac{0.1}{1.2} \times 100 = 8.3\%$

✔✔ Method of calculation and values are correct

(a) State a suitable instrument for measuring:

A micrometer measures to a precision of 0.01 mm.

A ruler can only measure to the nearest mm. A travelling microscope or other vernier scale is needed.

(i) d ·····················micrometer ✔····················· **(ii)** x ·····················30 cm ruler ✘····················· [2]

(b) Complete the table by calculating the percentage uncertainties in the measurements. [2]

(c) The Young modulus is found from the equation:

$$E = \frac{4F\ell_0}{\pi d^2 x}$$

✔ Correct substitutions including conversion of mm to m

(i) Calculate the value of E.

$$E = \frac{4F\ell_0}{\pi d^2 x} = \frac{4 \times 5.15 \times 0.953}{\pi \times (0.21 \times 10^{-3})^2 \times (1.2 \times 10^{-3})} = 1.18 \times 10^{11}$$

✔ Correct value.

$$E = \quad 1.18 \times 10^{11} \quad \text{N m}^{-2}$$

(ii) Calculate the percentage uncertainty in E.

The percentage uncertainty in d^2 is twice the percentage uncertainty in d. (If there had been a d^3 term the percentage uncertainty would be three times as much as that in d.)

Total percentage uncertainty $= 0.2 + 4.8 + 1.0 + 8.3 = 14.3\%$ ✘

The correct value is:

$0.2 + 2 \times 4.8 + 1.0 + 8.3 = 19.1\%$.

Percentage uncertainty = ·····14.3%·····

(iii) State the value of E and its uncertainty to the appropriate number of significant figures.

$$14.3\% \text{ of } 1.18 \times 10^{11} = 0.169 \times 10^{11}$$

✔ The value is correct allowing for 'error carried forward' from (c) (ii); only one decimal place is justified because of the large uncertainty.

$$E = (1.2 \pm 0.2) \times 10^{11} \text{ N m}^{-2} \quad [4]$$

(d) Suggest a possible cause of systematic error in the measurements made.

·····'zero error' in the micrometer····· ✔ A valid answer. ····················· [1]

1 What is the reading on the vernier scale?

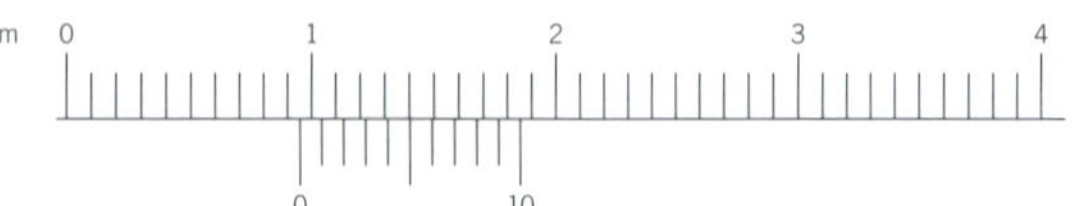

   A 0.94 cm **B** 0.95 cm **C** 1.04 cm **D** 1.05 cm [1]

2 A student is using a micrometer to measure the diameter of a resistance wire. This is the reading on the micrometer scale.

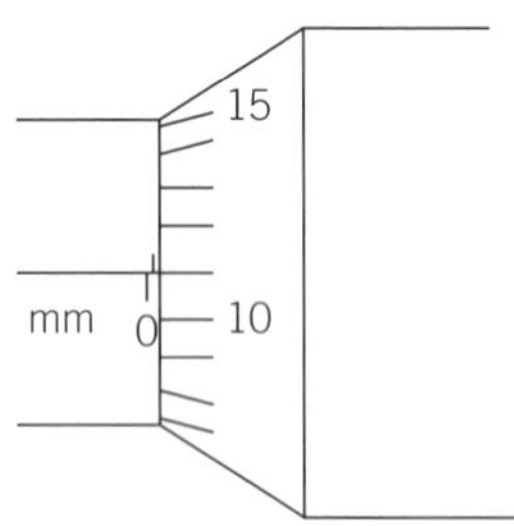

What is the diameter of the wire?

   A 0.11 mm   **B** 0.61 mm

   C 1.10 mm   **D** 1.11 mm   [1]

3 This is an electrical signal displayed on a cathode-ray oscilloscope (c.r.o.).

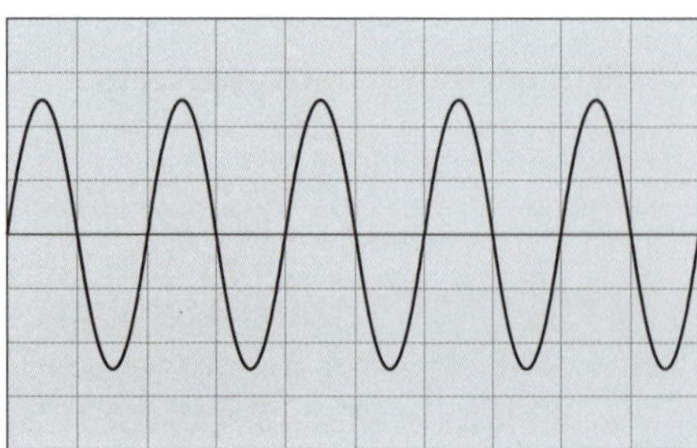

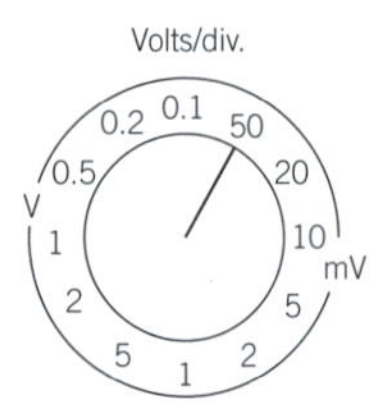

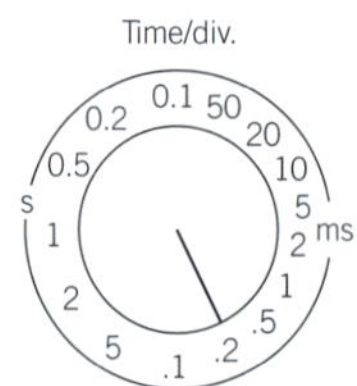

What are the frequency and amplitude of the signal?

	Frequency/kHz	Amplitude/mV
A	2.5	125
B	2.5	250
C	5.0	125
D	5.0	250

    [1]

4 The density of brass was found by measuring the diameter d and the mass m of a brass sphere.

d	16.5 ± 0.1 mm
m	19.7 ± 0.1 g

The density ρ can be found using the equation:

$$\rho = \frac{6m}{\pi d^3}$$

What is the percentage uncertainty in the value of the density?

   A 0.5%   **B** 0.6%   **C** 1.1%   **D** 2.3%   [1]

5 Several students in a class each measure the height of a bench using the same metre ruler. The results are shown in the chart below.

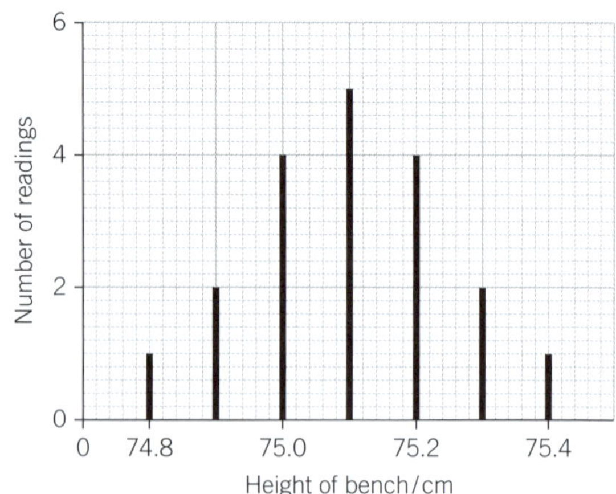

The true value of the height is 75.3 cm.

(a) State how the chart is evidence of:

   (i) random error

   (ii) systematic error.   [2]

(b) Describe how you would expect the chart to change if the measurements were more accurate.   [1]

6 A force of 40 ± 1 N is exerted by a gas on a piston of diameter 7.0 ± 0.5 cm.

(a) Calculate the pressure on the piston.

(b) Calculate the percentage uncertainty in the value of the pressure.

(c) State the value of the pressure and its absolute uncertainty to the appropriate number of significant figures.   [6]

3 Kinematics

Key points

- ☐ Define distance, displacement, speed (including average speed), velocity, and acceleration.
- ☐ Use graphs to represent distance, displacement, speed, velocity, and acceleration.
- ☐ Calculate velocity from the gradient of a displacement–time graph
- ☐ Calculate displacement from the area under a velocity–time graph
- ☐ Calculate acceleration from the gradient of a velocity–time graph.
- ☐ Derive and use the equations of motion for constant acceleration in a straight line.
- ☐ Describe an experimental method for measuring g, the acceleration of free fall.
- ☐ Solve problems about projectiles and objects falling in a uniform gravitational field without air resistance.

Describing motion

Kinematics is the study of motion and the relationship between quantities such as displacement, velocity, and acceleration. Here is a reminder of the key ideas.

Displacement s is the distance moved in a particular direction.

Velocity v is the rate of change of displacement with time:

$$v = \frac{\Delta s}{\Delta t}$$

Average speed is the **total** distance travelled divided by the time taken:

$$v_{av} = \frac{\text{total distance}}{\text{time taken}}$$

Acceleration a is the rate of change of velocity with time:

$$a = \frac{\Delta v}{\Delta t}$$

Displacement, velocity, and acceleration are all examples of **vectors**. Distance and speed are both **scalars**.

> **Remember**
>
> Δ means 'the change in'.
>
> **Vectors** have magnitude and direction; for example, displacement, velocity and acceleration.
>
> **Scalars** only have magnitude; for example, distance and speed.

Using graphs

Displacement–time graphs

The **gradient** of the displacement–time graph (see Figure 3.1) is the **velocity** $\left(\dfrac{\Delta s}{\Delta t}\right)$.

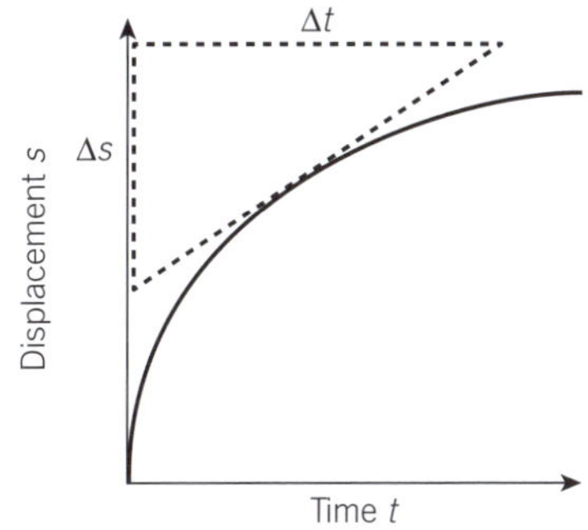

▲ **Figure 3.1** Velocity is the gradient of a displacement–time graph

> ★ **Exam tip**
>
> When calculating a velocity by finding the gradient, draw a tangent, and complete a large triangle as shown in Figure 3.1. Δs can be read directly from the axis of the graph.

Changes in the gradient of a displacement–time graph give information about acceleration as well as velocity (see Figure 3.2).

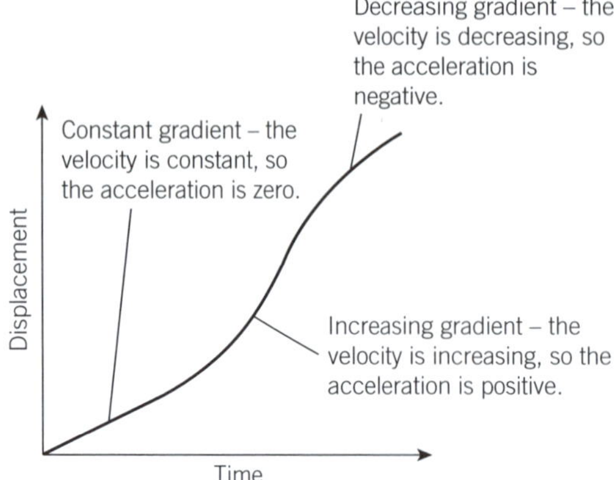

▲ **Figure 3.2** Acceleration in displacement–time graphs

Worked example

The graph shown in Figure 3.3 is the displacement–time graph for an athlete running a 100 m race.

a) Describe how the velocity of the athlete changes:

   i) during the first 2 s

   ii) between 5 s and 7 s

   iii) between 9 s and 11 s.

b) Use the graph to estimate the velocity of the athlete after 8.0 s.

Answer

a) i) The gradient of the line is increasing so the velocity of the athlete is increasing. The athlete is accelerating.

   ii) The gradient is constant, so the athlete is moving with constant velocity.

   iii) The gradient is decreasing so the athlete is decelerating.

b) Draw a large tangent on the graph at $t = 8.0$ s and complete the right-angled triangle (see dotted line in Figure 3.4).

$$v = \frac{\Delta s}{\Delta t} = \frac{116 - 4}{12 - 0} = 9.3\,\text{ms}^{-1}$$

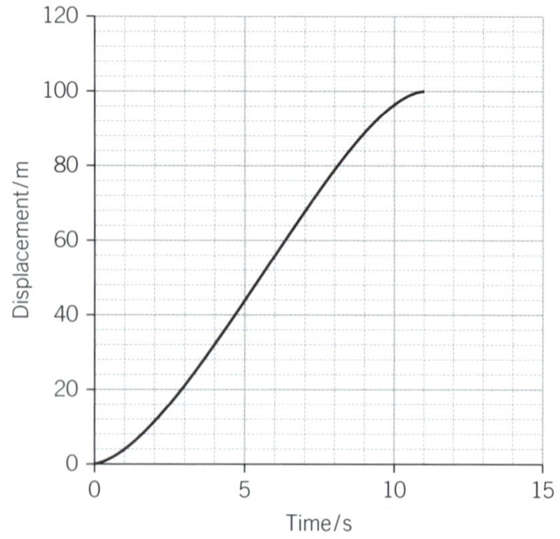

▲ **Figure 3.3**

Maths skills

For more on drawing tangents and calculating gradients see Appendix: Maths skills.

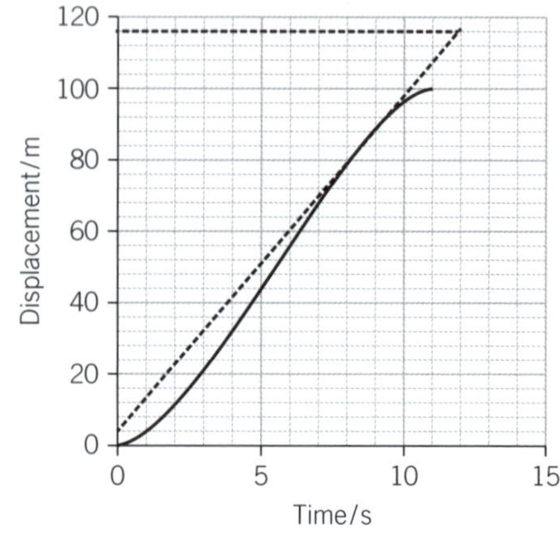

▲ **Figure 3.4**

Velocity–time graphs

Velocity–time graphs are very useful. The gradient of a velocity–time graph is the acceleration, and the area under a velocity–time graph between any two points is the displacement (see Figure 3.5).

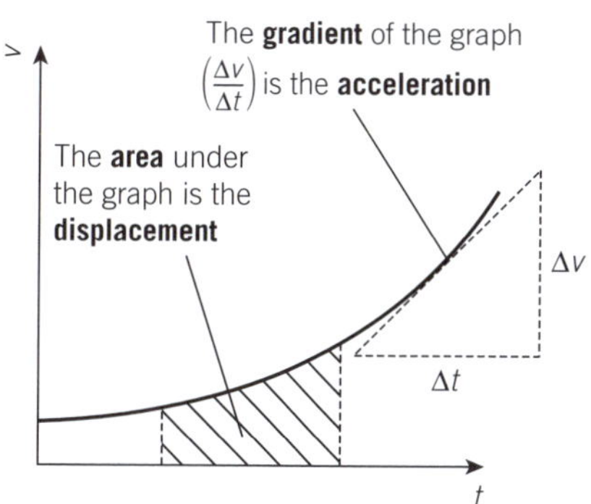

▲ **Figure 3.5** Acceleration and displacement can be calculated from a velocity–time graph

---

Worked example

Figure 3.6 shows the velocity–time graph for a car starting from rest.

a) Determine the car's acceleration at $t = 4\,s$.

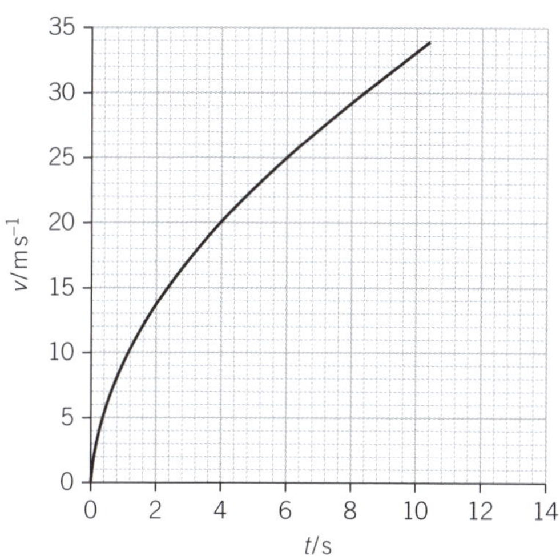

▲ **Figure 3.6**

b) Estimate the distance travelled in the first 6 s.

Answer

a) The acceleration at $t = 4\,s$ is the gradient at $t = 4\,s$ (see Figure 3.7).

$$a = \frac{\Delta v}{\Delta t} = \frac{32.0 - 11.0}{8.0} = 2.6\,\mathrm{m\,s^{-2}}$$

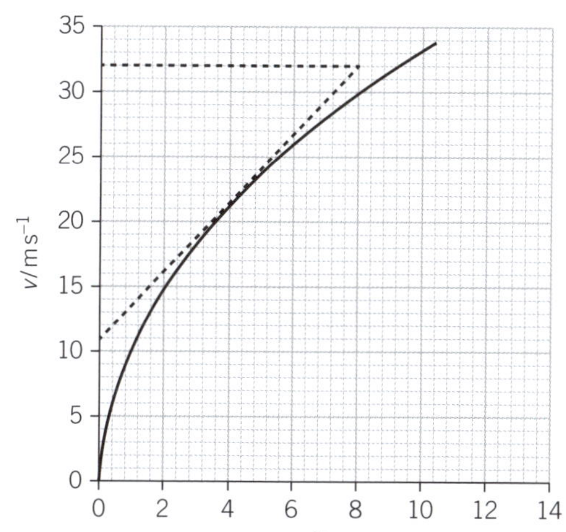

▲ **Figure 3.7**

> **Remember**
>
> Don't forget to include the units in your answer.

b) The distance travelled is the area under the graph (see Figure 3.8) from $t = 0\,s$ to $t = 6\,s$.

distance travelled $\approx \frac{1}{2} \times 6 \times 35 = 105\,m$

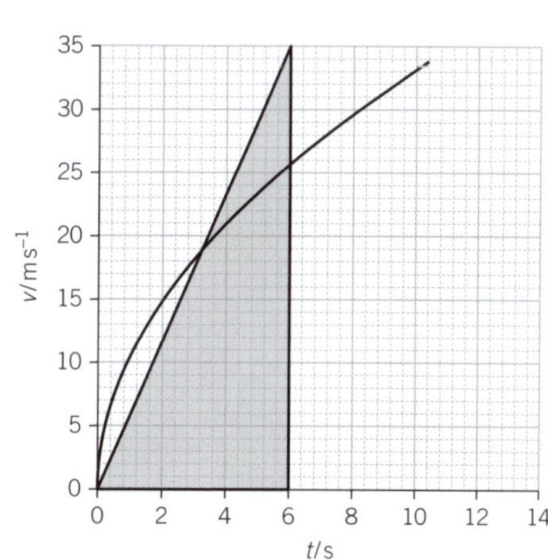

▲ **Figure 3.8**

> **Maths skills**
>
> For more on estimating areas see Appendix: Maths skills.

Equations of motion

If an object is travelling with **constant acceleration**, the **equations of motion** may be used to analyse its motion.

A velocity–time graph for an object moving with constant acceleration a, starting with velocity u, and reaching a velocity v in t seconds, shows how these equations arise.

From Figure 3.9:

$$\text{acceleration} = \text{gradient}$$

$$a = \frac{v - u}{t}$$

so $\qquad v = u + at \qquad$ (equation 1)

The displacement, s, of an object can be found by working out the area under the graph (see Figure 3.10):

$$\text{displacement} = \text{area under the graph}$$

$$s = ut + \tfrac{1}{2}(v - u)\,t$$

from equation 1, $v - u = at$ so:

$$s = ut + \tfrac{1}{2}(at)t$$

$$s = ut + \tfrac{1}{2}at^2 \qquad \text{(equation 2)}$$

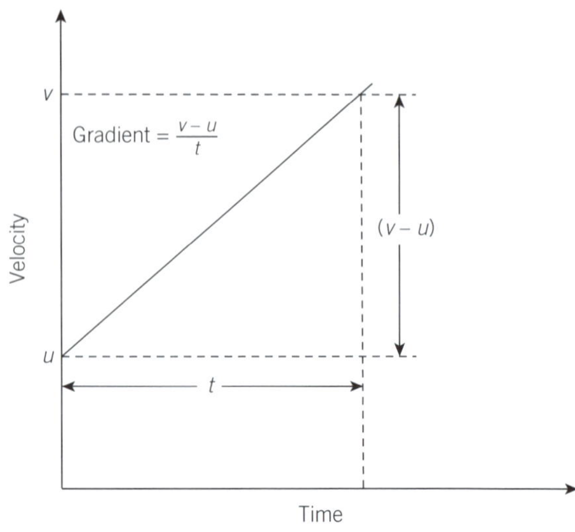

▲ **Figure 3.9** The acceleration is given by the gradient of the velocity–time graph

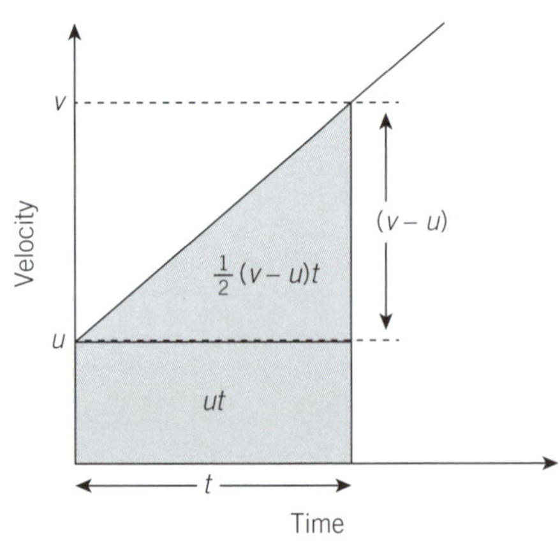

▲ **Figure 3.10** The displacement is the area under the graph

The third equation can be derived from equations 1 and 2:

First, 'square' equation 1:

$$v^2 = (u + at)^2 = u^2 + 2uat + a^2t^2$$

$$= u^2 + 2a(ut + \tfrac{1}{2}at^2)$$

so $\qquad v^2 = u^2 + 2as \qquad$ (equation 3)

> **Remember**
>
> These equations only apply when the acceleration or deceleration is **constant**, including zero acceleration (constant velocity).

> **Remember**
>
> The three equations of motion: $v = u + at$
>
> $$s = ut + \tfrac{1}{2}at^2$$
>
> $$v^2 = u^2 + 2as$$

> **Exam tip**
>
> You need to be able to recall these equations and you may be asked to derive them.

Using the equations of motion

Worked example

An aeroplane touches down at the end of a runway travelling at a speed of $72\,\text{m s}^{-1}$ (see Figure 3.11). It decelerates uniformly at a rate of $3\,\text{m s}^{-2}$.

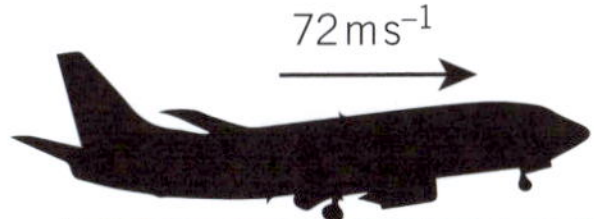

▲ **Figure 3.11**

Calculate:

a) the speed of the aeroplane 8 s after touchdown

b) the distance travelled along the runway before coming to rest.

Answer

a) Using $v = u + at$:
$$v = 72 + (-3) \times 8 = 48\,\text{m s}^{-1}$$

b) Using $v^2 = u^2 + 2as$ with $v = 0$ when the aeroplane comes to rest:
$$0^2 = 72^2 + 2 \times (-3) \times s$$
$$s = 864\,\text{m}$$

> The acceleration is negative because the aeroplane is decelerating.

Motion under gravity

At low speeds air resistance has a negligible effect on falling objects and can be ignored. The equations of motion can be used to solve problems where $a = g$, the acceleration of free fall.

> ★ **Exam tip**
>
> The value of g is provided in Exam Papers 1, 2, and 4.

Worked example

1 A stone is thrown vertically upwards with a velocity of $30.0\,\text{m s}^{-1}$ and falls back down to the ground (Figure 3.12).

Calculate:

a) the velocity of the stone after 4.0 s

b) the maximum height reached by the stone

Answer

a) Using $v = u + at$:

$$v = 30.0 + (-9.81) \times 4.0 = -9.24\,\text{m s}^{-1}$$

The velocity is $9.24\,\text{m s}^{-1}$ **downwards**.

b) Using $v^2 = u^2 + 2as$ with $v = 0$ when the stone reaches its highest point:

$$0^2 = 30.0^2 + 2 \times (-9.81) \times s$$

$$s = 45.9\,\text{m}$$

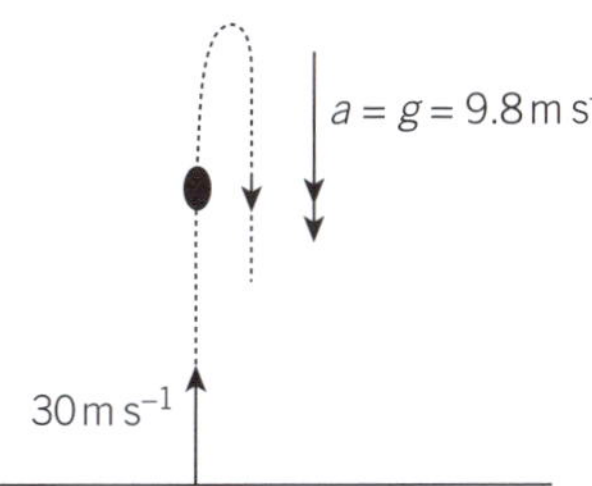

▲ **Figure 3.12**

> 💡 **Remember**
>
> Take care deciding on the **sign** of g. An object thrown into the air, for example, is always accelerating downwards, whether the object happens to be moving upwards or downwards.

2 A hot-air balloon (see Figure 3.13) is ascending at a constant speed of $3.0\,\text{m s}^{-1}$. A sandbag is dropped from the balloon and hits the ground after $5.0\,\text{s}$.

a) Calculate the height of the balloon when the sandbag was released.

b) Draw the graph of velocity against time for the sandbag, from the moment the sandbag is released until it hits the ground. Ignore air resistance.

Answer

a) Using $s = ut + \tfrac{1}{2}at^2$ from the moment the sandbag is released until it hits the ground:

$$s = 3.0 \times 5.0 + \tfrac{1}{2} \times (-9.81) \times 5.0^2$$

$$= -108\,\text{m}$$

b) See Figure 3.14.

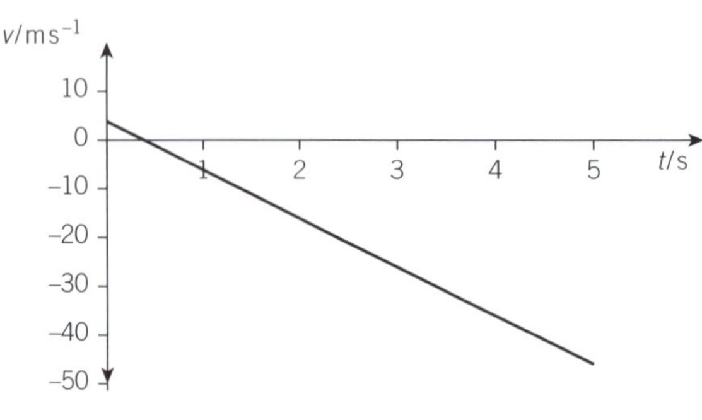

▲ **Figure 3.14**

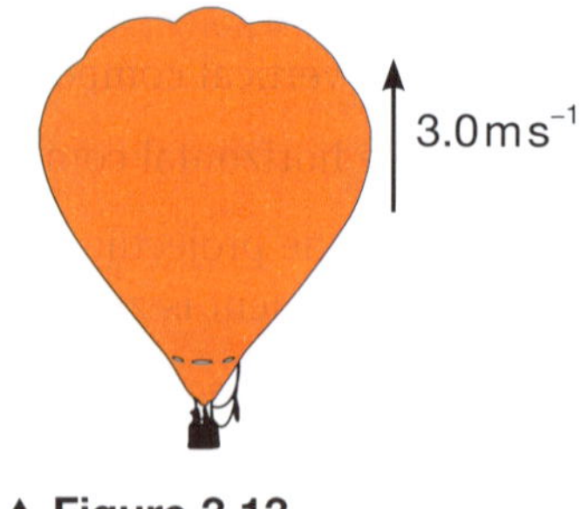

▲ **Figure 3.13**

> s is negative as the displacement is downwards.

Projectiles

The movement of any object travelling through the air (see Figure 3.15) can be described in terms of its horizontal and vertical components. If air resistance can be ignored, the components are:

- constant acceleration g vertically downwards

- constant velocity horizontally (as no forces act horizontally).

The equations of motion can be applied separately in the horizontal and vertical directions to solve problems.

If the initial velocity is V, making an angle θ to the horizontal (see Figure 3.16), then:

- the horizontal velocity is $V\cos\theta$

- the initial vertical velocity is $V\sin\theta$.

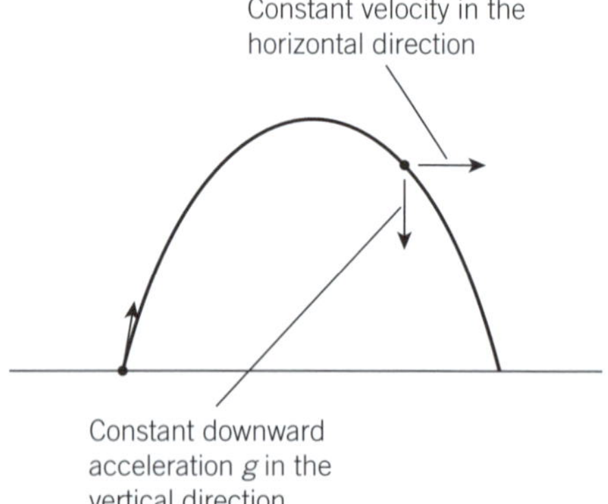

▲ **Figure 3.15** Projectile motion

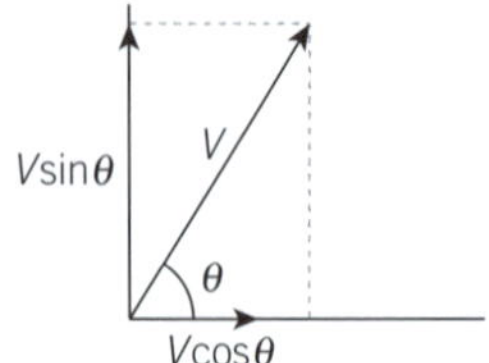

▲ **Figure 3.16** Horizontal and vertical components of velocity

> **Maths skills**
>
> For more on resolving vectors in two directions, see Appendix: Maths skills.

Figure 3.17 summarises some facts about the projectile's motion.

- At its maximum height H:
  - the vertical component of velocity is zero
  - the horizontal component of the velocity stays constant at $V\cos\theta$.
- When the projectile has travelled its full range R, the vertical displacement is zero (the net distance travelled vertically is 0).

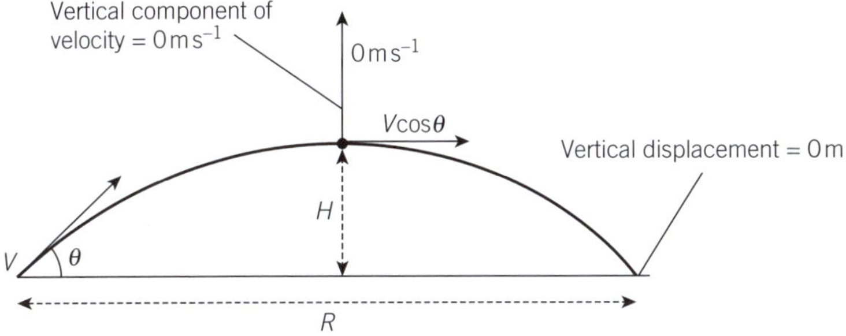

▲ **Figure 3.17** Projectile motion

Applying the equations of motion to projectiles, it can be shown that:
$$H = \frac{V^2 \sin^2 \theta}{2g}$$
and
$$R = \frac{V^2 \sin 2\theta}{g}$$

Worked example

A cannon fires a cannonball with an initial velocity of $12\,\mathrm{m\,s^{-1}}$ at an angle of $50°$ to the horizontal.

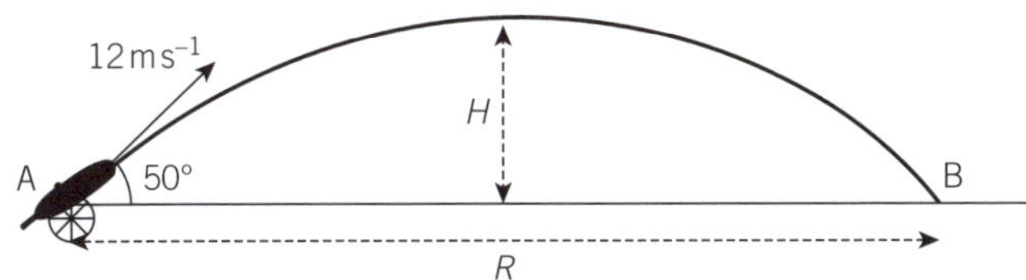

▲ **Figure 3.18**

Show that:

a) the initial horizontal component of the cannonball's velocity is $7.7\,\mathrm{m\,s^{-1}}$

b) the maximum height H reached is $4.3\,\mathrm{m}$.

Answer

a) Initial horizontal velocity $= 12\cos 50°$

$$= 7.7\,\mathrm{m\,s^{-1}}$$

b) Using $v^2 = u^2 + 2as$ applied vertically from the starting point A to the highest point:

$$0^2 = (12\sin 50)^2 + 2 \times (-9.81) \times H$$
$$H = \frac{(12\sin 50)^2}{2 \times 9.8} = 4.3\,\mathrm{m}$$

> ★ **Exam tip**
>
> 'Show' means you are expected to derive the answer given, showing all your working.

> The vertical component of the velocity at the highest point is zero.

 # ↑ Raise your grade

The graph shows how the velocity of a drag-racing car changes with time. The graph can be divided into three separate stages, as shown below.

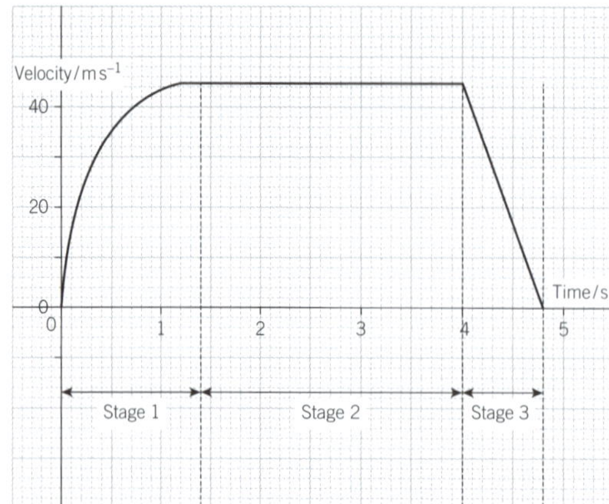

(a) Describe the motion of the car during the three separate stages.

The car accelerates at first, then travels at constant speed. After 4 seconds it starts to
slow down

✔ ✗ ✗ The statement is correct, but lacks sufficient detail. A better answer would be: *In stage 1 the car accelerates from rest. As the velocity increases, the acceleration decreases (the gradient of the graph decreases). In stage 2 the velocity is constant at 45 m s⁻¹. After 4 s, at the beginning of stage 3, the car decelerates quickly and at a uniform rate, coming to rest after 4.8 s.*

[3]

(b) Use the graph to find the acceleration of the car.

(i) after 0.5 s

$$\text{Acceleration} = \frac{velocity}{time} = \frac{35}{0.5} = 70$$

✗ ✗ Acceleration is **change** in velocity/time. A tangent should be drawn at $t = 0.5$ s and the gradient of the tangent calculated.

A good answer would be:

acceleration = gradient of graph at $t = 0.5$ s
$$= \frac{55 - 22}{1.2} = 27.5 \text{ m s}^{-2}$$

acceleration = 70 m s⁻² [2]

(ii) after 4.5 s.

$$\text{acceleration} = \frac{45}{5.0 - 4.0} = 45$$

✗ ✗ The candidate has misread the second time value (it should be 4.8 s not 5.0 s). The answer should also be negative (the car is decelerating).

A good answer would be $a = \dfrac{45 - 0}{4.0 - 4.8} = -56.3 \text{ m s}^{-2}$

acceleration = 45 m s⁻² [2]

(c) Estimate the total distance travelled by the car.

Total distance = area under graph = $45 \times 4 + \frac{1}{2} \times 0.8 = 198$

The correct method has been used to find the distance travelled for one mark, but the calculation hasn't taken into account the area under the curve of the graph between $t = 0$ s and $t = 1.4$ s.

A better answer: distance = $51 \times 5.0 \times 0.20 + 45 \times 2.6 + \frac{1}{2} \times 45 \times 0.80 = 186$ m

51 small squares under the curved part of the graph

distance = 198 m [2]

22

Exam-style questions

1 A cyclist travels from one town to the next at an average speed of $40\,\text{km}\,\text{h}^{-1}$. She completes the return journey at an average speed of $20\,\text{km}\,\text{h}^{-1}$.

What was her average speed for the whole journey?

A $25\,\text{km}\,\text{h}^{-1}$ **B** $27\,\text{km}\,\text{h}^{-1}$

C $30\,\text{km}\,\text{h}^{-1}$ **D** $33\,\text{km}\,\text{h}^{-1}$ [1]

2 The graph shows the distance travelled by a car in the first $20\,\text{s}$ of a journey.

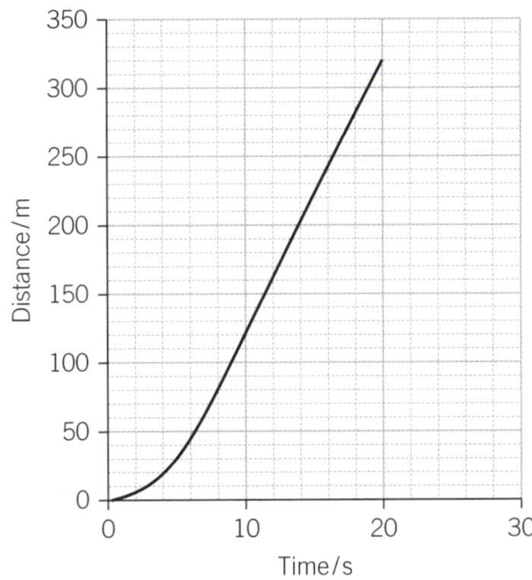

What is the best estimate of the speed of the car after $10\,\text{s}$?

A $8\,\text{m}\,\text{s}^{-1}$ **B** $12\,\text{m}\,\text{s}^{-1}$ **C** $16\,\text{m}\,\text{s}^{-1}$ **D** $20\,\text{m}\,\text{s}^{-1}$ [1]

3 A stone is thrown vertically upwards with a speed of $20\,\text{m}\,\text{s}^{-1}$ near the edge of a cliff and falls down to hit the beach below the cliff $6.0\,\text{s}$ later.

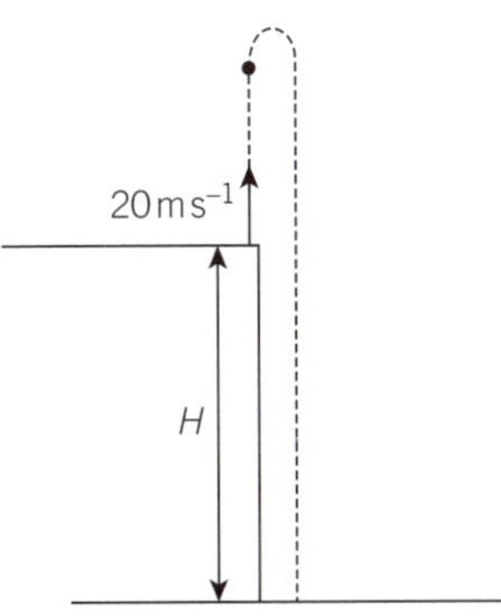

What is the height H of the cliff?

A $56.4\,\text{m}$ **B** $86.4\,\text{m}$ **C** $176\,\text{m}$ **D** $296\,\text{m}$ [1]

4 **(a)** Explain the difference between a *scalar* quantity and a *vector* quantity. [1]

(b) Underline the vector quantities in the list below:

speed   displacement   acceleration   velocity   [1]

(c) A tennis player hits a tennis ball horizontally with a speed of $60\,\text{m}\,\text{s}^{-1}$. The ball is initially at a height of $1.40\,\text{m}$ above the ground and $11.90\,\text{m}$ from the net. Air resistance is negligible.

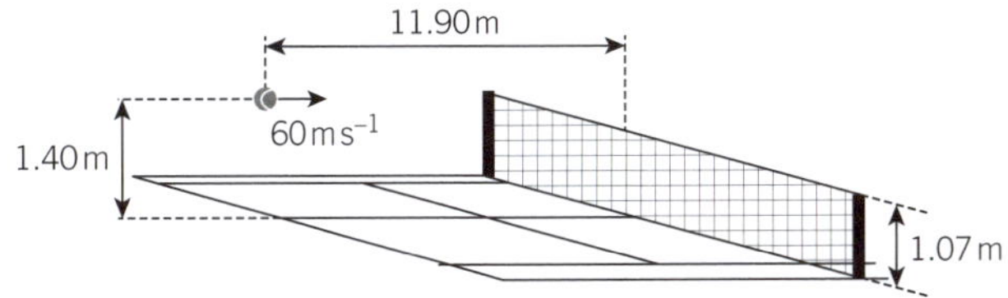

(i) Calculate the time it takes for the ball to reach the net.

(ii) Show that the ball passes over the net if the net is $1.07\,\text{m}$ high. [4]

(d) Calculate the distance the ball is from the net when it lands on the other side of the court. [2]

(e) The distance the ball moves is different from its displacement. Explain why. [1]

5 A basketball player throws a basketball into the hoop of the basket.

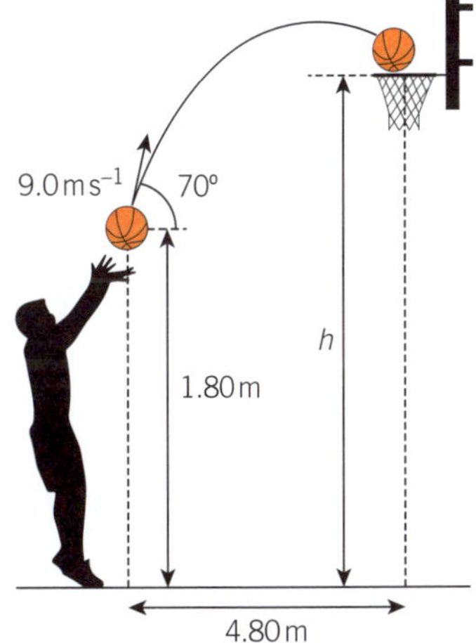

(a) Calculate:

(i) the initial horizontal component of the velocity of the basketball

(ii) the time it takes for the ball to reach the basketball

(iii) the height h of the basket above the ground. [5]

(b) Determine the velocity of the ball as it reaches the basket. [3]

Key points

- ☐ Understand that mass is the property of an object that resists change in motion.
- ☐ Recall $F = ma$ and solve problems using it; understand that acceleration and resultant force are always in the same direction.
- ☐ Understand that linear momentum is given by mv, and that force is the rate of change of momentum.
- ☐ State and apply Newton's laws of motion.
- ☐ Understand the concept of weight as the effect of a gravitational field on a mass and recall that the weight of a body is mg.
- ☐ State the principle of conservation of momentum and apply it to solve problems, including elastic and inelastic collisions between objects in one and two dimensions.
- ☐ Know that, for elastic collisions, the relative speed of approach is equal to the relative speed of separation.
- ☐ Understand that the momentum of a system is always conserved in collisions between objects, but that some change in kinetic energy may occur.
- ☐ Describe qualitatively the effect of air resistance on the motion of objects falling in a uniform gravitational field.

Dynamics is concerned with the affect forces have on the movement of masses as described by Newton's three laws motion.

Momentum and Newton's laws of motion

Mass and inertia

When a car brakes suddenly we feel we are 'thrown forward' – in reality we are just trying to carry on moving in a straight line at constant speed. The reluctance of an object to change its motion (to speed up, slow down or change direction) is called its **inertia**. The mass of an object is an indication of its inertia and is measured in kilograms (kg).

The greater the mass of an object, the greater the resistance to change (see Figure 4.1).

Newton's first law of motion

Newton's first law of motion states that a **resultant force** is needed to accelerate or decelerate an object.

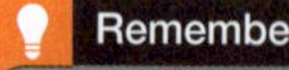

Remember

Newton's first law: An object will remain stationary, or continue at constant speed in a straight line, unless acted on by an external force.

a Large mass – difficult to start moving

b Large mass – difficult to stop moving

▲ **Figure 4.1** Large masses – difficult to start and stop

AS Level

The first law appears to contradict our daily experience of forces and movement – to make something move at constant speed it has to be continually pushed. But this ignores the friction forces that oppose the motion. A parachutist falling at constant speed has balanced forces of weight acting downwards and air resistance acting upwards – the resultant force on the parachutist is zero (see Figure 4.2).

In Figure 4.3, the driving force D from the engine is initially greater than the air resistance and friction forces F, and so the car accelerates. At a certain speed the friction forces will equal the driving force and the car will travel at constant speed. If the brakes are applied, the friction forces are bigger than the driving force and the car decelerates.

▲ **Figure 4.2** The forces are balanced ($F = W$) so the parachutist falls at a constant speed

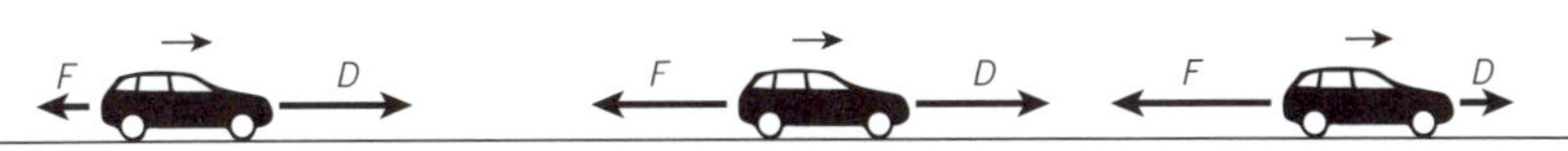

a $D > F$: accelerates as forces are unbalanced

b $D = F$: moves at a constant speed as forces are balanced

c $D < F$: decelerates as forces are unbalanced

▲ **Figure 4.3** Balanced and unbalanced forces

Momentum

The **momentum** p of an object is its mass m multiplied by its velocity v.

$$p = mv$$

> **Remember**
>
> Momentum is a vector. It has units of kg m s^{-1}.

Newton's second law of motion

If a resultant force acts on an object, it speeds up, slows down or changes direction; that is, its momentum changes. Newton's second law links the size of the force applied to the change in momentum.

> **Remember**
>
> **Newton's second law:** The rate of change of momentum of an object is proportional to the resultant force on it.

Consider an object of mass m moving with speed u acted on by a constant force F for a time t (Figure 4.4):

▲ **Figure 4.4** Force = rate of change of momentum

Using Newton's second law:

$$F \propto \frac{\text{change in momentum}}{\text{time taken}} = \frac{mv - mu}{t} = \frac{m(v - u)}{t} = ma$$

$$F \propto ma$$

> **Remember**
>
> Force = rate of change of momentum
>
> $$F = ma$$

where a is the acceleration of the object. By defining the unit of force (the newton) as that force which gives a mass of 1 kg an acceleration of 1 m s^{-2}, we can write:

$$F = ma$$

Mass and weight

When an object of mass m is held above the Earth's surface and then released, it accelerates downwards at $9.81\,\mathrm{m\,s^{-2}}$ (ignoring air resistance) because there is an unbalanced force acting on it (its weight W). Using $F = ma$:

$$W = mg$$

where g is the acceleration of free fall.

Newton's third law of motion

When you push a door, the door pushes you – forces always act in pairs. Newton's third law is often quoted as 'action = reaction', but it can be stated more formally.

Newton's third law: When two objects interact, they exert equal and opposite forces on each other.

Action and reaction forces are often misunderstood. The key points to remember are:

- Forces do not exist individually, but in pairs.
- The forces are of the same type; for example, both gravitational.
- The two forces act on different objects.
- The third law applies to every situation.

Think of a book resting on a table, as shown in Figure 4.5.

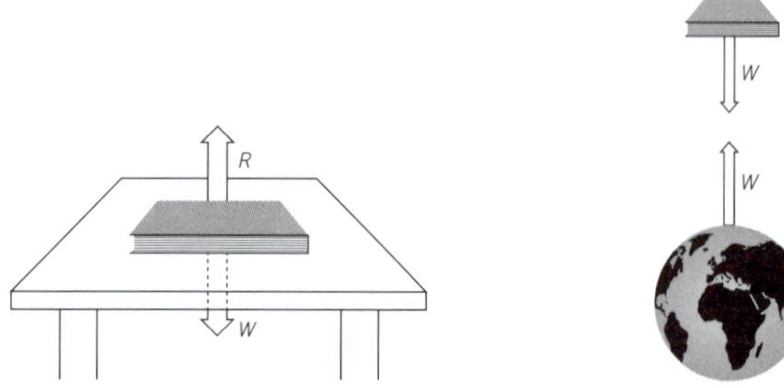

▲ **Figure 4.5** Newton's third law

- The weight W of the book acts downwards.
- The push up from the table, R, is equal to W because the book is in equilibrium (if it wasn't the book would fall through the table, or take off!).

However, this is not the 'reaction' force in Newton's third law. If the table were to disappear there would still be a 'reaction' force.

The two forces referred to in Newton's third law act on **different** objects.

- The force causing the book to have weight is the gravitational pull of the Earth.
- The 'reaction' force is the gravitational pull upwards on the Earth by the book.

Conservation of linear momentum

Newton's second law can be written:

$$F = \frac{\Delta(mv)}{\Delta t}$$

so

$$F\Delta t = \Delta(mv)$$

When two objects collide, the same force F acts for the same time Δt on both objects, and so the magnitude of the change in momentum will be the same for both objects.

In Figure 4.6 the momentum of one of the masses will decrease, but the momentum of the other mass increases by an equal amount, and so the **total** momentum of the two masses is the same as before the collision. There is no change in the total overall momentum. This is an example of the principle of conservation of momentum.

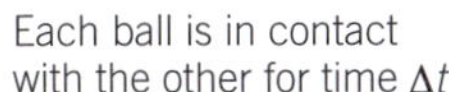

▲ **Figure 4.6** Impulse = $F\Delta t$

> 💡 **Remember**
>
> **Principle of conservation of momentum:** For a system of interacting objects, the total momentum remains constant provided no external resultant force acts on the system.

This is why momentum is such a useful quantity to calculate. Although the momentum of individual objects changes, the total momentum in any interaction (e.g., a collision or explosion) remains constant provided no external force acts.

> **Worked examples**
>
> 1 A ^{235}U nucleus has a mass of 3.9×10^{-25} kg. It decays by emitting an alpha particle, of mass 6.6×10^{-27} kg, with a speed of 1.6×10^{7} m s^{-1}. What is the recoil velocity of the nucleus?
>
> **Answer**
>
> The total momentum before and after the alpha particle is emitted is zero. Let the velocity of the nucleus after the alpha particle is emitted be v.
>
> $$\begin{array}{c}\text{momentum of} \\ \text{decayed nucleus}\end{array} + \begin{array}{c}\text{momentum of alpha} \\ \text{particle}\end{array} = 0$$
>
> $$(3.9 \times 10^{-25} - 6.6 \times 10^{-27})v + 6.6 \times 10^{-27} \times 1.6 \times 10^{7} = 0$$
>
> so
>
> $$v = \frac{-6.6 \times 10^{-27} \times 1.6 \times 10^{7}}{(3.9 \times 10^{-25} - 6.6 \times 10^{-27})} = -2.8 \times 10^{5}\ \text{m s}^{-1}$$

The quantity $F\Delta t$ is called the **impulse**.

The minus sign shows that the nucleus recoils; it moves in the opposite direction to the alpha particle.

2 A physics student holds a ball above his head. He drops the ball and it falls to the ground, bounces a few times and eventually comes to rest. He says the ball initially has no momentum, gains momentum as it falls but has negative momentum when it bounces back up, and so momentum cannot be conserved. What would you say to the student in reply?

Answer

- Before the ball is released, the total momentum of the ball and the Earth is zero.

- As the ball descends, the Earth is ascending upwards with the same magnitude of momentum, but in the opposite direction due to the gravitational pull of the ball on the Earth (see Figure 4.7). We don't notice the Earth moving up because its mass is so large compared to the ball, making its speed too small to measure. The total momentum of the ball and Earth 'system' is still zero.

- When the ball hits the Earth, it exerts a sudden force on the Earth which causes its momentum to change direction.

- As the ball moves up the Earth moves down with an equal magnitude of momentum.

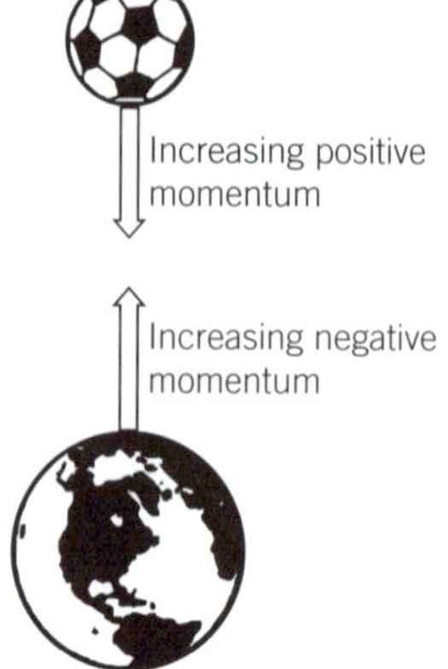

▲ **Figure 4.7** Conservation of momentum

Collisions

Elastic collisions

When objects collide, the total momentum of the objects remains constant. If kinetic energy is also conserved in a collision, it is called an **elastic collision**. The collisions of molecules in an ideal gas are considered elastic, and the collisions of snooker balls are almost elastic.

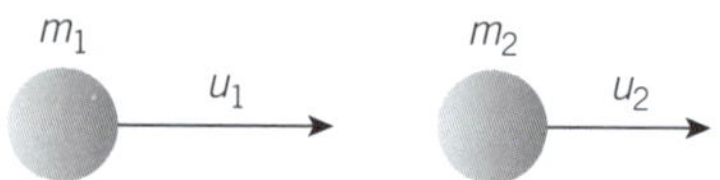

a before the collision ($u_1 > u_2$)

In Figure 4.8, two hard spheres are both travelling in the same direction: m_1 is moving with speed u_1 and m_2 is moving more slowly with speed u_2. m_1 has a head-on collision with m_2. After the collision, m_2 is travelling faster than m_1.

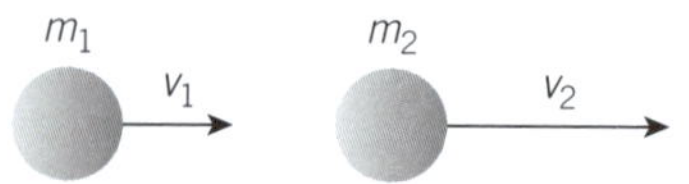

b after the collision ($v_2 > v_1$)

▲ **Figure 4.8** Elastic collisions

Applying conservation of total momentum:

$$m_1 u_1 + m_2 u_2 = m_1 v_1 + m_2 v_2$$

Simplifying:

$$m_1(u_1 - v_1) = -m_2(u_2 - v_2) \qquad \text{(equation 1)}$$

Applying conservation of kinetic energy:

$$\frac{1}{2}m_1 u_1^2 + \frac{1}{2}m_2 u_2^2 = \frac{1}{2}m_1 v_1^2 + \frac{1}{2}m_2 v_2^2$$

Simplifying this equation:

$$m_1(u_1^2 - v_1^2) = -m_2(u_2^2 - v_2^2) \qquad \text{(equation 2)}$$

Dividing equation 2 by 1:

$$\frac{m_1(u_1^2 - v_1^2)}{m_1(u_1 - v_1)} = \frac{-m_2(u_2^2 - v_2^2)}{-m_2(u_2 - v_2)}$$

Simplifying this equation: $\quad u_1 + v_1 = u_2 + v_2$

Rearranging: $\quad v_2 - v_1 = u_1 - u_2$

So velocity of separation = velocity of approach

> **Maths skills**
>
> 'The difference of two squares'
>
> $a^2 - b^2 = (a - b)(a + b)$

> **Remember**
>
> **Elastic collision:** momentum and kinetic energy are conserved.
>
> The velocity of separation equals the velocity of approach.

Inelastic collisions

Most collisions are **inelastic collisions** – some of the initial kinetic energy is 'lost' (usually transferred as heat or sound energy). A car colliding with a tree, or the collisions of electrons in a gas in which the atoms of the gas become excited, are both examples of inelastic collisions. The principle of conservation of momentum still applies to inelastic collisions.

> **Remember**
>
> **Inelastic collision:** total momentum is conserved but total kinetic energy is not.

Worked example

A dynamics trolley of mass 0.90 kg, travelling at a speed of $2.50\,\mathrm{m\,s^{-1}}$, collides head–on with another trolley of mass 1.80 kg, travelling in the opposite direction with a speed of $1.40\,\mathrm{m\,s^{-1}}$, as shown in Figure 4.9. The two trolleys stick together after the collision.

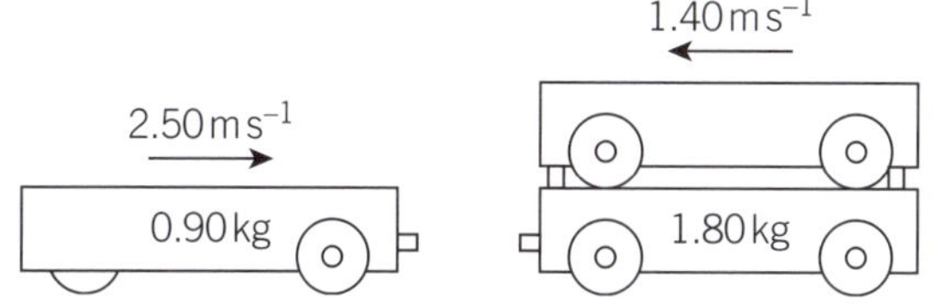

▲ **Figure 4.9** Inelastic collisions

a) Calculate their common speed, v, after the collision.

b) Determine in which direction they both move after the collision.

c) Calculate the total kinetic energy E_k:

    i) before the collision   **ii)** after the collision.

d) Why are your answers to **c) i)** and **c) ii)** not the same?

Answer

a) Total momentum before collision = total momentum after collision:

$$0.90 \times 2.50 - 1.80 \times 1.40 = (1.80 + 0.90) \times v$$

$$v = \frac{(0.90 \times 2.50 - 1.80 \times 1.40)}{(1.80 + 0.90)} = -0.10\,\mathrm{m\,s^{-1}}$$

b) The final velocity is negative; the trolleys move in the same direction as the 1.80 kg trolley is moving before the collision.

c) **i)** E_k before collision $= \frac{1}{2} \times 0.90 \times 2.50^2 + \frac{1}{2} \times 1.80 \times (-1.40)^2$

$$= 4.58\,\mathrm{J}$$

    ii) E_k after collision $= \frac{1}{2} \times 2.7 \times (-0.1)^2 = 0.014\,\mathrm{J}$

d) Kinetic energy is lost (as heat and sound). As the trolleys 'stick' there is a friction force acting on each trolley accelerating/decelerating it (the friction forces are doing work).

Collisions in two dimensions

The principle of conservation of momentum can be applied to collisions in two dimensions. The total momentum in any direction must remain constant.

In Figure 4.10 a mass M is travelling at speed u towards a mass m which is at rest. When they collide, the line joining their centres makes an angle θ with the original direction of M.

Applying the principle of conservation of momentum:

In the x direction: $Mu = Mv_1\cos\phi + mv_2\cos\theta$

In the y direction: $Mv_1\sin\phi = mv_2\sin\theta$

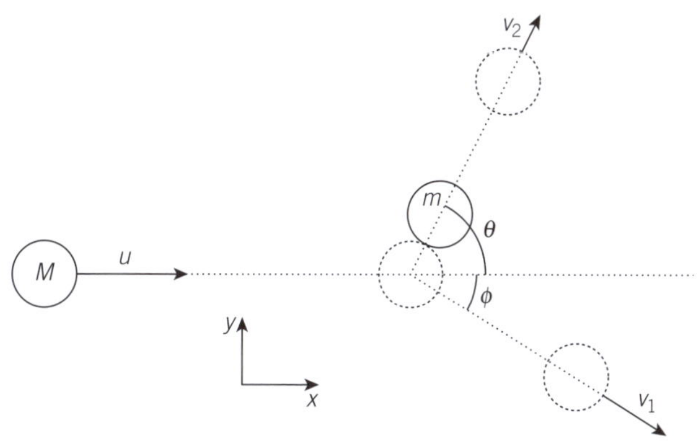

▲ **Figure 4.10** Conservation of momentum in two dimensions

Worked example

A snooker ball, travelling at a speed of $2.0\,\text{m s}^{-1}$, has an off-centre collision with an identical ball which is at rest, as shown in Figure 4.11. After the collision the first ball moves at an angle of $60°$ to its original direction at a speed of $0.8\,\text{m s}^{-1}$.

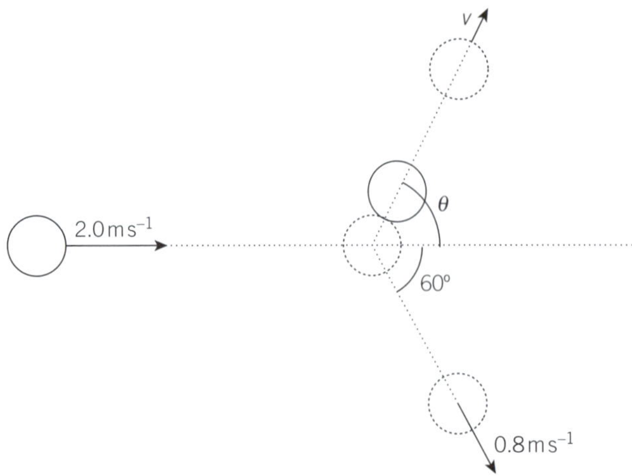

▲ **Figure 4.11** Conservation of momentum in two dimensions

a) Calculate the speed v and direction θ of the second ball after the collision.

b) State whether the collision is elastic or inelastic. Justify your answer.

Answer

a) Let the mass of each ball be m. Using the principle of conservation of momentum:

In the initial direction of travel of the first ball:

$$m \times 2.0 = (m \times 0.8 \cos 60) + (mv \cos \theta)$$

$$v \cos \theta = 1.6 \qquad \text{(equation 1)}$$

At $90°$ to the initial direction of the first ball:

$$0 = mv \sin \theta - m \times 0.8 \times \sin 60$$

$$v \sin \theta = 0.693 \qquad \text{(equation 2)}$$

Equation 2 divided by equation 1:

$$\frac{\sin \theta}{\cos \theta} = \frac{0.693}{1.6} = 0.4333$$

so $\qquad \theta = \tan^{-1}(0.433) = 23.4°$

Substituting this value into equation 1:

$$v = \frac{1.6}{\cos 23.4°} = 1.74\,\text{m s}^{-1}$$

b) Change in $E_k = \tfrac{1}{2}\,m \times 2.0^2 - \tfrac{1}{2}\,m \times 0.8^2 - \tfrac{1}{2}\,m \times 1.74^2 = 0.166\,m$

The collision is inelastic as some kinetic energy is lost.

★ **Exam tip**

'Justify' means 'give some evidence for' – in this case some calculation to show whether the collision is elastic or not.

∟ **Maths skills**

$$\frac{\sin \theta}{\cos \theta} = \tan \theta$$

Non-uniform motion

Free fall and air resistance

Figure 4.12 shows different stages after a skydiver jumps from an aeroplane, and Figure 4.13 summarises her motion in a graph.

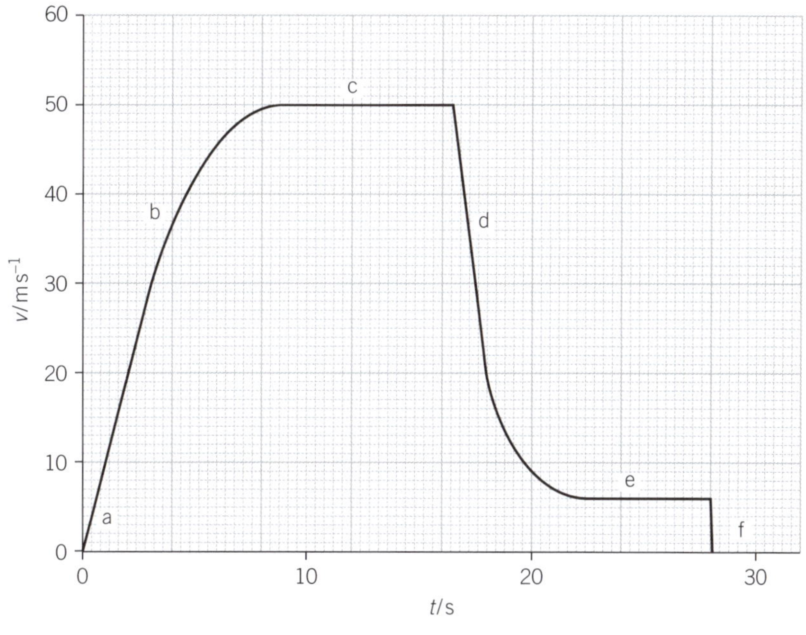

▲ **Figure 4.13** Velocity–time graph for a skydiver

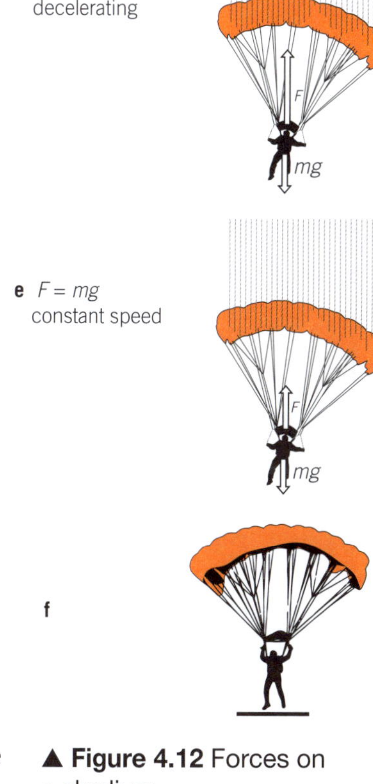

▲ **Figure 4.12** Forces on a skydiver

When she first leaves the aeroplane (Figure 4.12a) there is only one force acting on her – her weight mg, and so she accelerates towards the ground at $9.81\,\mathrm{m\,s}^{-2}$.

As her speed increases, she is hitting more air molecules every second, and so the upward air resistance force F increases (Figure 4.12b). The net, or resultant, force is still downward, and so she continues to accelerate downwards, but at a slower rate.

Eventually the skydiver is falling at such a high speed (about $50\,\mathrm{m\,s}^{-1}$) that her weight mg and the air resistance force F exactly balance each other (Figure 4.12c), and so there is no resultant force on her. She continues to fall at this speed, known as the terminal velocity.

When the skydiver opens her parachute, the air resistance force is suddenly greatly increased (Figure 4.12d). The upwards force is now greater than the weight force, and so she begins to decelerate.

After a short time she will reach a much slower speed (about $6\,\mathrm{m\,s}^{-1}$), where the air resistance force and the weight force balance again (Figure 4.12e), and she will continue to fall at this slower speed until she hits the ground (Figure 4.12f).

Raise your grade

(a) State the principle of conservation of momentum.

The total momentum of a system remains constant. ✔ ✘ [2]

> The statement is correct but insufficient – it is only true if there is **no resultant force** acting on the system.

(b) Explain what is meant by an *elastic* collision.

Energy is conserved. ✘ [1]

> An elastic collision is one in which the total **kinetic** energy remains constant.

(c) A ball of mass m is travelling with speed u in a straight line when it collides elastically with a stationary ball of mass $2m$, as shown below. After the collision, the smaller mass is moving with a velocity v_1 and the larger mass with a velocity v_2.

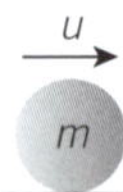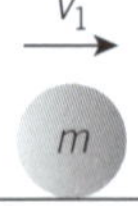

(i) Show that $v_2 = \dfrac{2u}{3}$.

Conservation of momentum: $mu = mv_1 + 2mv_2 \rightarrow u = v_1 + 2v_2$ (1) ✔

Elastic collision: velocity of separation = velocity of approach

$$v_2 - v_1 = u \qquad (2) ✔$$

Adding equations (1) and (2): $\quad 3v_2 = 2u \quad \rightarrow \quad v_2 = \dfrac{2u}{3}$ ✔

(ii) Calculate v_1.

From equation (1): $\quad v_1 = u - 2v_2 = u - 2 \times \dfrac{2u}{3} = -\dfrac{u}{3}$ ✔ [4]

(d) The ball of mass $2m$ then collides with a wall and bounces back with a speed of $\dfrac{u}{3}$. The collision with the wall lasts t seconds.

Calculate: **(i)** the change in momentum of the ball

Change in momentum $= 2m \times \dfrac{2u}{3} - 2m \times \dfrac{u}{3} = \dfrac{2mu}{3}$

> ✘ Momentum is a vector – the 'rebound' momentum is negative. The change in momentum is:
> $$2m \times \dfrac{2u}{3} - 2m \times \left(-\dfrac{u}{3}\right) = \dfrac{6mu}{3} = 2mu$$

(ii) the average force acting on the wall during the collision.

$$F = \dfrac{\Delta(mv)}{\Delta t} = \dfrac{\frac{2mu}{3}}{t} = \dfrac{2mu}{3t}$$

> ✔ Method is correct (error carried forward) (correct answer is $\dfrac{2mu}{t}$) [2]

(e) The momentum of the ball has changed. Explain how the principle of conservation of momentum still applies.

The wall (and the rest of the Earth it is connected to) gains momentum in the opposite direction – the total momentum of the ball and wall 'system' remains the same. ✔✔

> A good answer. [2]

? Exam–style questions

1 An electron in an electron 'gun' is accelerated from rest to a speed of $4.2 \times 10^7\,\text{m s}^{-1}$ over a distance of 15 mm. What is the force on the electron?

$[m_e = 9.11 \times 10^{-31}\,\text{kg}]$

A $5.4 \times 10^{-17}\,\text{N}$

B $1.1 \times 10^{-16}\,\text{N}$

C $5.4 \times 10^{-14}\,\text{N}$

D $1.1 \times 10^{-13}\,\text{N}$ [1]

2 A lift of mass $1.4 \times 10^3\,\text{kg}$ is ascending with an acceleration of $1.6\,\text{m s}^{-2}$. What is the tension in the cables supporting the lift?

A 2.2 kN **B** 11.5 kN

C 13.7 kN **D** 16.0 kN [1]

3 A box of mass 40 kg is pulled by a force of 600 N at an angle of 40° to the horizontal, as shown below. The friction force F is equal to half the weight of the box.

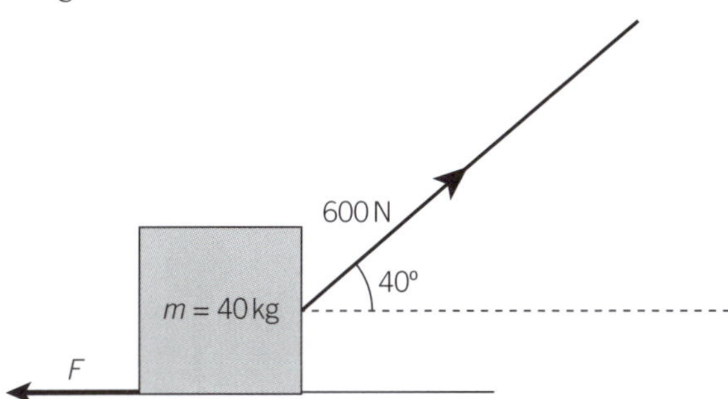

What is the acceleration of the box?

A $4.7\,\text{m s}^{-2}$ **B** $6.6\,\text{m s}^{-2}$

C $9.2\,\text{m s}^{-2}$ **D** $11.0\,\text{m s}^{-2}$ [1]

4 A mass of 1.4 kg moving with a velocity of $0.7\,\text{m s}^{-1}$ collides with a mass of 0.7 kg moving in the opposite direction with a speed of $0.2\,\text{m s}^{-1}$. After the collision, the 1.4 kg mass is moving with a speed of $0.3\,\text{m s}^{-1}$ in the same direction as before.

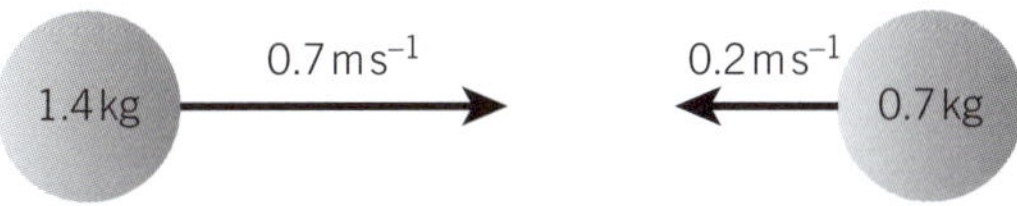

What is the speed of the 0.7 kg mass after the collision?

A $0.6\,\text{m s}^{-1}$ **B** $1.0\,\text{m s}^{-1}$

C $1.8\,\text{m s}^{-1}$ **D** $2.2\,\text{m s}^{-1}$ [1]

5 Two dynamics trolleys P and Q are initially at rest. The mass of trolley P is 0.90 kg; the mass of the trolley Q is unknown. They are 'exploded apart' by the release of a spring–loaded plunger at the front of one of the trolleys, and move off in opposite directions as shown below.

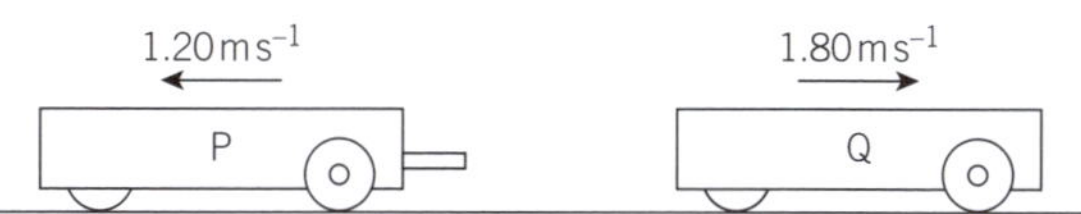

What is the mass of trolley Q?

A 0.60 kg **B** 0.74 kg **C** 1.35 kg **D** 1.67 kg [1]

6 (a) State the principle of conservation of momentum. [2]

(b) Curling is a sport in which players slide circular 'stones' on a sheet of ice towards a target. The stones each have a mass of 18 kg and a diameter of 19.0 cm.

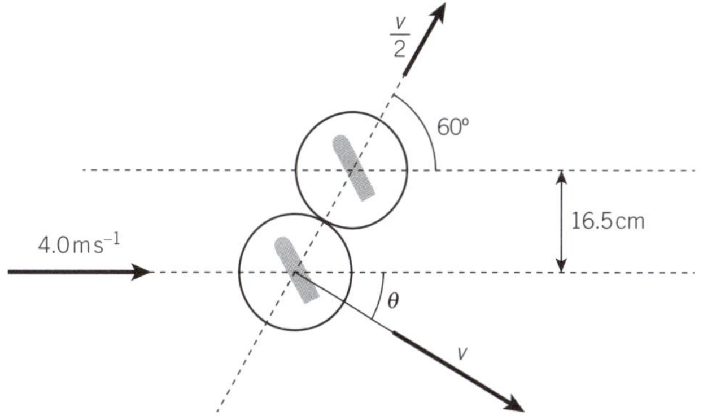

A stone is travelling at a speed of $4.0\,\text{m s}^{-1}$ when it collides with a second identical but stationary stone, as shown above.

(i) Calculate the momentum of the first stone before the collision.

(ii) Show that the stationary stone moves off at an angle of 60° to the original direction of travel of the first stone. [3]

(c) The speed v of the first stone after the collision is twice the speed of the second stone.

Calculate:

(i) θ, the angle the first stone is deflected

(ii) v. [4]

Key points

- ☐ Describe the force on a mass in a uniform gravitational field and on a charge in a uniform electric field.
- ☐ Understand the origin of the upthrust acting on a body in a fluid.
- ☐ Show a qualitative understanding of frictional forces and viscous forces including air resistance.
- ☐ Understand that the weight of a body may be taken as acting at a single point known as its centre of gravity.
- ☐ Define and apply the moment of a force.
- ☐ Understand that a couple is a pair of forces that only produces rotation.
- ☐ Define and apply the torque of a couple.
- ☐ State and apply the principle of moments.
- ☐ Understand that, when there is no resultant force and no resultant torque, a system is in equilibrium.
- ☐ Use a vector triangle to represent coplanar forces in equilibrium.
- ☐ Define and use density and pressure.
- ☐ Derive, from the definitions of pressure and density, the equation $\Delta p = \rho g \Delta h$ and use it to solve problems.

Types of force

Gravitational forces

Any mass in a gravitational field experiences a force. The **gravitational field strength** g is defined as the force on unit mass (1 kg). The force F on a mass m is:

$$F = mg$$

Near the Earth's surface the gravitational field is **uniform** – constant in both **magnitude** and **direction** – and has a value of $9.81\,\mathrm{N\,kg^{-1}}$. As $F = ma$, an object dropped near the surface of the Earth will have an acceleration of $9.81\,\mathrm{m\,s^{-2}}$ ignoring air resistance (see Figure 5.1). On the Moon $g = 1.6\,\mathrm{N\,kg^{-1}}$ so the acceleration of free fall on the Moon is $1.6\,\mathrm{m\,s^{-2}}$.

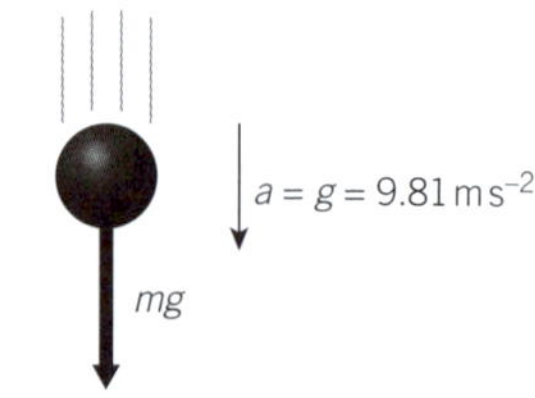

▲ **Figure 5.1** Gravitational field strength

Electrical forces

An electrically charged object in an electric field experiences a force. The **electric field strength** E is defined as the force per unit **positive** charge (1 C) on a stationary point charge.

The force on a charge q (see Figure 5.2) is:

$$F = Eq$$

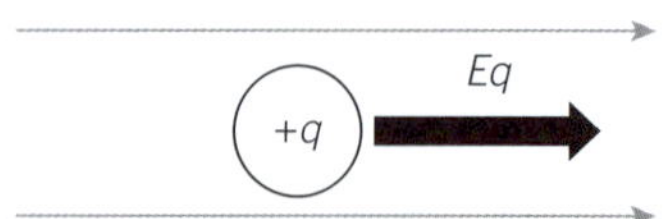

▲ **Figure 5.2** Electric field strength

Upthrust (buoyancy) forces

An object immersed in a fluid (liquid or gas) experiences an upward force, an **upthrust**, equal to the weight of the fluid that has been displaced. For example, a ball of radius r submersed in a liquid of density ρ has displaced a volume $V = \frac{4}{3}\pi r^3$ of liquid.

Volume of a sphere of radius $r = \frac{4}{3}\pi r^3$

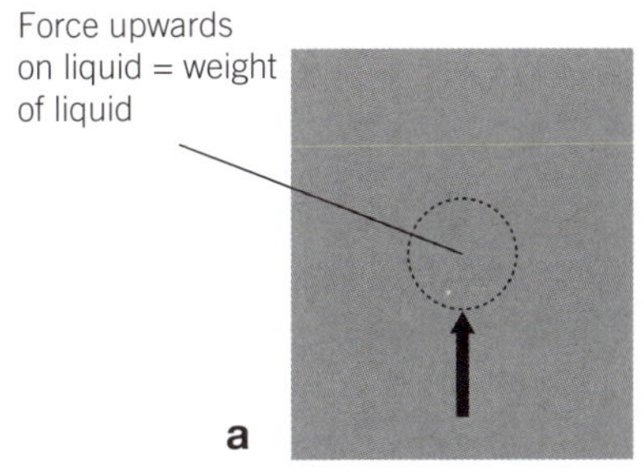

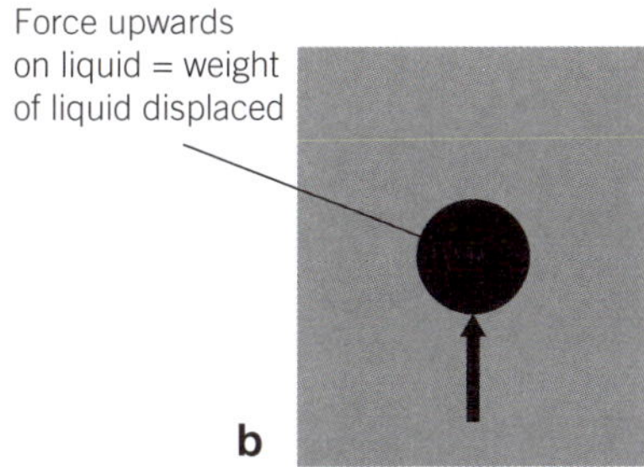

▲ **Figure 5.3** Upthrust (buoyancy) forces

The upward force on the liquid must have been equal to the weight of the liquid (see Figure 5.3a), as it was in equilibrium. This same force now acts on the ball (see Figure 5.3b) – the upthrust will be $V = \frac{4}{3}\pi r^3 \rho g$.

Frictional and viscous forces

Friction is a force which always opposes motion. If a book resting on a table is gently pushed one way, there is a friction force in the opposite direction and the book will remain at rest (see Figure 5.4a).

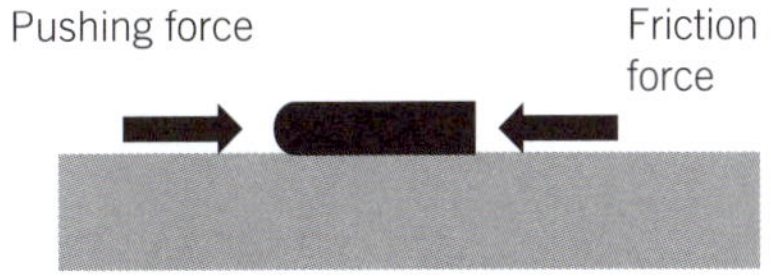

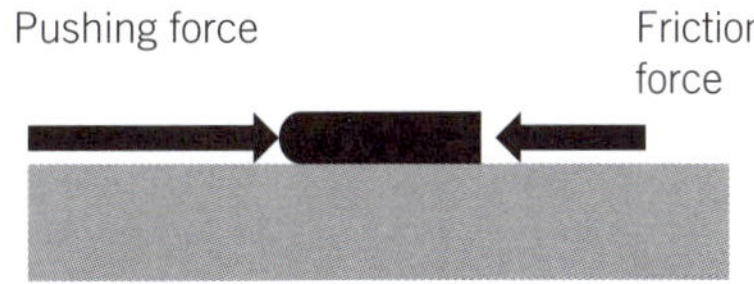

a In equilibrium (at rest)   **b** Beginning to accelerate

▲ **Figure 5.4** Frictional forces on a book

If the pushing force is gradually increased, the book will eventually slide along the table as the friction force reaches a maximum value (sometimes referred to as the **limiting** friction force). Once the book is moving, the friction force is called the **dynamic** or **sliding** friction force, and is usually less than the limiting friction force (see Figure 5.4b).

Objects moving through liquids or gases also experience resistive forces due to friction – these are usually referred to as **viscous (drag)** forces. A swimmer swimming in the sea, or a raindrop falling through the air, experiences viscous (drag) forces.

The size of the viscous drag depends on a range of factors including the **viscosity** of the fluid (its 'thickness' or 'stickiness') and the speed and shape of the object moving through the fluid.

The viscosity of gases is generally much less than the viscosity of liquids, but can still have significant effects. A skydiver falling from an aeroplane initially accelerates (Figure 5.5a) but as the speed of the skydiver increases so does the drag force. Eventually the drag force equals the weight of the skydiver, who then continues to fall at a constant speed, called the **terminal velocity** (Figure 5.5b).

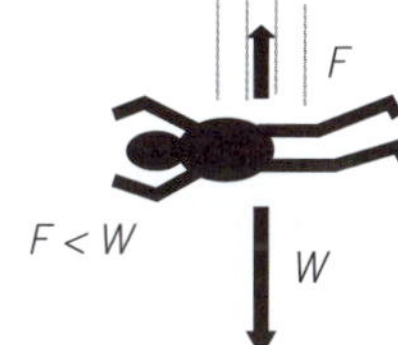

a acceleration

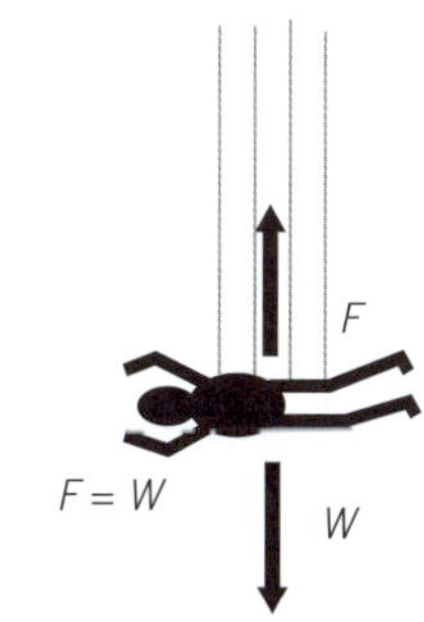

b constant speed

▲ **Figure 5.5** Forces in free fall

> ## Worked example
>
> A hot-air balloon ascends at constant speed (Figure 5.6).
>
> **a)** State the forces acting on the balloon and their directions.
>
> **b)** Derive an equation relating the forces you have identified.
>
> **Answer**
>
> **a)** The weight of the balloon W and the drag or friction force F both act downwards. The upthrust or buoyancy force B acts upwards.
>
> **b)** As the balloon is rising at constant speed, the net force on the balloon must be zero.
>
> $$W + F = B$$

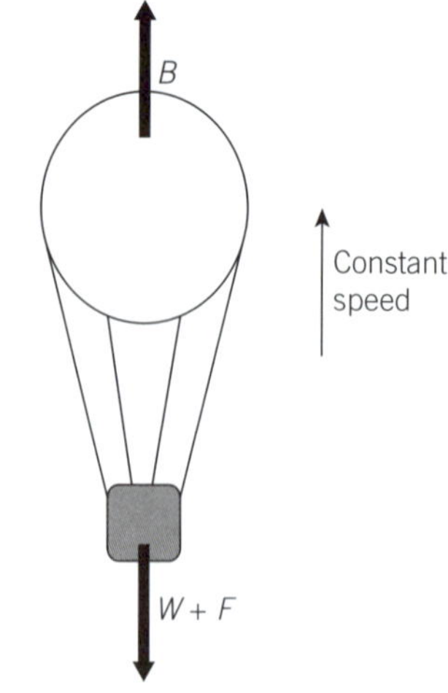

▲ **Figure 5.6** Drag and buoyancy

Centre of gravity

The **centre of gravity** of a body (an **object**) is the point where all the weight of the body can be considered to act. Knowledge of the position of the centre of gravity is helpful in assessing the stability of an object.

In Figure 5.7a the object is stable – the weight force is trying to rotate the object anticlockwise, returning it to an upright position.

In Figure 5.7b the object has been tilted further so that the line of action of the weight force is outside the right-hand edge of the object. The weight force is trying to rotate the object clockwise, causing it to topple over.

The turning effects of forces

The turning effect of a force is called the **moment** of the force.

> **Remember**
>
> The moment of a force about a point is the force multiplied by the **perpendicular** distance from the line of action of the force to the point.
>
> The SI units of the moment of a force are N m.

For a force F, a perpendicular distance d from a point P, the moment about P is Fd (see Figure 5.8).

The principle of moments

If an object is subjected to a number of forces, but is in equilibrium, the turning effects (moments) of each of the forces must balance out. This statement is known as the **principle of moments**.

> **Remember**
>
> Sum of the clockwise moments = sum of the anticlockwise moments

▲ **Figure 5.7** Centre of gravity

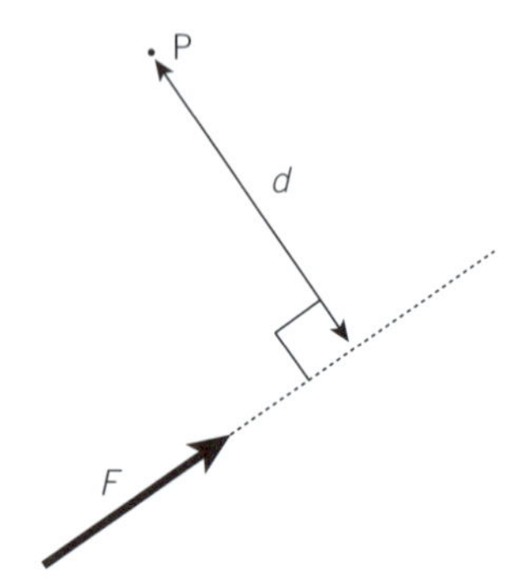

▲ **Figure 5.8** Moment of a force

Worked example

A horizontal beam is hinged at one end and supports a light fitting. The beam is held in place by a rope, as shown in Figure 5.9. Calculate the tension T in the rope.

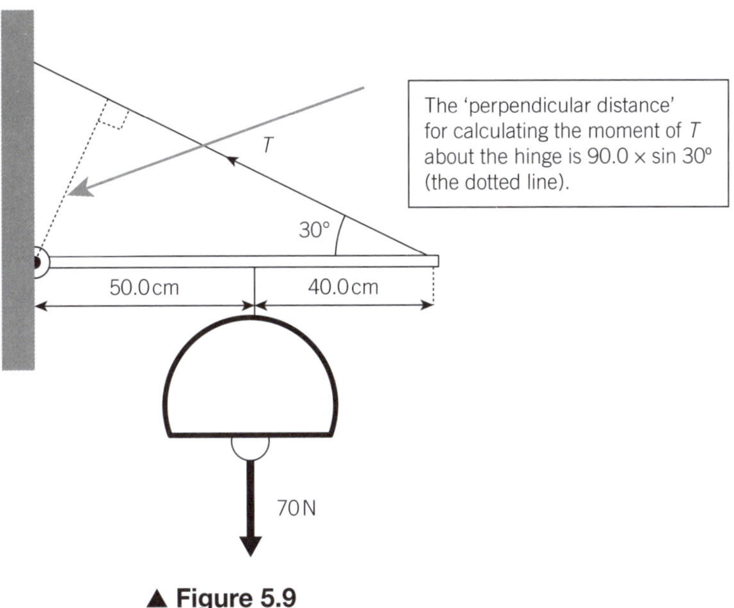

▲ **Figure 5.9**

Answer

Using the principle of moments about the hinge:

sum of the anticlockwise moments = sum of the clockwise moments

$$T \times (90.0 \times \sin 30°) = 70 \times 50.0$$

$$T = 77.8 \, \text{N}$$

Torque and couple

A **couple** is a pair of equal and opposite forces acting on a body, but not along the same line. A couple can only cause a body to rotate.

The **torque** of a couple is the total moment of the couple, and so has the same units as moments (N m).

The torque about point P is $F \times d$ (see Figure 5.10). (The torque is the same value, regardless of the position of P).

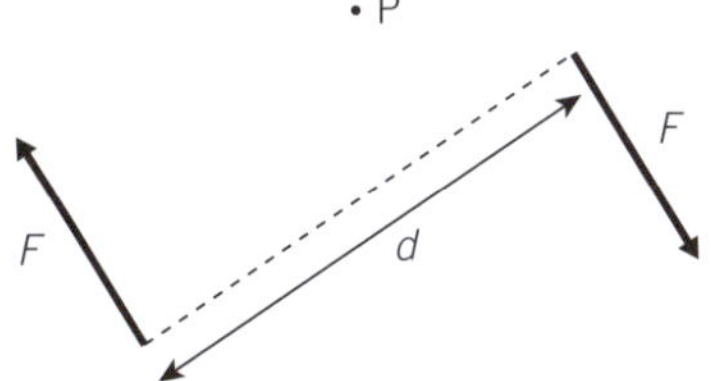

▲ **Figure 5.10** Torque of a couple

Equilibrium of forces

For a body to be in **equilibrium**, two conditions must be satisfied:

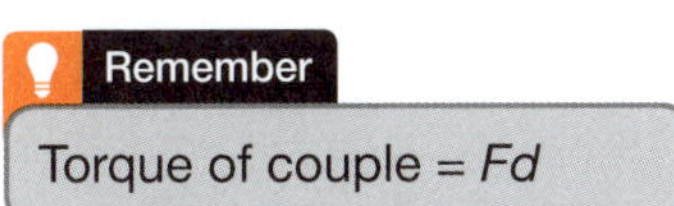

- The resultant force acting in any direction must be zero.
- The resultant torque about any point must be zero.

Worked example

A uniform ladder AB, of length 6.0 m and weight 100 N, rests against a smooth wall. The base of the ladder rests on the floor and is 2.0 m from the wall, as shown in Figure 5.11.

Calculate:

a) the angle the ladder makes with the wall

b) the force on the ladder from the wall

c) the size and direction of the force on the ladder from the floor.

Answer

The ladder is uniform, so the centre of gravity of the ladder is halfway along the ladder.

a) $\quad \sin\theta = \dfrac{2.0}{6.0} = \dfrac{1}{3} \qquad$ so $\quad \theta = \sin^{-1}\left(\dfrac{1}{3}\right) = 19.5°$

b) Taking moments about point B:

$$S \times 6\cos\theta = 100 \times 3\sin\theta$$

$$S = 50\tan\theta = 50\tan 19.5 = 17.7\,\text{N}$$

c) Resolving vertically: $\quad P = 100\,\text{N}$

   Resolving horizontally: $F = S = 17.7\,\text{N}$

The force at B has two components (see Figure 5.12):

The resultant force $R = \sqrt{(100^2 + 17.7^2)}$

$$= 101.6\,\text{N}$$

$$\text{angle } \phi = \tan^{-1}\left(\dfrac{17.7}{100}\right) = 10.0°$$

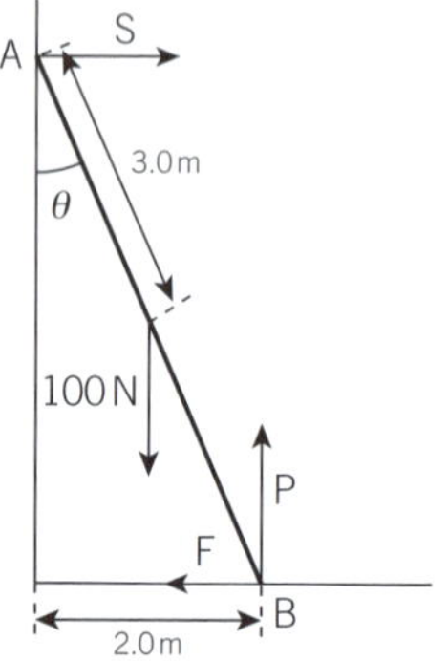

▲ Figure 5.11

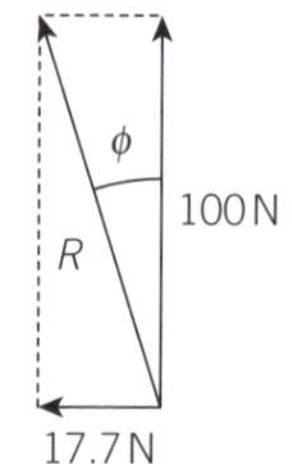

▲ Figure 5.12

Equilibrium under three forces

If an object acted on by three forces is in equilibrium, as shown in Figure 5.13a, the resultant of the three forces must be zero. The three forces, drawn as vectors, must form a triangle (see Figure 5.13b).

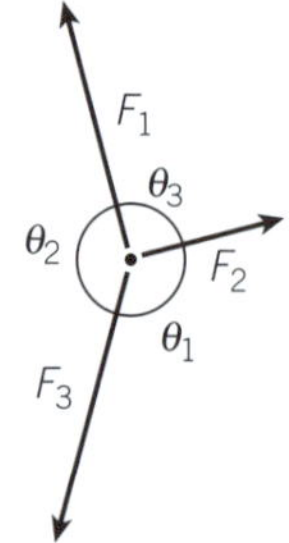

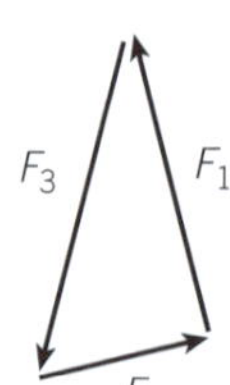

a Free-body diagram   **b** Vector triangle of forces

▲ **Figure 5.13** Equilibrium under three forces

'Three force' questions can be approached in a number of ways:

* by resolving forces in two directions and taking moments about a suitable point

* using Lami's theorem (see Figure 5.14)

* drawing a scale diagram of the triangle of forces.

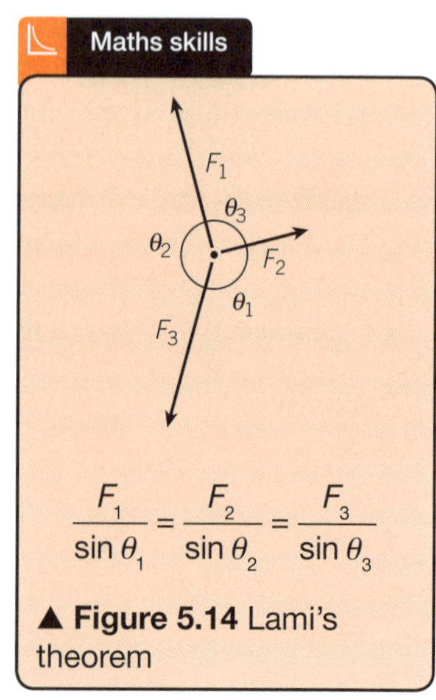

Worked example

A block of weight W rests on a rough slope making an angle θ with the horizontal, as shown in Figure 5.15. Calculate:

a) the friction force F **b)** the normal force N.

Method 1: Resolving forces

a) Resolving forces along the slope: $F = W \sin \theta$

Resolving forces perpendicular to the slope: $N = W \cos \theta$

Method 2: Using Lami's theorem

$$\frac{F}{\sin(180 - \theta)} = \frac{N}{\sin(90 + \theta)} = \frac{W}{\sin 90°}$$

so

a) $F = W \sin \theta$ **b)** $N = W \cos \theta$

Method 3: Drawing a scale diagram of the triangle of forces

See Figure 5.16.

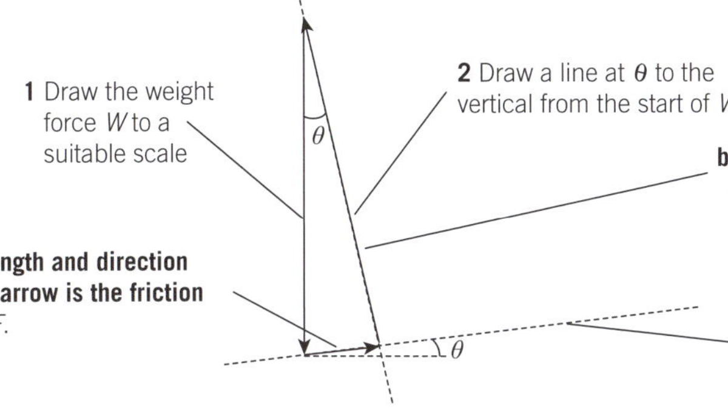

▲ **Figure 5.16** Drawing a scale diagram

▲ **Figure 5.15** Equilibrium under three forces

Math skills

$\sin(180 - \theta) = \sin \theta$

$\sin(90 + \theta) = \cos \theta$

Density and pressure

Density

The **density** ρ of a material is defined as the mass per unit volume. For an object of mass M and volume V: $\rho = \dfrac{M}{V}$

Worked example

The density of air is $1.29\,\text{kg m}^{-3}$. Estimate the mass of air in a school hall.

Answer

Estimated volume of school hall $= 35\,\text{m} \times 20\,\text{m} \times 8\,\text{m} = 5600\,\text{m}^3$

Mass M of air $= \rho V = 1.29 \times 5600 = 7200\,\text{kg}$

Remember

$$\text{density} = \frac{\text{mass}}{\text{volume}}$$

SI units are kg m^{-3}

Some densities that are useful to know:

air: $1.29\,\text{kg m}^{-3}$

water: $1.0 \times 10^3\,\text{kg m}^{-3}$

1 ml of water ($1\,\text{cm}^3$) has a mass of $1.0\,\text{g}$.

Pressure

The **pressure** p is defined as the normal force per unit area (see Figure 5.17):

$$p = \frac{F}{A}$$

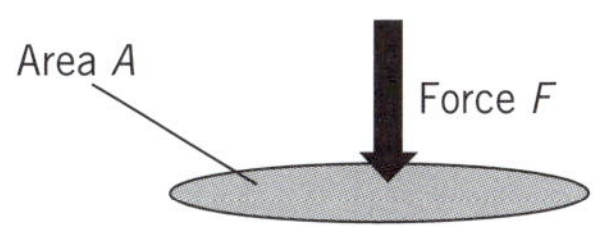

▲ **Figure 5.17** Pressure

Remember

$$p = \frac{F}{A}$$

SI units for pressure are N m^{-2} or **pascal** (Pa)

Worked example

A bar of gold has dimensions $16.0\,\text{cm} \times 5.0\,\text{cm} \times 2.5\,\text{cm}$ and a mass of $3.86\,\text{kg}$.

a) What is the density of gold?

b) What is the **maximum** pressure the bar can exert when placed on a table?

Answer

a) $\rho = \dfrac{M}{V} = \dfrac{3.86}{0.160 \times 0.050 \times 0.025} = 19.3 \times 10^3\,\text{kg m}^{-3}$

b) See Figure 5.18 for the orientation for maximum pressure.

$$p_{\text{max}} = \frac{F}{A} = \frac{3.86 \times 9.81}{0.050 \times 0.025} = 30.3\,\text{kPa}$$

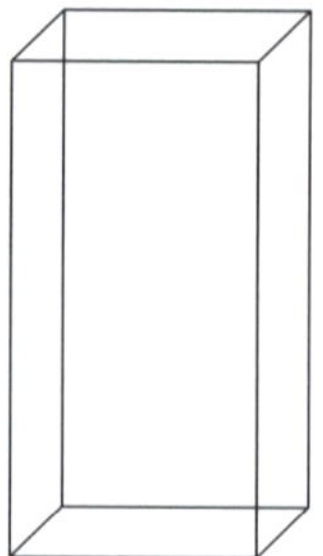

▲ **Figure 5.18** Orientation for maximum pressure

Pressure in liquids and gases

The pressure on the surface of a column of liquid is the atmospheric pressure p_0. Below the surface, the pressure increases with depth due to the weight of liquid above, as shown in Figure 5.19. For a column of liquid of depth Δh, cross-sectional area A and density ρ, the increase in pressure is Δp. For equilibrium:

$$\Delta p\, A = \rho(A\,\Delta h)\,g$$

$$\Delta p = \rho g \Delta h$$

The net upwards force due to the increased pressure at depth Δh is the buoyancy force discussed earlier.

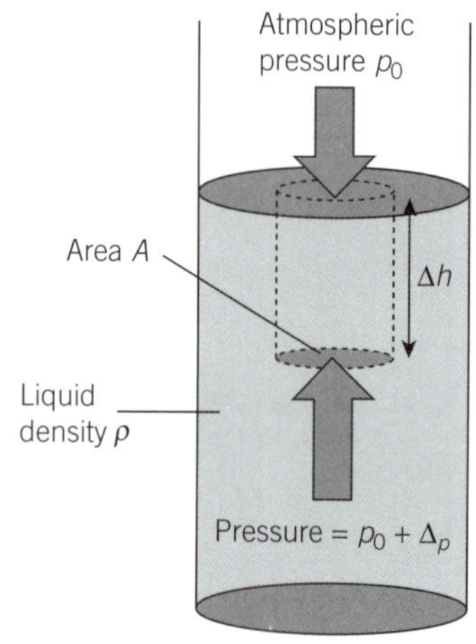

▲ **Figure 5.19** Pressure in liquids

Worked examples

1 Submersibles used for exploring the deepest oceans can dive to a depth of $6000\,\text{m}$. What is the **total** pressure on the outside of them?

[density of water $= 1.00 \times 10^3\,\text{kg m}^{-3}$, atmospheric pressure $= 1.0 \times 10^5\,\text{Pa}$]

Answer

Hydrostatic pressure p due to the weight of water:

$$p = \rho g \Delta h = 1.00 \times 10^3 \times 9.81 \times 6000 = 5.89 \times 10^7\,\text{Pa}$$

$$\text{total pressure} = \text{hydrostatic pressure} + \text{atmospheric pressure}$$

$$= 5.89 \times 10^7 + 1.0 \times 10^5 = 5.90 \times 10^7\,\text{Pa}$$

[Note: the atmospheric pressure is very small compared to the hydrostatic pressure, and can be ignored for most calculations.]

2 The atmospheric pressure at a height of $10\,\text{km}$ is $2.8 \times 10^4\,\text{Pa}$.

   a) A window of an aeroplane has an area $900\,\text{cm}^2$. Calculate the force on the window when the aeroplane is flying at this height, assuming that the air pressure inside the aeroplane is $1.0 \times 10^5\,\text{Pa}$.

   b) In which direction is this force?

Answer

a) Net force exerted on window $= \Delta p \times A$

$$= (1.0 \times 10^5 - 2.8 \times 10^4) \times 900 \times 10^{-4}$$

$$= 6.5\,\text{kN}$$

b) The force acts **outwards** from inside the aeroplane.

💡 **Remember**

$$\Delta p = \rho g \Delta h$$

You also need to be able to derive this equation.

⬆ Raise your grade

(a) State the two conditions necessary for a body to be in equilibrium.

The overall force acting on the body in any direction must be zero. ✔ ✘

> The candidate has stated the first condition for equilibrium correctly, but should also have stated that the total moment (or torque) acting on the body must be zero.

[2]

(b) A uniform beam AB, of length 1.400 m and weight 34 N, is attached by a hinge to a wall at A, as shown in the diagram. The beam is kept horizontal by a wire attached to the beam at C and the wall at D. A lantern of weight 26 N hangs from end B.

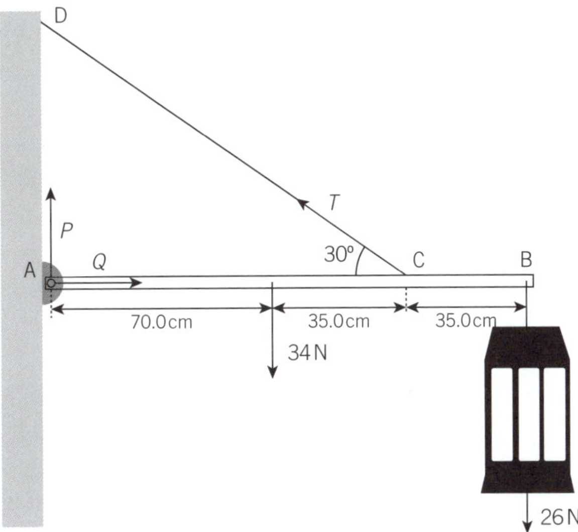

(i) Take moments about A to show that the tension T in the wire is 115 N.

$$T \times 1.05 \sin 30° = 34 \times 0.70 + 26 \times 1.40$$

✔ ✔

$$T = 115\,N$$

> The candidate has calculated the clockwise and anticlockwise moments correctly, and found the correct value for T.

[2]

(ii) Calculate the vertical force P exerted by the hinge.

Resolving forces vertically: $P + T \sin 30° = 34 + 26$ ✔

$$P = 60 + 115 \sin 30° ✘$$

$$P = 117.5\,N$$

vertical force P = ...117.5.... N   [2]

> The method for finding P is correct for the first mark, but the calculation of P is incorrect – the value of P should be $60 - 115 \sin 30 = 2.5\,N$.

(iii) Show that the horizontal force Q exerted by the hinge is 100 N.

Resolving forces horizontally: $Q = T \cos 30°$ ✔

$$= 115 \cos 30° = 99.6\,N$$

> A valid method for finding Q. An alternative method is to 'take moments' about point D.

[1]

(iv) Find the size and direction of the resultant force exerted by the hinge.

$$R^2 = 117.5^2 + 99.6^2 ✔$$

$$R = 154.0\,N$$

$$\tan \theta = \frac{117.5}{99.6} = 1.18$$ ✔

$$\theta = 49.7°$$

> A valid <u>method</u> for finding R, allowing error carried forward (e.c.f.) – the correct value is 99.6 N.

> A valid <u>method</u> for finding θ, again allowing for e.c.f.– the correct value is 1.4°.

e.c.f. (should be 2.5 N)

117.5 N
R
θ
99.6 N

[2]

1 Four forces act on a point.

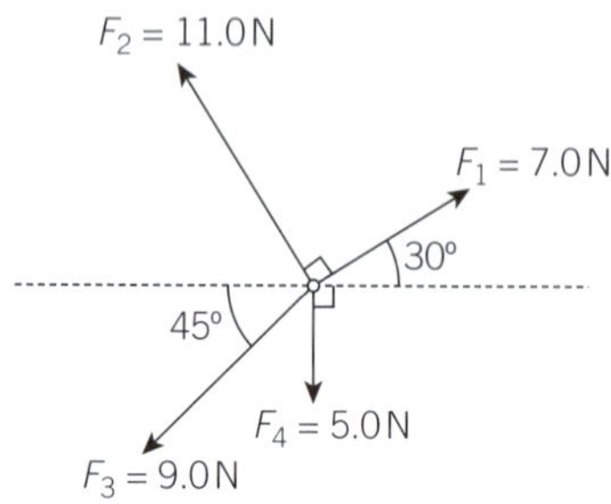

Which statement is **incorrect**?

A The vertical component of F_1 is 3.5 N.

B The horizontal component of F_2 is 9.5 N.

C The vertical component of F_3 is 6.4 N.

D The horizontal component of F_4 is 0.0 N.　　[1]

2 A load of 400 N is supported by two cables, as shown below. One cable is pulled horizontally with a force of 170 N.

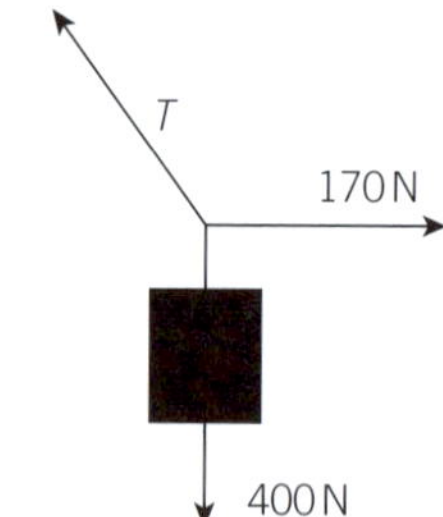

What is the best estimate of the tension in the other cable?

A 165 N　**B** 230 N　**C** 400 N　**D** 435 N　　[1]

3 A book is held upright by gripping it in the corner, between thumb and forefinger, as shown below.

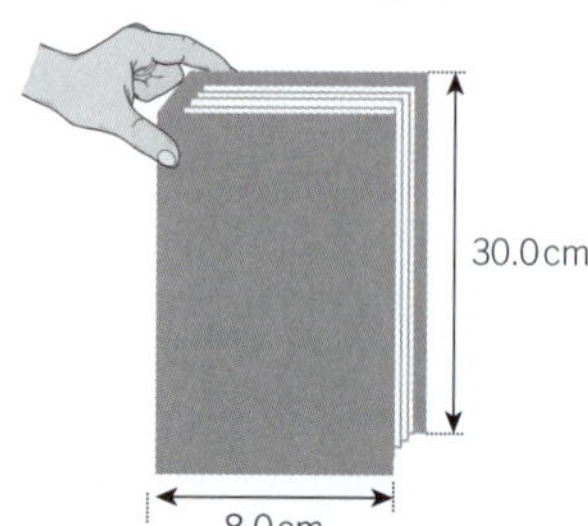

The book weighs 12.0 N. What is the torque applied by the hand to the book?

A 48.0 N cm clockwise

B 48.0 N cm anticlockwise

C 180.0 N cm clockwise

D 180.0 N cm anticlockwise　　[1]

4 A measuring cylinder contains oil to a depth of 6 cm floating on water. The depth of the water is 9 cm. The density of the oil is $900\,\mathrm{kg\,m^{-3}}$ and the density of water is $1000\,\mathrm{kg\,m^{-3}}$.

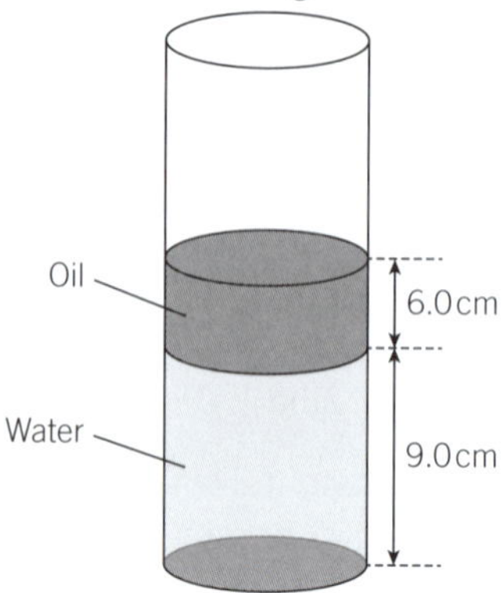

What is the pressure at the bottom of the cylinder due to the liquids?

A 1.38 kPa　**B** 1.41 kPa　**C** 1.87 kPa　**D** 2.21 kPa　　[1]

5 **(a)** Define the *moment* of a force.　　[1]

(b) State the *principle* of moments.　　[1]

(c) The drawing shows a beam QT, of length 1.5 m and negligible mass, supporting a load of 4.8 kN. The beam is held in equilibrium by cord PR. The cord is at an angle of 70° to the vertical and length TR is 1.0 m.

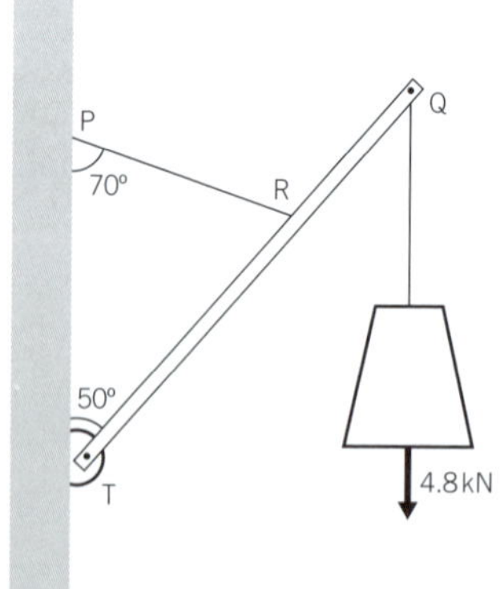

Calculate the tension in the cord.　　[3]

6 A cable-car is at rest supported by a cable, as shown below.

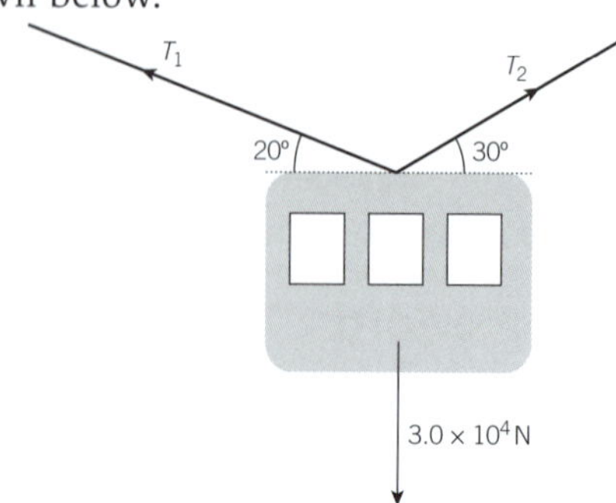

By drawing the triangle of forces to a suitable scale, find the tensions T_1 and T_2 in the cable.　　[3]

6 Work, energy, and power

Key points

- [ ] Give examples of energy in different forms, its conversion and conservation, and apply the principle of conservation of energy to simple examples.
- [ ] Understand 'work' as the product of a force and displacement in the direction of the force.
- [ ] Calculate the work done in different situations, including the work done by a gas expanding against a constant external pressure: $W = p\Delta V$.
- [ ] Recall and understand that the efficiency of a system is $\dfrac{\text{useful energy output}}{\text{total energy input}}$
- [ ] Appreciate the implications of energy losses in devices and use the concept of efficiency to solve problems.
- [ ] Derive, from the equations of motion, the equation for kinetic energy $E_k = \dfrac{1}{2}mv^2$ and apply the equation.
- [ ] Distinguish between gravitational potential energy and elastic potential energy.
- [ ] Understand and use the relationship between force and potential energy in a uniform field to solve problems.
- [ ] Derive, from the equation $W = Fs$, the formula $\Delta E_p = mg\Delta h$ for potential energy changes near the Earth's surface.
- [ ] Recall and use the formula $\Delta E_p = mg\Delta h$ for potential energy changes near the Earth's surface.
- [ ] Define power as the work done per unit time $\left(P = \dfrac{W}{t}\right)$ and derive $P = Fv$.
- [ ] Solve problems using the relationships $P = \dfrac{W}{t}$ and $P = Fv$.

Energy conversion and conservation

Energy

Energy is needed to make objects move, to lift them up, or to make them warmer. It can be described as the ability to do work or change temperature, and can take many forms including heat, light, sound, movement (kinetic energy), and electrical energy. There are various forms of **potential energy** (energy which is stored), including elastic potential energy (e.g., of a stretched rubber band), gravitational potential energy (the energy a mass has by virtue of its position in a gravitational field), and electrical potential energy (energy a charged particle has because of its position in an electric field).

Principle of conservation of energy

 **Remember**

Energy cannot be created or destroyed, only converted from one form to another.

An electric motor converts electrical energy to movement, together with some heat and sound energy. A candle converts chemical energy into light and heat. A solar cell converts light energy into electrical energy.

Work and energy types

Work

Work is done when a force moves in the direction of the force (Figure 6.1). When work is done, energy is transferred, perhaps as kinetic energy, gravitational potential energy, heat or sound.

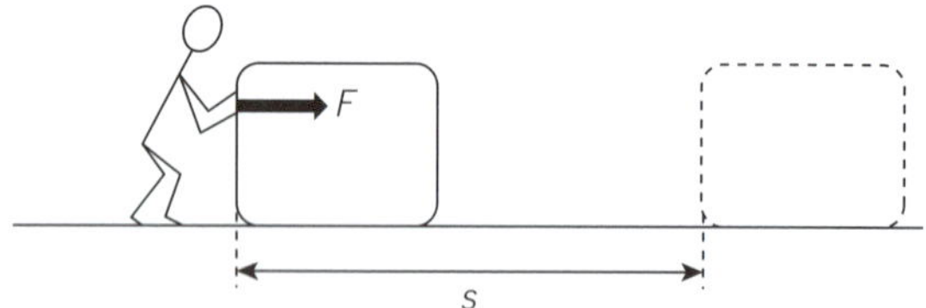

▲ **Figure 6.1** Work done and energy transferred

The work done (in joules) is the force (in newtons) multiplied by the distance moved in the direction of the force (in metres).

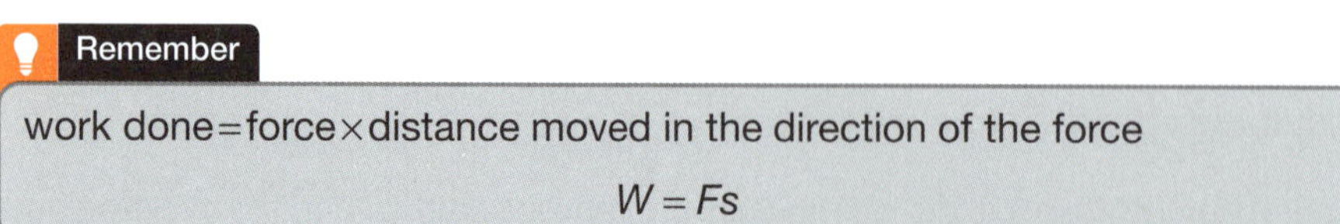

Remember

work done = force × distance moved in the direction of the force

$$W = Fs$$

Types of energy

Table 6.1 shows different types of energy and their descriptions.

▼ **Table 6.1** Forms of energy

Energy	Description
Gravitational potential energy	Energy a mass has due to its position in a gravitational field (work must be done to 'lift' a mass against a gravitational field)
Electrical potential energy	Energy a charged object has due to its position in an electric field (work must be done to move a positively charged object in the opposite direction to an electric field)
Elastic potential (strain) energy	Energy stored in an object or material due to deformation (e.g., stretching or compressing a spring)
Kinetic energy	The energy a mass has due to its speed
Internal energy	The combined kinetic and potential energies of all the particles in a body
Chemical and nuclear energy	The energy that can be released during chemical or nuclear reactions

Link

See Unit 9 *Deformation of solids* for more about elastic potential energy.

See Unit 12 *Thermal properties of materials* for more on internal energy and the first law of thermodynamics.

See Unit 17 *Electric fields* for more on electrical potential energy.

Potential energy and kinetic energy

Gravitational potential energy

When a load is lifted, work is done on the load, and the load gains gravitational potential energy. The force needed to just lift a mass m is mg. If the mass is lifted a vertical height Δh (see Figure 6.2):

change in gravitational potential energy of the mass ΔE_p = work done
$$= mg \times \Delta h$$

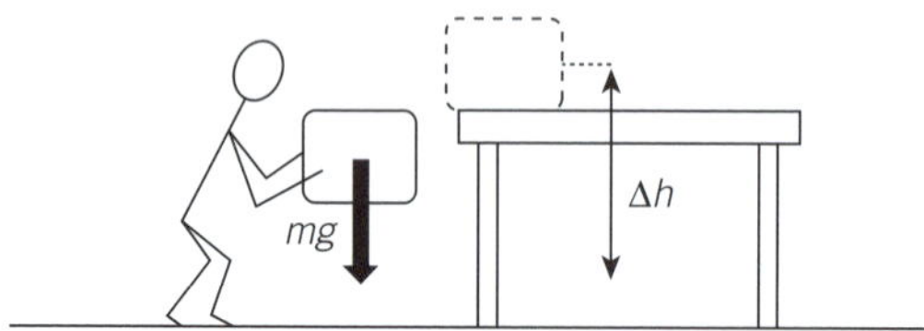

▲ **Figure 6.2** Gravitational potential energy

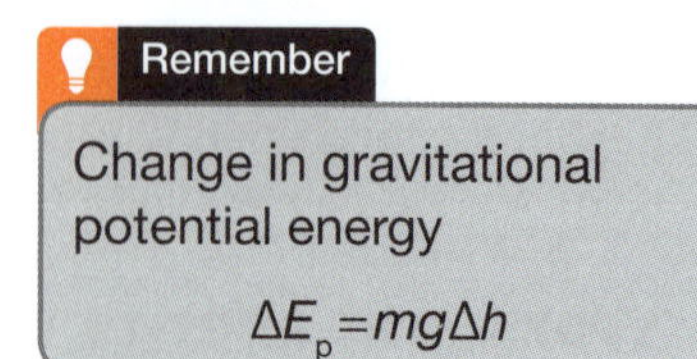

Remember

Change in gravitational potential energy

$$\Delta E_p = mg\Delta h$$

Kinetic energy

If work is done on an object that is free to move, the object will accelerate and gain kinetic energy E_k (Figure 6.3).

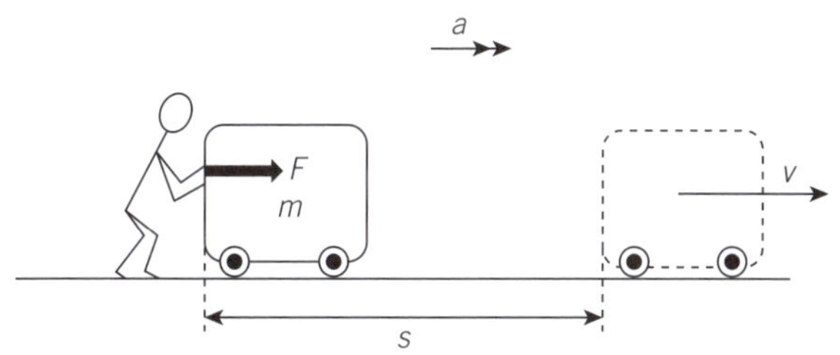

▲ **Figure 6.3** Kinetic energy

$$\text{work done} = E_k = Fs$$

using Newton's second law, $F = ma$:

$$E_k = (ma)s$$

For uniform acceleration:

$$v^2 = u^2 + 2as.$$

So for an object starting from rest ($u = 0$):

$$as = \frac{v^2}{2}$$

so

$$E_k = m\left(\frac{v^2}{2}\right) = \frac{1}{2}mv^2$$

> **Remember**
>
> When calculating work done, kinetic energy, or change in gravitational potential energy, remember that quantities such as mass m, change in height Δh and speed v must all be in SI units for the calculated value to be in joules.

> **Remember**
>
> $$E_k = \frac{1}{2}mv^2$$
>
> You also need to be able to derive this equation.

Worked examples

[Take g, the acceleration of free fall, as $9.81\,\text{m}\,\text{s}^{-2}$.]

1 A child pulls a sledge a horizontal distance of 50 m with a force of 240 N at an angle of 30° to the horizontal, as shown in Figure 6.4.

How much work does the child do?

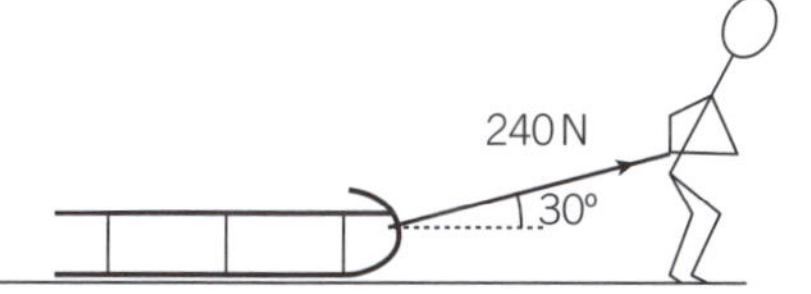

▲ **Figure 6.4** Child pulling a sledge

Answer

$$\text{work done} = (240\cos 30°) \times 50 = 10.4\,\text{kJ}$$

2 A human flea has a mass of 0.45 mg and can leap vertically with an initial velocity of $90\,\text{cm}\,\text{s}^{-1}$, reaching a height of 3.5 cm (see Figure 6.5).
  a) Calculate:
    i) the initial kinetic energy of the flea
    ii) the change in gravitational potential energy of the flea.
  b) Why are the answers to **a) i)** and **a) ii)** not the same?

> 240 cos 30° is the horizontal component of the force – only this force does work. The vertical component does no work as the sledge does not move vertically.

Answer

a) **i)** $E_k = \dfrac{1}{2}mv^2 = \dfrac{1}{2} \times (0.45 \times 10^{-6}) \times (90 \times 10^{-2})^2 = 1.8 \times 10^{-7}\,\text{J}$

  ii) $\Delta E_p = mg\,\Delta h = (0.45 \times 10^{-6}) \times 9.81 \times (3.5 \times 10^{-2}) = 1.5 \times 10^{-7}\,\text{J}$

b) Air resistance causes some energy to be 'lost' as heat.

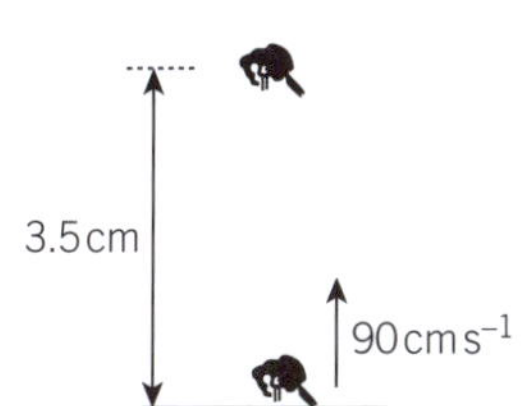

▲ **Figure 6.5** Human flea jumping

Work done by an expanding gas

When a gas expands, it is applying a force against its surroundings and doing work because the force applied by the gas is moving. Figure 6.6 shows a gas at pressure p inside a cylinder pushing against a piston of surface area A.

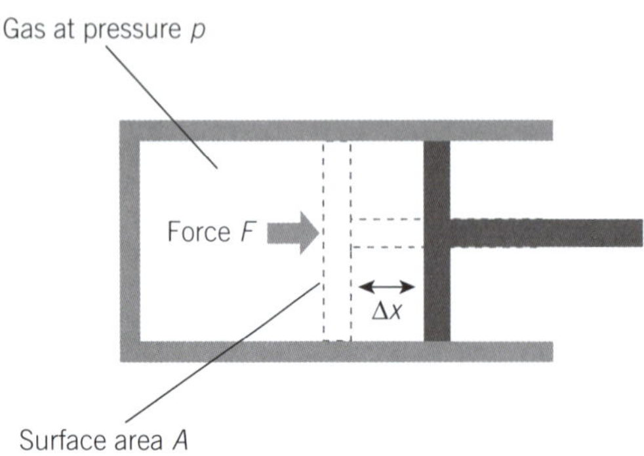

▲ **Figure 6.6** Work done by a gas

If the piston moves a small distance Δx (small enough for the pressure not to alter significantly) the work done by the gas ΔW is:

$$\Delta W = F\Delta x = pA\,\Delta x = p\,\Delta V$$

where ΔV is the change in volume of the gas.

> ### Worked example
>
> A gas at a constant pressure of 2.0×10^5 Pa is cooled so that its volume decreases from 2.5 m³ to 1.8 m³. Calculate the work done by the gas.
>
> **Answer**
>
> $\Delta W = p\,\Delta V = 2.0\times10^5\times(1.8 - 2.5) = -1.4\times10^5$ J

Efficiency

A light bulb is designed to convert electrical energy into light energy, a car engine to convert chemical energy into kinetic energy, and a nuclear power station to change nuclear (atomic) energy into electricity. None of these processes are 100% efficient. In each example, some energy is 'lost' ('wasted') as unwanted forms of energy such as heat or sound energy (see Figure 6.7).

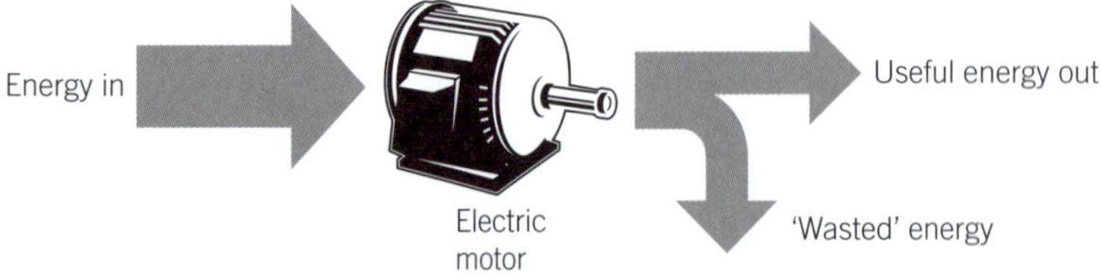

▲ **Figure 6.7** Efficiency of an electric motor

Worked example

A hydraulic jack lifts one wheel of a car, applying a force of 3.6 kN to to one of the wheels when the handle of the jack is pressed down with a force of 150 N. Each time the handle is moved down 20.0 cm, the car wheel rises 0.5 cm, as shown in Figure 6.8.

a) Calculate the efficiency of the hydraulic jack.

b) Explain why the hydraulic jack is not 100% efficient.

Answer

a) $\text{Efficiency} = \dfrac{\text{useful energy output}}{\text{total energy input}} \times 100\%$

$$= \frac{3.6 \times 10^3 \times 0.5 \times 10^{-2}}{150 \times 20.0 \times 10^{-2}} \times 100\%$$

$$= 60\%$$

b) Energy is 'lost' as heat and sound energy caused by friction between the moving parts of the jack. The platform of the hydraulic jack has mass so gains some potential energy when the wheel is lifted.

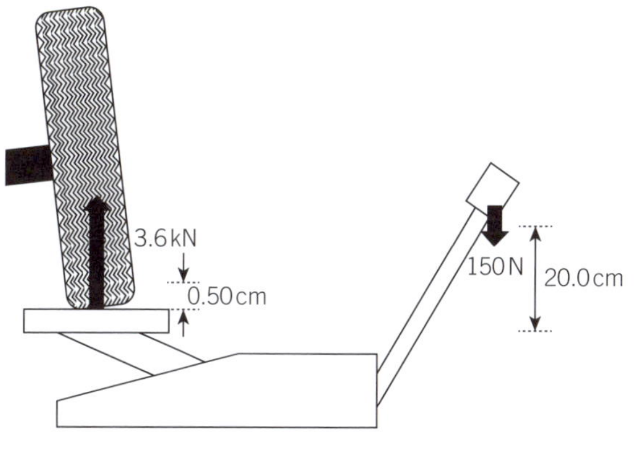

▲ **Figure 6.8** Hydraulic jack

Change in potential energy in a gravitational field

A mass m experiences a force mg when it is in a gravitational field, where g is the gravitational field strength, as shown in Figure 6.9.

In order to move the mass a small distance Δx in the opposite direction to the gravitational field (small enough for the gravitational field strength g not to change significantly), an amount of work $mg\Delta x$ must be done on the mass. The gravitational potential energy of the mass increases by the same amount:

$$\Delta E_p = mg\,\Delta x$$

If the gravitational field is **uniform** (g is constant):

$$\Delta E_p = mg\,\Delta h$$

where Δh is the total distance moved in the opposite direction to the gravitational field.

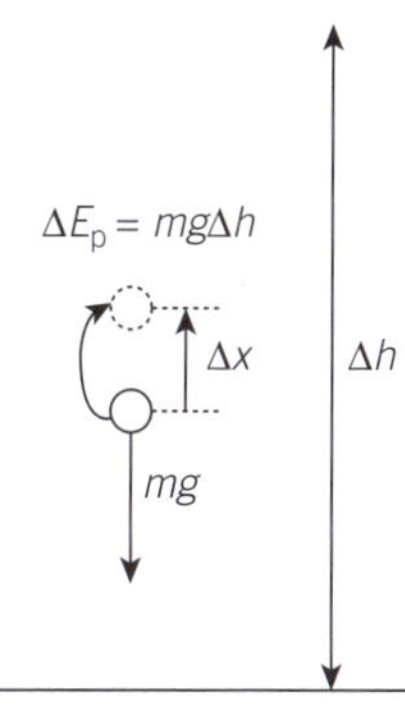

▲ **Figure 6.9** Gravitational field and potential energy

Change in potential energy in an electric field

A charge $+q$ experiences a force Eq when placed in an electric field of strength E, as shown in Figure 6.10.

In order to move the charge a small distance Δx in the opposite direction to the electric field (small enough for the electric field strength E not to change significantly), an amount of work $Eq\Delta x$ must be done on the charge. The electrical potential energy of the charge increases by the same amount.

$$\Delta E_p = Eq\,\Delta x$$

If the electric field is uniform (E is constant):

$$\Delta E_p = Eqd$$

where d is the total distance moved in the opposite direction to the electric field.

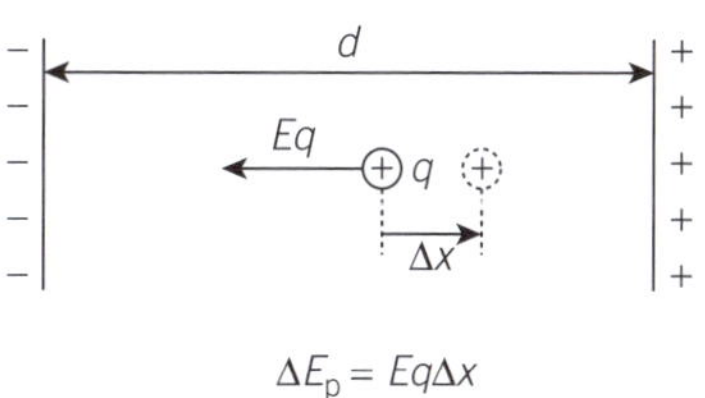

▲ **Figure 6.10** Electric field and potential energy

★ **Exam tip**

Notice that the symbol E represents electric field strength, but E_p represents potential energy.

Power

Power is the rate of doing work (or transferring energy).

$$\text{power (W)} = \frac{\text{work done (J)}}{\text{time taken (s)}}$$

Using the equation for work done: $W = Fs$, gives:

$$\text{power} = \frac{Fs}{t} = F\left(\frac{s}{t}\right) = Fv$$

Worked examples

1 Gravel falls onto a conveyor belt at the rate of $5.0\,\text{kg}\,\text{s}^{-1}$ as shown in Figure 6.11. The conveyor belt is moving at a constant speed of $3.0\,\text{m}\,\text{s}^{-1}$.

  a) What is the power of the electric motor needed to drive the conveyor belt?

  b) Why is the answer to **a)** an underestimate of the true value?

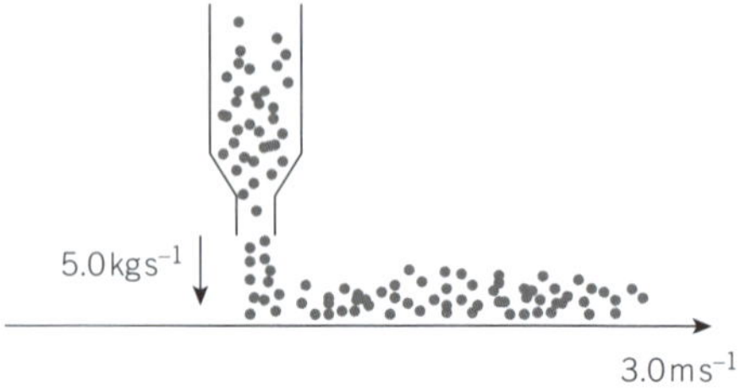

▲ **Figure 6.11** Conveyor belt

Answer

  a) Each second $5.0\,\text{kg}$ of gravel is accelerated to a speed of $3.0\,\text{m}\,\text{s}^{-1}$. The minimum power required is the kinetic energy gained by the gravel each second.

  $$\text{power} = \frac{1}{2} \times 5 \times 3^2 = 22.5\,\text{W}$$

  b) The conveyor belt is not 100% efficient. Energy is lost as heat and sound energy caused by the frictional forces acting on the moving belt.

2 A $1\,\text{kW}$ electric motor is used to lift a load of $300\,\text{N}$ with the aid of a simple pulley system, as shown in Figure 6.12. A force of $200\,\text{N}$ applied by the motor will just lift the load.

  a) Explain why the force needed to lift the load is less than $300\,\text{N}$.

  b) Calculate the efficiency of the pulley system.

  c) The motor lifts the load $8.0\,\text{m}$ in $5.0\,\text{s}$. What is the overall efficiency of the motor and pulley system? State any assumptions you make.

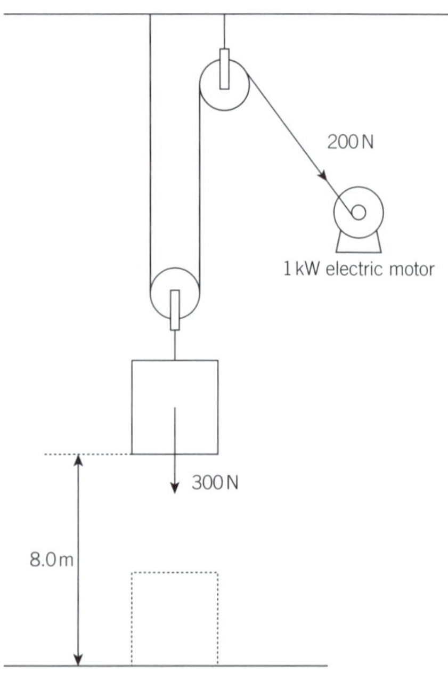

▲ **Figure 6.12** Pulley system

Answer

  a) To lift the load $1.0\,\text{m}$ the motor has to pull the rope a distance of $2.0\,\text{m}$. The work done by the motor is $200 \times 2.0 = 400\,\text{J}$; the work done on the load is only $300 \times 1.0 = 300\,\text{J}$.

  b) For every $1.0\,\text{m}$ the load is raised:

  $$\text{efficiency} = \frac{\text{useful energy output}}{\text{total energy input}} \times 100\% = \frac{300}{400} \times 100 = 75\%$$

  c) $\text{output power} = \text{useful energy out per second} = \dfrac{300 \times 8.0}{5.0} = 480\,\text{W}$

  $\text{input power} = 1 \times 10^3\,\text{W}$

  $$\text{overall efficiency} = \frac{\text{useful power output}}{\text{total power input}} = \frac{480}{1 \times 10^3} \times 100\% = 48\%$$

↑ **Raise your grade**

(a) Explain what is meant by *work done*.

work done = force × distance moved ✗

> Work done = force × distance moved **in the direction of the force.**

.... [1]

(b) Distinguish between *gravitational potential energy* and *elastic potential energy*.

Gravitational P.E. is the stored energy a mass has due to its position ✗

> Answer should include reference to **gravitational field** e.g add '…its position in a gravitational field'.

Elastic P.E. is energy stored because something has been stretched ✔

.... [2]

(c) The picture shows part of a rollercoaster ride.

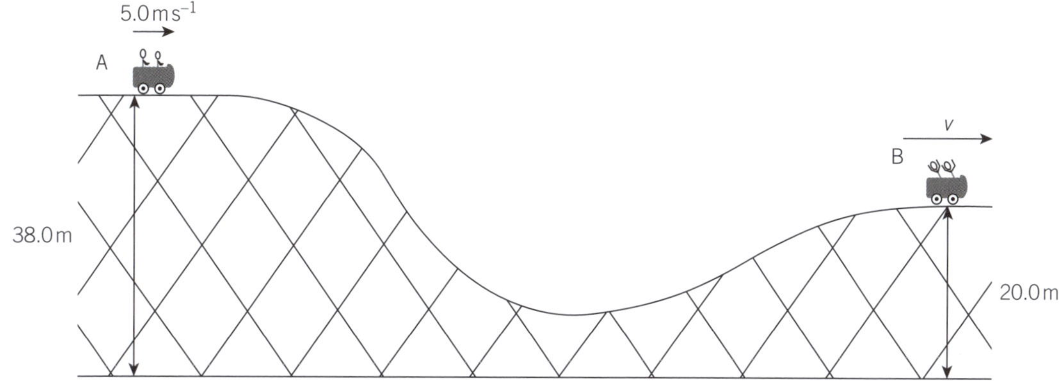

The carriage and passengers, of total mass 500 kg, is moving horizontally at A with a speed of 5.0 m s⁻¹. Air resistance is negligible and other friction forces can be ignored.

Show that the change in gravitational potential energy of the carriage and passengers, as the carriage moves from A to B, is 88 kJ.

$\Delta E_p = mg\Delta h = 500 \times 9.81 \times (38.0 - 20.0) = 88$ kJ ✔   Method.

.... [1]

(d) Calculate:

> The kinetic energy the carriage already had at A should be added to this value.

(i) the kinetic energy of the carriage and passengers at B.

Potential energy lost = kinetic energy gained = 88 kJ ✗ ✔

.... [2]

> Uses principle of conservation of energy.

kinetic energy = ...88... kJ

> (Correct value is 95 kJ.)

(ii) the speed v of the carriage and passengers at B.

$$\frac{1}{2}mv^2 = 88 \times 10^3 \rightarrow v = \sqrt{\frac{2 \times 88 \times 10^3}{500}} = 18.8 \text{ m s}^{-1}$$ ✔ ✔

speed = ...19... m s⁻¹ [2]

Correct method used. ················· Correct calculation allowing for e.c.f. ··············

(e) In fact, air resistance and other frictional forces are significant. The speed of the carriage at B is 13 m s⁻¹. The length of the track from A to B is 30 m. Calculate the average frictional force acting on the carriage and passengers as it moves from A to B.

$Fs = \Delta E_p - \Delta E_k = 88 \times 10^3 - \frac{1}{2} \times 500 \times (13^2 - 5^2) = 5.2 \times 10^4$ ✔   Correct method.

$$F = \frac{5.2 \times 10^4}{30} = 1.7 \text{ kN}$$ ✔   Correct calculation.

average frictional force = ...1.7×10^3... N [2]

1 A uniform square paving stone, of dimensions 40.0 cm × 40.0 cm × 8.0 cm, has a mass of 30 kg and is lying flat on the ground.

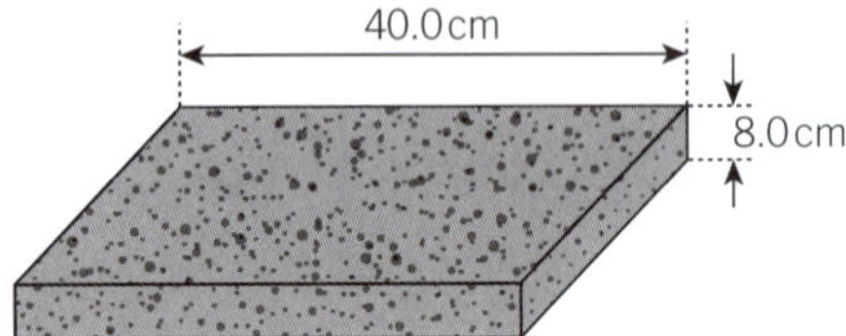

How much work is needed to stand the paving stone on its end?

A 47 J   **B** 59 J

C 106 J   **D** 118 J   [1]

2 Starting from rest, a skier skis down a 30° slope of length 25 m. Ignoring frictional forces, what is her speed at the end of the slope?

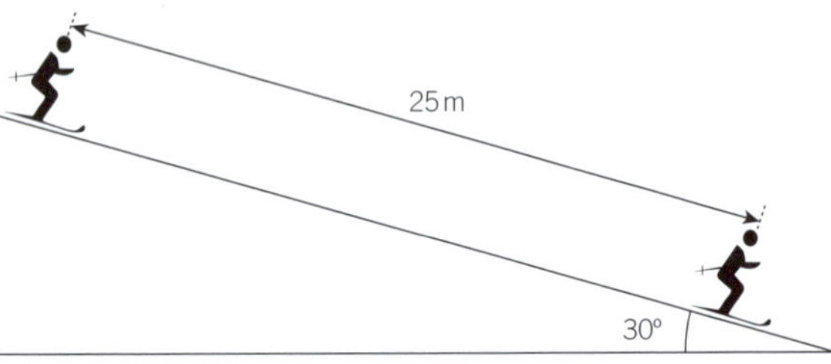

A 16 m s^{-1} **B** 21 m s^{-1}

C 24 m s^{-1} **D** 31 m s^{-1} [1]

3 A constant force F is applied to an object which is initially at rest.

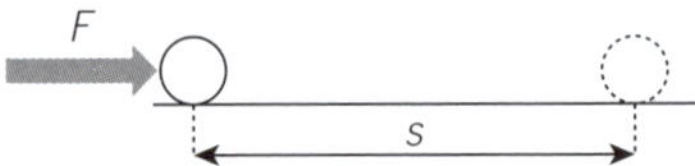

Which graph shows the variation of W, the work done by the force, against s, the distance travelled by the object?

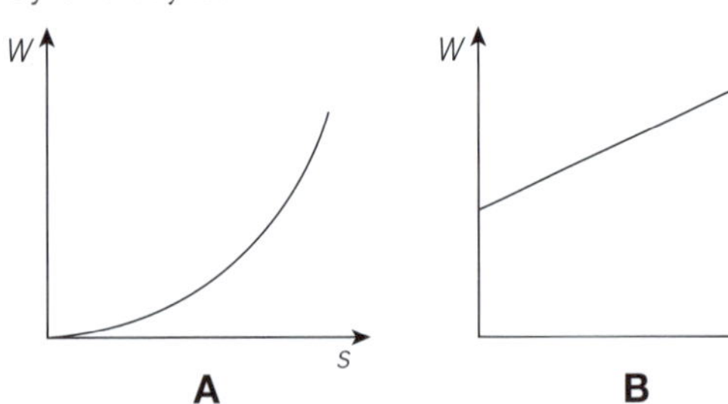

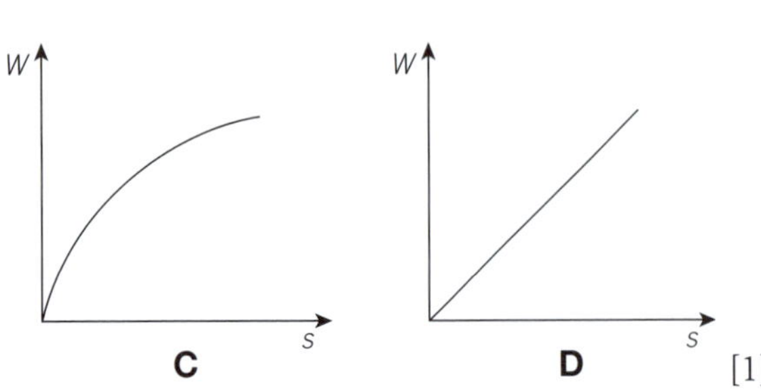

[1]

4 (a) Distinguish between *gravitational potential energy* and *electrical potential energy*.   [2]

(b) A tennis ball of mass 60 g is dropped from a height of 4.0 m onto level ground.

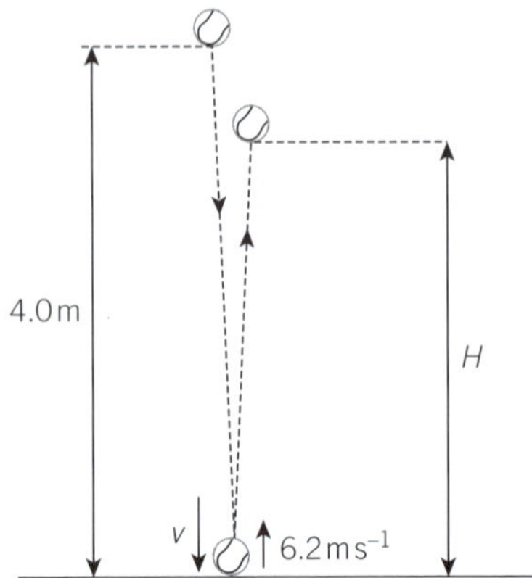

Ignoring air resistance, calculate:

(i) the gravitational potential energy lost by the ball when it reaches the ground

(ii) the speed v of the ball just before it hits the ground.   [3]

(c) The ball rebounds with a speed of 6.2 m s^{-1}. Calculate:

(i) the fraction of energy 'lost' in the collision

(ii) the maximum height H reached by the ball after the first bounce.   [3]

5 (a) (i) Define *power*.

(ii) Use your definition to show that:
power = force × velocity   [3]

(b) A lorry of mass 3000 kg moves at a speed of 15 m s^{-1} along a horizontal road. A resistive force of 2.4 kN acts on the lorry.

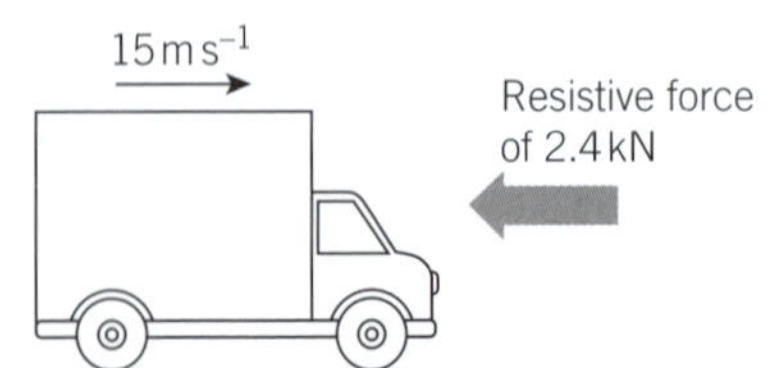

The lorry accelerates at 0.50 m s^{-2}. Calculate:

(i) the driving force produced by the lorry's engine

(ii) the output power of the engine.   [3]

Key points

- ☐ Define the radian and express angular displacement in radians.
- ☐ Understand and use angular speed to solve problems.
- ☐ Recall and use $v = r\omega$ to solve problems.
- ☐ Understand the centripetal acceleration for motion in a circle at constant speed.
- ☐ Recall and use the centripetal acceleration equations $a = r\omega^2$ and $a = \dfrac{v^2}{r}$.
- ☐ Describe how motion in a curved path needs a perpendicular force.
- ☐ Recall and use the centripetal force equations $F = mr\omega^2$ and $F = \dfrac{mv^2}{r}$.

Kinematics of uniform circular motion

Radians and angular displacement

The **angular displacement** θ of an object moving in a circular path is usually measured in **radians** (Figure 7.1).

For a complete circle (360°), $l = 2\pi r$, so $\theta = 2\pi$ radians:

$$1 \text{ radian} = \frac{360}{2\pi} = 57.3°$$

Angular velocity

The **angular velocity** ω of an object is the rate of change of angular displacement with time:

$$\omega = \frac{\Delta\theta}{\Delta t}$$

where $\Delta\theta$ is the angle 'swept' by the radius in time Δt and is measured in radians per second (rad s^{-1} or just s^{-1}). The linear speed v is related to the angular speed ω by the equation:

$$v = r\omega$$

For an object moving with constant angular speed ω around a circle of radius r (see Figure 7.2), the time for one complete revolution is T, where:

$$T = \frac{2\pi r}{v} = \frac{2\pi}{\omega}$$

> **Remember**
>
> $$\theta = \frac{\text{arc length}}{\text{radius}} = \frac{l}{r}$$

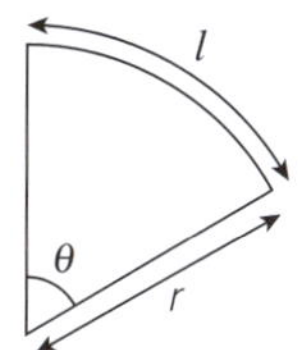

▲ **Figure 7.1** Radian measure

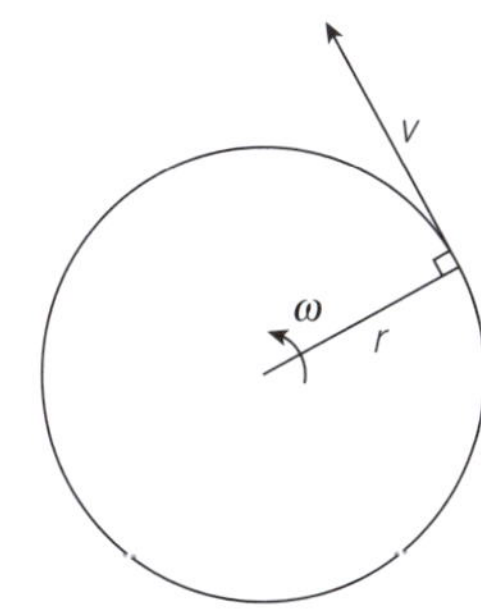

▲ **Figure 7.2** Angular velocity

Worked example

The approximate distance from the Earth to the Moon is 3.84×10^8 m; the Moon takes 27.3 days to orbit the Earth.

Calculate:

a) the angular velocity ω of the Moon around the Earth

b) the average speed v of the Moon.

Answer

a) $\omega = \dfrac{2\pi}{T} = \dfrac{2\pi}{27.3 \times 24 \times 60 \times 60} = 2.66 \times 10^{-6} \text{ rad s}^{-1}$

b) $v = r\omega = 3.84 \times 10^8 \times 2.66 \times 10^{-6} = 1.02 \text{ km s}^{-1}$

> **Remember**
>
> $$v = r\omega$$
> $$T = \frac{2\pi}{\omega}$$

Centripetal acceleration and centripetal force

Centripetal acceleration

Velocity is a **vector** quantity – it has both magnitude and direction. Since an object moving in a circle is constantly changing direction, its velocity is constantly changing. It is **accelerating**.

From Figure 7.3, for an object moving a small angular displacement $\Delta\theta$ in a small time Δt, the change in velocity Δv is:

$$\Delta v = 2v\sin\left(\frac{\Delta\theta}{2}\right) = v\,\Delta\theta \qquad \text{(as } \Delta\theta \text{ is very small)}$$

So the **centripetal acceleration** a is:

$$a = \frac{\Delta v}{\Delta t} = \frac{v\Delta\theta}{\Delta t} = v\omega$$

Substituting $v = r\omega$, the centripetal acceleration a can also be written:

$$a = r\omega^2 = \frac{v^2}{r}$$

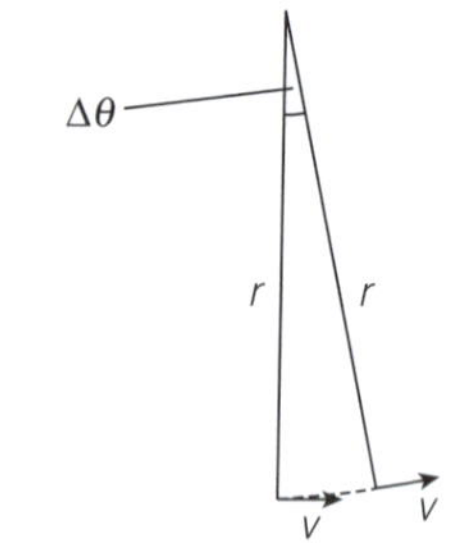

▲ **Figure 7.3** Centripetal acceleration

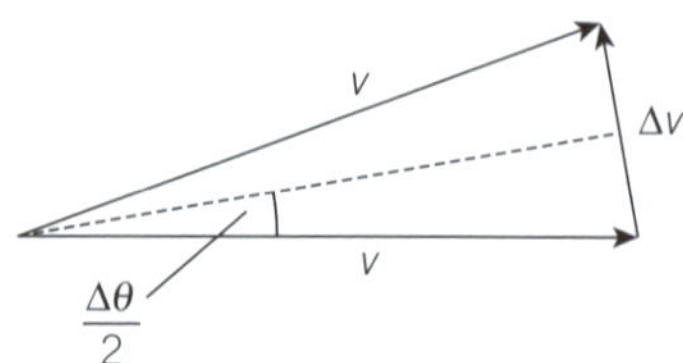

Maths skills

For small angles:

$$\sin\theta = \theta$$

where the angle θ is measured in **radians**.

Remember

$$a = r\omega^2 = \frac{v^2}{r}$$

Worked example

A wind turbine (see Figure 7.4) rotates at 14 revolutions per minute (r.p.m.). The diameter of the turbine is 70 m.

    a) Calculate the angular velocity of the turbine in rad s⁻¹

    b) Determine the centripetal acceleration of a point:
      i) at the end of one of the turbine blades
      ii) at the midpoint of one of the turbine blades.

Answer

    a) $\omega = 14 \text{ r.p.m.} = \dfrac{14 \times 2\pi}{60} = 1.47 \text{ rad s}^{-1}$

    b) **i)** $a = r\omega^2 = 35 \times 1.47^2 = 75.6 \text{ m s}^{-2}$

        ii) $a = r\omega^2 = 17.5 \times 1.47^2 = 37.8 \text{ m s}^{-2}$

Centripetal force

From $F = ma$, a force is needed to accelerate an object. If an object is moving along a curved path it must be accelerating because its velocity (a vector) is changing direction, even if its speed is constant.

The force needed to keep an object of mass m moving in a circle of radius r with constant speed v is:

$$F = ma = mr\omega^2 = \frac{mv^2}{r}$$

The direction of the acceleration is towards the centre of the circle so the force must act towards the centre of the circle (Figure 7.5).

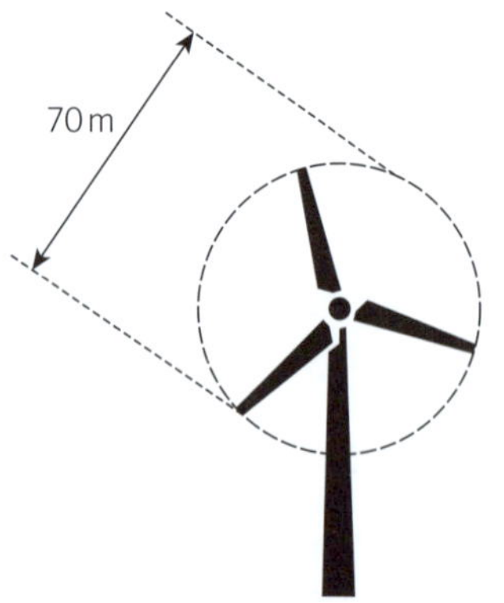

▲ **Figure 7.4** Wind turbine

Remember

$$F = mr\omega^2 = \frac{mv^2}{r}$$

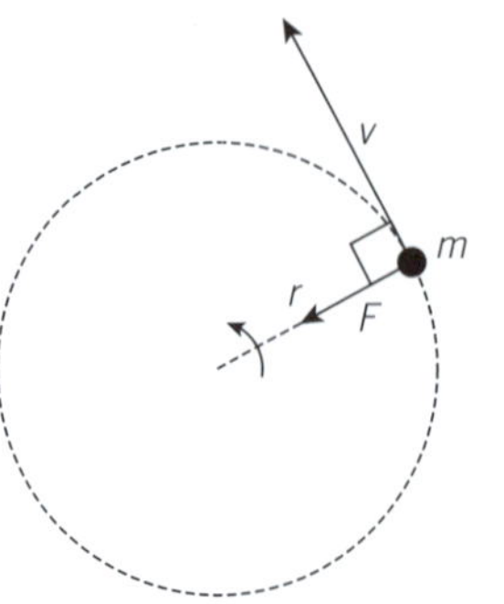

▲ **Figure 7.5** Centripetal force

Worked examples

1 The international space station ISS (Figure 7.6) orbits the Earth at a height of 400 km above the Earth's surface, taking 92 minutes to complete one orbit. The radius of the Earth is 6370 km.

   a) Calculate:

      i) the angular velocity of the space station

      ii) the centripetal acceleration of the space station.

   b) The mass of the space station is 4.2×10^5 kg.

      i) Calculate the centripetal force acting on the space station.

      ii) What provides this force?

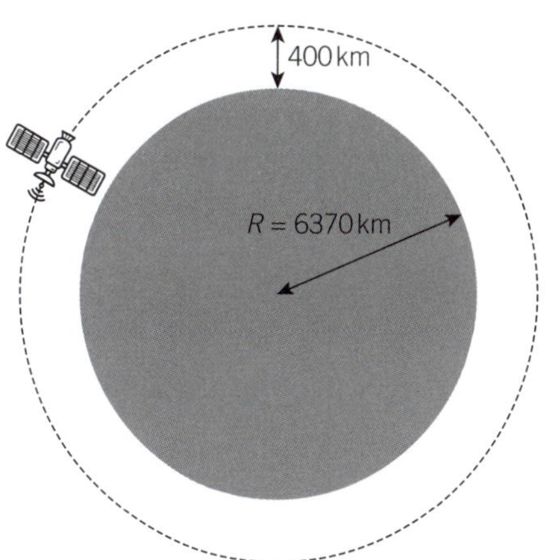

▲ **Figure 7.6** International Space Station orbiting the Earth

Answer

   a) **i)** $\omega = \dfrac{2\pi}{T} = \dfrac{2\pi}{92 \times 60} = 1.14 \times 10^{-3}$ rad s^{-1}

      ii) $a = r\omega^2 = (6.37 \times 10^6 + 4 \times 10^5) \times (1.14 \times 10^{-3})^2 = 8.80$ m s^{-2}

   b) **i)** $F = ma = 4.2 \times 10^5 \times 8.80 = 3.70 \times 10^6$ N

      ii) The gravitational pull of the Earth on the space station keeps it in orbit around the Earth.

2 A ball of mass m connected to a string of length l is whirled in a vertical circle at a constant speed v, as shown in Figure 7.7.

   a) Explain why the ball is accelerating even though it is travelling at constant speed.

   b) Calculate the tension in the string:

      i) when the ball is at its lowest point

      ii) when the ball is at its highest point.

   c) Determine the minimum velocity needed for the ball to continue to travel in a circular path.

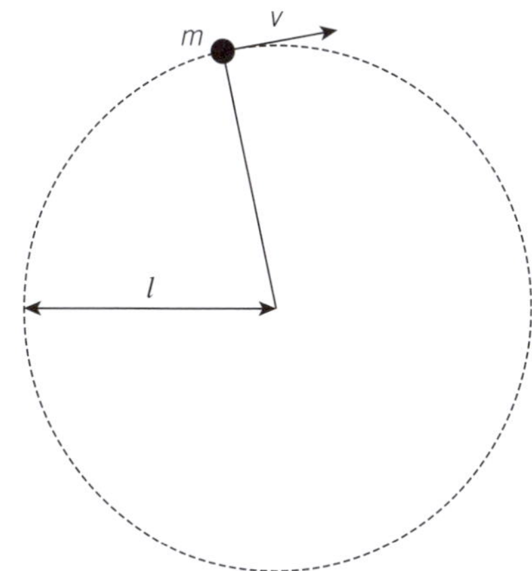

▲ **Figure 7.7** Vertical rotation of a ball

Answer

   a) Acceleration is the rate of change of velocity – a vector quantity, which has both magnitude and direction. As the direction of the ball is changing continuously, the velocity is also changing. It is accelerating.

   b) **i)** At the lowest point (see Figure 7.8):

$$T - mg = \frac{mv^2}{r}$$

$$T = mg + \frac{mv^2}{r}$$

      ii) At the highest point (see Figure 7.9):

$$T + mg = \frac{mv^2}{r}$$

$$T = \frac{mv^2}{r} - mg$$

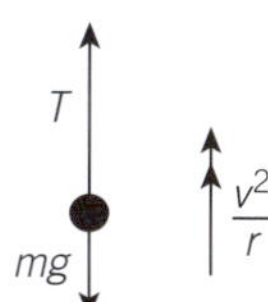

▲ **Figure 7.8**

   c) From b) ii), the tension in the string cannot be less than zero (otherwise the string will be slack and the ball will be in free fall).

$$T = \frac{mv^2}{r} - mg > 0 \quad \text{so} \quad v^2 > gr$$

$$v > \sqrt{(gr)}$$

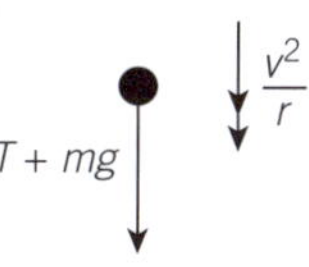

▲ **Figure 7.9**

 # Raise your grade

A conical pendulum consists of a small ball of mass 50 g suspended on a string and rotated in a circle of radius 0.40 m at a constant speed of 90 r.p.m., as shown. [The value of g is 9.81 m s^{-2}.]

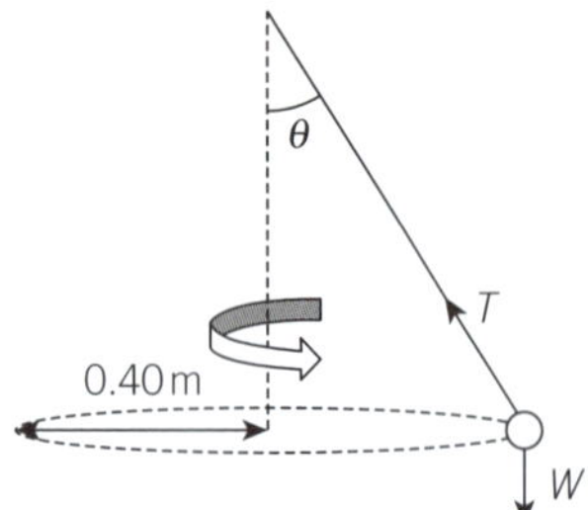

(a) (i) Show that the angular velocity ω of the sphere is 9.4 rad s^{-1}.

$$\omega = 90 \text{ r.p.m.} = \frac{90 \times 2\pi}{60} = 9.4 \text{ rad s}^{-1} \checkmark$$

Correct method and calculation. [1]

(ii) Hence show that the centripetal acceleration of the ball is 35 m s^{-2}.

$$a = r\omega^2 = 0.40 \times 9.4^2 = 35.3 \text{ m s}^{-2} \checkmark$$

Correct method and calculation. [1]

(b) (i) Calculate the centripetal force acting on the ball.

$$F = ma = (50 \times 10^{-3}) \times 35.3 = 1.77 \checkmark \times$$

Correct method and calculation, but units omitted (should be N), so loses second mark.

centripetal force = 1.77 [2]

(ii) State the direction of this force.

Outwards from the centre of the circle $\times$

Incorrect statement – 'centripetal' means 'towards the centre'. [1]

(c) Show that the angle θ is approximately 75°.

$$\rightarrow : \quad T\sin\theta = mr\omega^2 = 50 \times 10^{-3} \times 0.40 \times 9.4^2$$

$F = ma$ applied in horizontal direction correctly.

$$= 1.77 \checkmark$$

$$\uparrow : \quad T\cos\theta = W = mg = 50 \times 10^{-3} \times 9.81 \checkmark$$

Forces in vertical direction resolved correctly.

$$= 0.491$$

$$\tan\theta = \frac{T\sin\theta}{T\cos\theta} = \frac{1.77}{0.491} = 3.60 \checkmark$$

$$\theta = 74.5° \checkmark$$

Correct value for θ. [4]

? Exam-style questions

[The value of g is $9.81\,\mathrm{m\,s^{-2}}$.]

1 (a) Express the following angles in radians:

(i) 60°   (ii) 250°   (iii) 95°   [3]

(b) Express the following radian measures in degrees:

(i) $\dfrac{\pi}{6}$ (ii) $\dfrac{3\pi}{4}$ (iii) $\dfrac{9\pi}{7}$ [3]

2 An electric fan rotates at 800 r.p.m. The distance from the tip of a blade of the fan to the centre is 30.0 cm.

Calculate:

(a) the angular speed of the fan in $\mathrm{rad\,s^{-1}}$

(b) the linear speed of the tip of a blade.   [2]

3 A simple pendulum consists of a small ball of mass 80.0 g tied to a thin string of length 0.60 m.

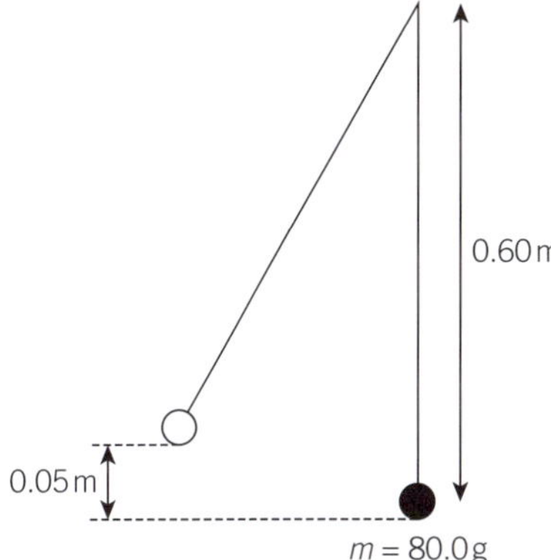

The ball is pulled to one side so that it is raised 0.05 m above its lowest position and released.

Calculate:

(a) the speed of the ball when it reaches its lowest position

(b) the maximum tension in the string.

4 A ball of mass m is connected to a string of length l and whirled around to form a conical pendulum moving with angular velocity ω in a horizontal circle of radius r.

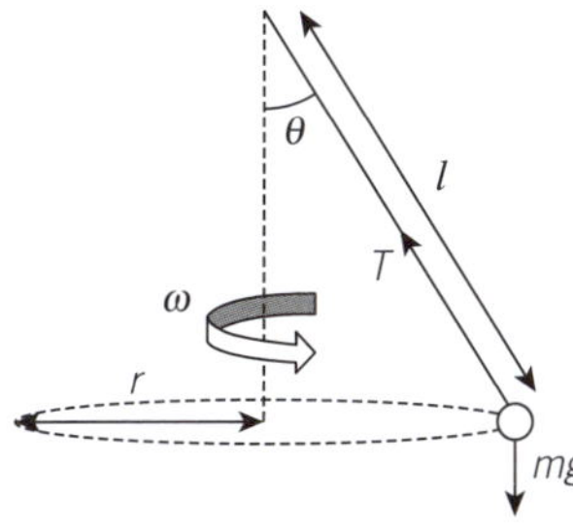

(a) State an expression, in terms of m, r and ω, for:

(i) the centripetal acceleration of the ball

(ii) the centripetal force acting on the ball.   [2]

(b) Hence show that:

(i) $T = ml\omega^2$

(ii) $\tan\theta = \dfrac{r\omega^2}{g}$ [2]

5 A ball of mass 0.20 kg is connected to a light, inextensible string and rotates in a vertical circle of radius 0.40 m. The speed of the ball at its lowest point is $6.0\,\mathrm{m\,s^{-1}}$.

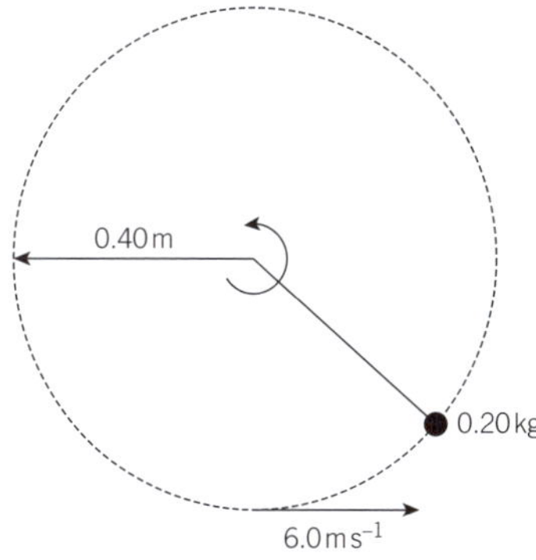

(a) Use the principle of conservation of energy to show that the speed of the ball at its highest point is $4.5\,\mathrm{m\,s^{-1}}$.   [2]

(b) Hence calculate the tension in the string:

(i) at its lowest point

(ii) at its highest point.   [3]

(c) Describe and explain what would happen if the speed of the ball at its lowest point was $3.0\,\mathrm{m\,s^{-1}}$.   [2]

6 A cyclist travels in a circle of radius 50.0 m at a speed of $20\,\mathrm{m\,s^{-1}}$, as shown below. The combined mass of the motorcyclist and motorcycle is 250 kg.

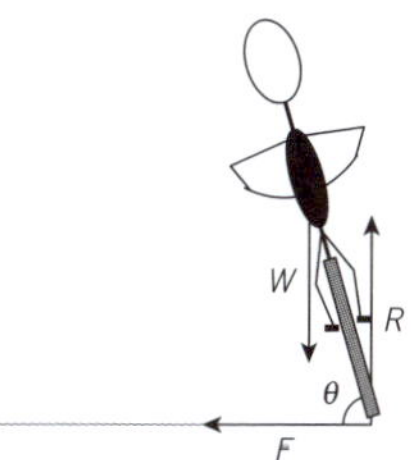

(a) Calculate:

(i) the combined weight W of the motorcyclist and the motorcycle

(ii) the vertical reaction force R.   [2]

(b) (i) Calculate the centripetal acceleration of the motorcycle and motorcyclist.

(ii) Hence show that the centripetal force F is 2.0 kN.   [3]

(c) Explain what provides the force F.   [2]

(d) Show that the angle θ is 51°. State any assumptions you make.   [2]

Key points

- ☐ Understand the idea of gravitational field as an example of field of force, and define gravitational field strength as force per unit mass.
- ☐ Understand that, outside the sphere, the mass of a sphere can be treated as a point mass at its centre.
- ☐ Recall and use Newton's law of gravitation in the form $F = \dfrac{Gm_1 m_2}{r^2}$.
- ☐ Analyse circular orbits in inverse square law fields, including geostationary orbits, by relating the gravitational force to the centripetal acceleration it causes.
- ☐ Derive, from Newton's law of gravitation and the definition of gravitational field strength, the equation $g = \dfrac{GM}{r^2}$ for the gravitational field strength of a point mass.
- ☐ Recall and solve problems using the equation $g = \dfrac{GM}{r^2}$ for gravitational field strength of a point mass.
- ☐ Appreciate that, on the surface of the Earth, g is approximately constant.
- ☐ Define gravitational potential at a point as the work done in bringing unit mass from infinity to the point.
- ☐ Solve problems using the equation $\phi = -\dfrac{GM}{r}$ for the potential in the field of a point mass.

Newton's law of gravitation

Any two masses will attract each other. The size of the force of attraction depends on the product of the two masses and varies inversely as the square of their distance apart. This can be summarised in **Newton's law of gravitation**:

$$F = \frac{Gm_1 m_2}{r^2}$$

where m_1 and m_2 are the masses, and r the distance between the centres of the two masses (Figure 8.1). G is a constant, called the universal gravitational constant, and is equal to $6.67 \times 10^{-11}\,\text{N}\,\text{m}^2\,\text{kg}^{-2}$.

For objects such as planets and moons, each mass can be treated as if all the mass is acting at the centre of the object (in the same way as a charged sphere can be treated as if all the charge is at the centre of the sphere).

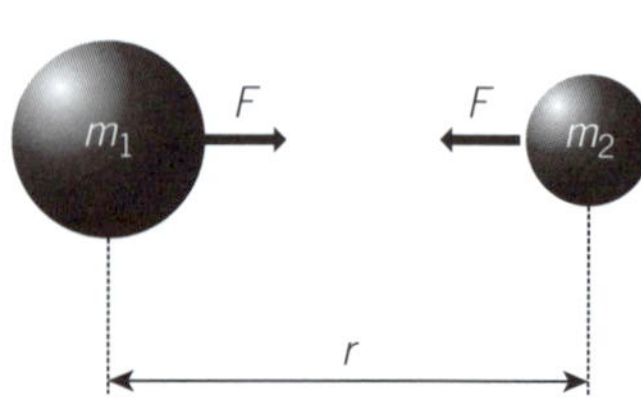

▲ **Figure 8.1** Newton's law of gravitation

> 💡 **Remember**
>
> $$F = \frac{Gm_1 m_2}{r^2}$$

> ⭐ **Exam tip**
>
> The value of G is provided in Exam Papers 1, 2, and 4.

> 💡 **Remember**
>
> Spherical masses can be treated as if all the mass is concentrated at the centre.

Worked examples

The Earth has a mass of $6.0 \times 10^{24}\,\text{kg}$, and the Moon has a mass of $7.4 \times 10^{22}\,\text{kg}$. The average distance between the centre of the Earth, and the centre of the Moon is $3.9 \times 10^5\,\text{km}$. Calculate the gravitational force exerted on the Moon by the Earth.

> This is also the force exerted by the Moon on the Earth.

Answer

$$F = \frac{Gm_1 m_2}{r^2} = \frac{6.67 \times 10^{-11} \times (6.0 \times 10^{24}) \times (7.4 \times 10^{22})}{(3.9 \times 10^5 \times 10^3)^2} = 1.95 \times 10^{20}\,\text{N}$$

Gravitational field strength

The **gravitational field strength** g at a point is the force per unit mass on a small test mass at that point (see Figure 8.2). It is a vector (having both magnitude and direction) and has units of $N\,kg^{-1}$.

From Newton's law, for a mass M the gravitational field a distance r from the centre of the mass is:

$$g = \frac{GM}{r^2}$$

The direction of the force is towards the centre of mass M. Gravitational field lines can be drawn to show the direction of the gravitational field.

Figure 8.3 illustrates that the Earth's gravitational field is **radial** (it varies as $1/r^2$), but on a smaller scale, such as near the Earth's surface, the field strength is constant in both magnitude and direction – a **uniform** field.

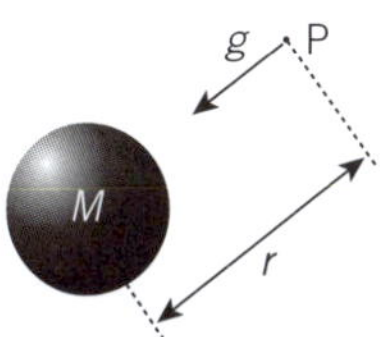
▲ **Figure 8.2** Gravitational field strength g

> Near the Earth's surface the gravitational field g is constant and has a value of $9.81\,N\,kg^{-1}$.

Worked example

[GM for Earth is $4.0\times10^{14}\,N\,m^2\,kg^{-1}$; radius of the Earth is $6.4\times10^3\,km$.]

Calculate the gravitational field strength:

a) on the Earth's surface

b) 1.00 km above the Earth's surface

Answer

a) $g = \dfrac{GM}{r^2} = \dfrac{4.0\times10^{14}}{(6.4\times10^6)^2} = 9.8\,N\,kg^{-1}$

b) $g = \dfrac{GM}{r^2} = \dfrac{4.0\times10^{14}}{(6.4\times10^6 + 10^3)^2} = 9.8\,N\,kg^{-1}$

> These values illustrate why g near the Earth's surface can be considered a constant.

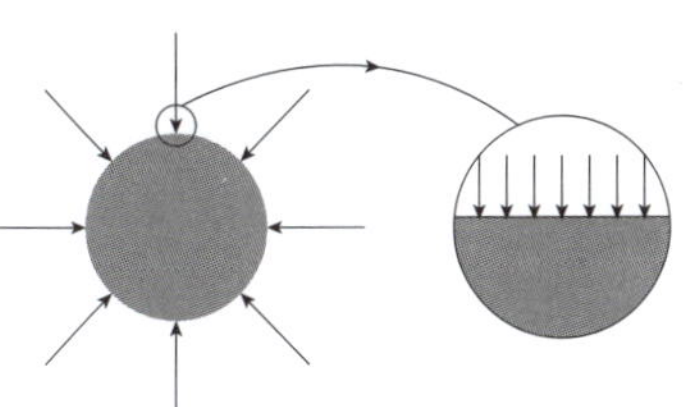
▲ **Figure 8.3** Radial gravitational field

> ★ **Exam tip**
>
> Take care to be consistent with units. In this question, GM is in S.I. units, so r must be converted to metres before substituting the value for r into the equation.

Orbital motion

Any object moving round in a circle of radius r at a constant speed v is accelerating because it is constantly changing direction. Acceleration is the rate of change of velocity (a vector), and an object moving in a circle is changing direction. The acceleration is a where

$$a = \frac{v^2}{r}.$$

Using $F = ma$, the force F needed to keep an object of mass m moving in a circle of radius r with constant speed v (see Figure 8.4) is:

$$F = \frac{mv^2}{r}$$

In the case of a satellite orbiting the Earth, or a planet orbiting the Sun, this force is the gravitational force of attraction.

For a mass m orbiting a mass M at a distance r:

$$\frac{GMm}{r^2} = \frac{mv^2}{r} \qquad \text{so} \qquad v = \sqrt{\frac{GM}{r}}$$

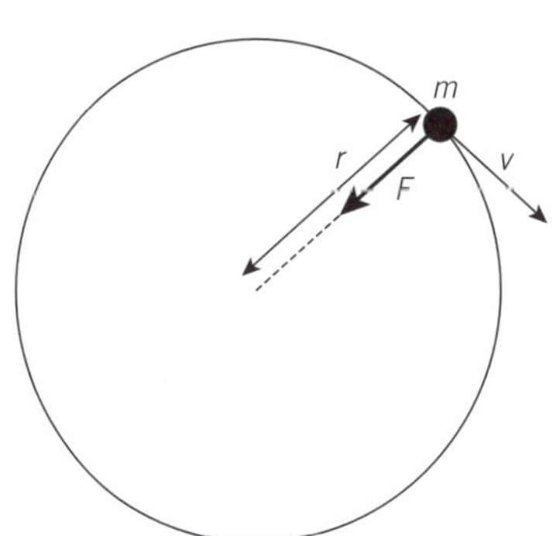
▲ **Figure 8.4** Orbital motion

The time T for one orbit is:

$$T = \frac{2\pi r}{v} = 2\pi\sqrt{\frac{r^3}{GM}}$$

Worked example

[GM for Earth $= 4.0 \times 10^{14}$ N m^2 kg^{-1}, radius of the Earth $= 6.4 \times 10^3$ km]

A satellite is orbiting in a **low Earth** orbit, 1000 km above the Earth's surface. Calculate:

a) the speed of the satellite

b) the time taken for one orbit of the Earth.

Answer

a) $\quad v = \sqrt{\dfrac{GM}{r}} = \sqrt{\dfrac{4.0 \times 10^{14}}{\left(6.4 \times 10^6 + 10^6\right)}} = 7.4 \times 10^3$ m s^{-1}

b) $\quad T = \dfrac{2\pi r}{v} = \dfrac{2\pi \times \left(6.4 \times 10^6 + 10^6\right)}{7.4 \times 10^3} = 6.28 \times 10^3$ s (about 105 minutes)

Geostationary satellites

Satellites in geostationary orbit complete one rotation of the Earth in 24 hours, as shown in Figure 8.5. Viewed from Earth, the satellites appear stationary.

Re-arranging $T = 2\pi \sqrt{\dfrac{r^3}{GM}}$

the radius R of a geostationary orbit is:

$$R = \sqrt[3]{\dfrac{GMT^2}{4\pi^2}} = \sqrt[3]{\dfrac{4 \times 10^{14} \times (24 \times 60 \times 60)^2}{4\pi^2}}$$

$$R = 4.23 \times 10^7 \text{ m}$$

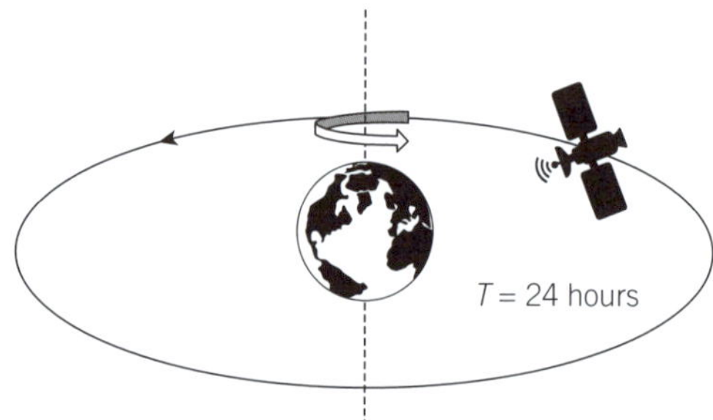

▲ **Figure 8.5** Geostationary orbit

The height h of a geostationary satellite **above the Earth's surface** is:

$$h = 4.23 \times 10^7 - 6.4 \times 10^6 = 3.59 \times 10^7 \text{ m}$$

Gravitational potential

The **gravitational potential** ϕ at a point P is defined as the work done in bringing unit mass (1 kg) from infinity to that point (see Figure 8.6). As work would need to be done in taking unit mass **from** the point **to** infinity, gravitational potential is always negative.

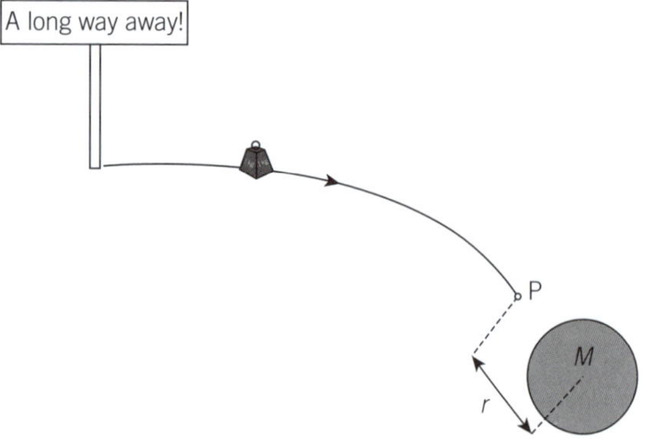

▲ **Figure 8.6** Gravitational potential

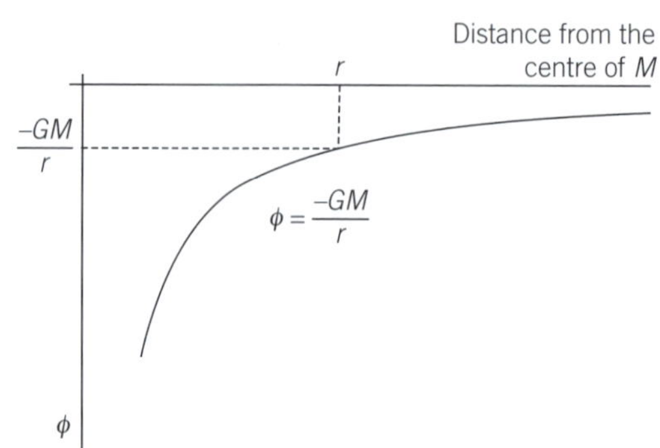

▲ **Figure 8.7** Variation of gravitational potential

For point P, a distance r from a mass M, as shown in Figures 8.6 and 8.7, the gravitational potential is:

$$\phi = -\frac{GM}{r}$$

This means that (GM/r) joules of energy are needed to move 1 kg from P to a long way away from M. The closer point P is to mass M (the smaller the value of r), the greater the value of (GM/r), and so the greater the energy needed to move 1 kg to a point far away from M.

> **Remember** 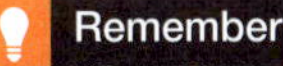
>
> $$\phi = \frac{GM}{r}$$
>
> The units of gravitational potential are $J\,kg^{-1}$.

Worked example

Ignoring air resistance how fast must an object be thrown up in the air to not come down again? (What is its **escape velocity**?)

Answer

On the Earth's surface the gravitational potential is $-\dfrac{GM}{R}$ where R is the radius of the Earth, and M its mass.

An amount of energy equal to $\dfrac{GM}{R}$ is needed for 1 kg to 'escape',

so a mass m would need $\dfrac{GMm}{R}$ joules of energy.

The kinetic energy given to the object must be at least as large as this for the object to escape:

$$\frac{1}{2}mv^2 \geq \frac{GMm}{R}$$

$$v \geq \sqrt{\frac{2GM}{R}}$$

For Earth, $GM = 4.0 \times 10^{14}\,N\,m^2\,kg^{-1}$, $R = 6.4 \times 10^3\,km$, so

$$v_{escape} = \sqrt{\frac{2 \times 4.0 \times 10^{14}}{6.4 \times 10^6}}$$

$$= 1.12 \times 10^4\,m\,s^{-1}$$

> The **escape velocity** is the velocity an object needs to completely break free from the gravitational field of a mass and 'escape' to infinity.

> The 'escape velocity' from Earth is about $11\,km\,s^{-1}$

Gravitational field and gravitational potential difference

The gravitational potential difference between two points is the work done in moving unit mass (1 kg) from one point to the other. If the two points are close enough together, such that the gravitational field strength g does not change significantly, the change in gravitational potential $\Delta\phi$ in moving a small distance Δh is:

$$\Delta\phi = g\Delta h$$

For a mass m moving a distance Δh, the change in gravitational potential energy ΔE_p is $mg\Delta h$ (the equation for gravitational potential energy calculations on, or near, the Earth's surface).

> **Link** 
>
> $$\Delta E_p = mg\Delta h$$
>
> See Unit 6 *Work, energy and power* for examples of the use of this equation.

 # Raise your grade

The graph shows how the gravitational potential ϕ due to the planet Mars varies with distance r from the centre of the planet. The mean radius of Mars is $3.4 \times 10^6\,$m.

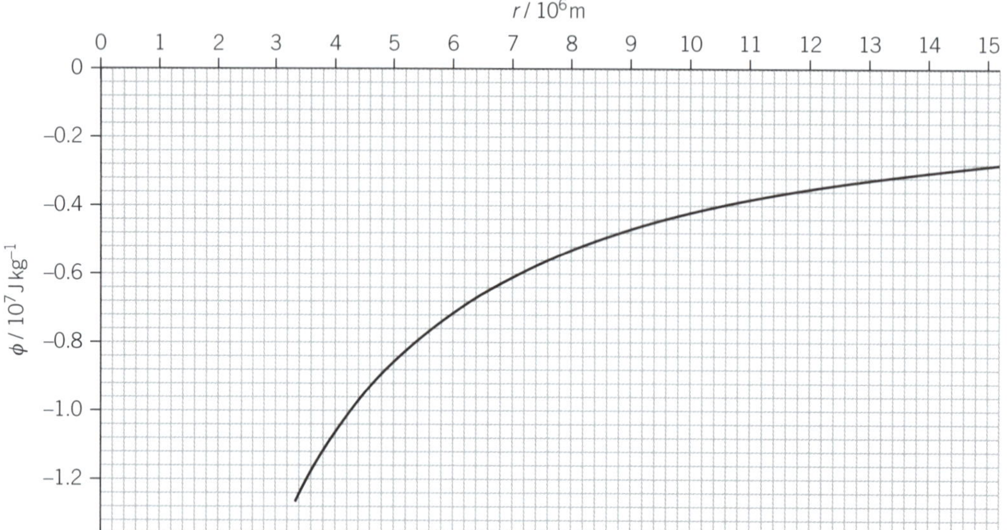

(a) (i) Define *gravitational potential*.

The gravitational potential at a point is the energy needed to move a mass from infinity to that point. ✔ ✗

> The statement is correct, but incomplete. Gravitational potential is the work done transferring **unit mass** (i.e., 1 kg) from infinity to the point concerned.

[2]

(ii) Explain why gravitational potential is negative.

Energy is needed to transfer a mass <u>from</u> a point <u>to</u> infinity – energy is gained moving a mass the other way. ✔

> A good answer.

[1]

(b) Use the graph to find the gravitational potential:

(i) on the surface of Mars

1.3×10^7 J kg⁻¹ ✔

> Good estimate from the graph.

[1]

gravitational potential = 1.3×10^7... J kg⁻¹

(ii) at a point P, $6.0 \times 10^6\,$m from the centre of Mars.

7.1×10^6 J kg⁻¹ ✔

[1]

gravitational potential = 7.1×10^6... J kg⁻¹

(c) Calculate the energy needed to move 100 kg from the surface of Mars to the point P.

Energy needed = $100 \times (1.3 \times 10^7 - 7.1 \times 10^6) = 5.9 \times 10^8$ J ✔ ✔

> Correct method, correct calculation with unit.

energy needed = 5.9×10^8 J [2]

❓ Exam-style questions

[Use $G = 6.67 \times 10^{-11}\,\mathrm{N\,m^2\,kg^{-2}}$; radius of the Earth $= 6.4 \times 10^3\,\mathrm{km}$]

1 Two average-sized adults are standing approximately 1.0 m apart.

  (a) Estimate the mass of one adult. [1]

  (b) Hence estimate the gravitational force between the two adults. [1]

  (c) If both adults are standing on a frictionless surface, and in the absence of air resistance, estimate the size of the acceleration towards each other. [1]

2 The Moon orbits the Earth once in 27.3 days. Its mean orbital radius is $3.8 \times 10^5\,\mathrm{km}$. Use this information to calculate the mass of the Earth. [3]

3 Use g on the Earth's surface as $9.81\,\mathrm{N\,kg^{-1}}$. The Earth's radius is $6.37 \times 10^3\,\mathrm{km}$.

  (a) Calculate the mass of the Earth.

  (b) Justify the number of significant figures in your answer. [3]

4 A planet is in a circular orbit around a star of mass M. The radius of the orbit is R, and the time for one orbit is T.

  (a) Show that the velocity v of the planet is:

$$v = \sqrt{\frac{2GM}{R}}$$

  (b) Hence show that:

$$\frac{R^3}{T^2} = \text{a constant} \quad [3]$$

5 Use g on the surface of the Moon as $1.6\,\mathrm{m\,s^{-2}}$. The radius of the Moon is $1.74 \times 10^3\,\mathrm{km}$.

Calculate:

  (a) the mass of the Moon

  (b) the gravitational potential on the surface of the Moon

  (c) the *escape velocity* from the Moon's surface. [4]

6 Europa, one of Jupiter's moons, orbits the planet once every 85 hours at a radius of $6.7 \times 10^5\,\mathrm{km}$.

Calculate:

  (a) the speed of Europa

  (b) the mass of Jupiter. [3]

7 **(a)** Define *gravitational potential*. [1]

  (b) A spacecraft, of mass $6.0 \times 10^4\,\mathrm{kg}$, is in orbit round the Earth at a height of 200 km.

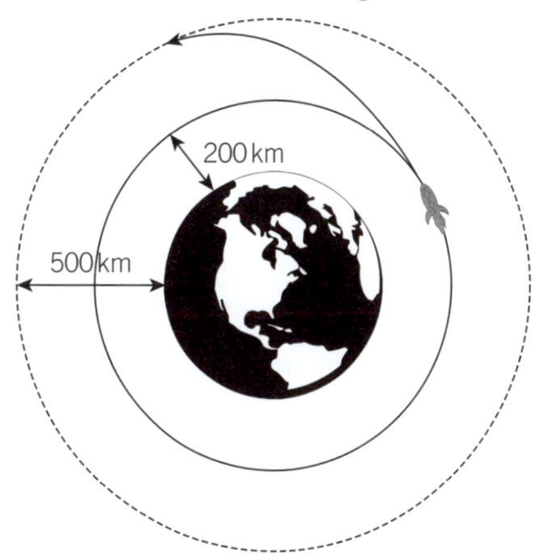

Calculate:

  (i) the gravitational field strength at this height

  (ii) the gravitational potential at this height. [3]

  (c) The spacecraft moves to an orbit 500 km above the Earth's surface.

Calculate:

  (i) the gravitational potential at this new height

  (ii) the increase in gravitational potential energy of the spacecraft. [3]

8 A weather satellite, of mass 320 kg, is to be placed in geostationary orbit, $3.56 \times 10^7\,\mathrm{m}$ above the Earth's surface.

  (a) **(i)** State the time taken for one orbit.

    (ii) Show that the speed of the satellite is $3.1 \times 10^3\,\mathrm{m\,s^{-1}}$.

    (iii) Calculate the kinetic energy of the satellite. [4]

  (b) Calculate the gravitational potential:

    (i) on the Earth's surface

    (ii) at the orbit height of $3.59 \times 10^7\,\mathrm{m}$. [3]

  (c) Calculate the increase in gravitational potential energy of the satellite in being lifted from the Earth's surface to its orbit height. [1]

  (d) Suggest one reason why the energy needed to launch the satellite into orbit is much more than the sum of your answers to (a)(iii) and (c). [1]

9 Deformation of solids

Key points

- ☐ Understand that forces cause materials to deform and that the deformation can be tensile or compressive.
- ☐ Describe the behaviour of springs in terms of load, extension, elastic limit, Hooke's law, and the spring constant.
- ☐ Define and use the terms stress, strain, and Young modulus.
- ☐ Describe an experiment to find the Young modulus of a metal wire.
- ☐ Understand that the area under the force–extension graph is the work done.
- ☐ Distinguish between elastic and plastic deformation of a material.
- ☐ Find the strain energy in a deformed material from the area under the force–extension graph.

Force and solid materials

When a pair of forces is applied to a solid material it **deforms**; that is, it changes shape. Forces that stretch a material are called **tensile** forces (see Figure 9.1a); forces which compress a material are called **compressive** forces (see Figure 9.1b).

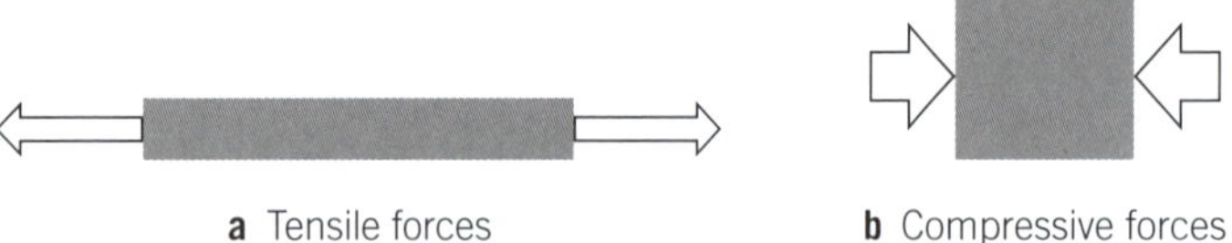

▲ **Figure 9.1** Forces on a solid material

The deformation is **elastic** if the material returns to its original shape once the forces have been removed. If there is some permanent deformation (e.g., compression or extension) when the forces have been removed, **plastic** deformation has occurred.

Stretching springs

The graph in Figure 9.2 shows the extension (stretch) of a spring as different loads are placed on it.

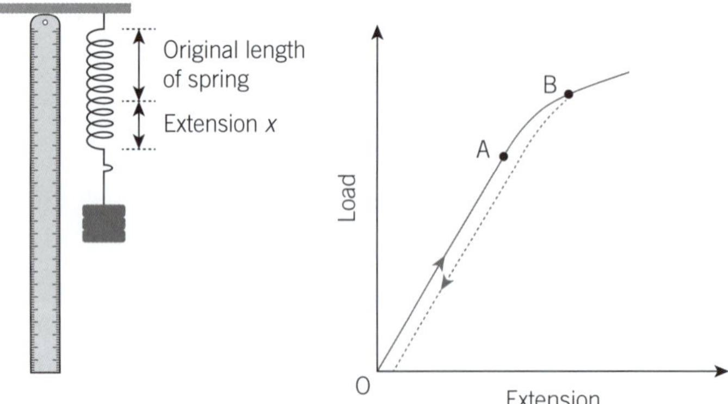

▲ **Figure 9.2** Stretching a spring

AS Level

Key terms

Deformation: change of shape.

Tensile: forces which stretch.

Compressive: forces which compress or squash.

Elastic: returns to original shape when forces are removed.

Plastic: some permanent deformation when forces are removed.

- **O–A:** The graph is a straight line, through the origin. Doubling the load doubles the stretch (extension) – the spring obeys **Hooke's law**:

$$F = kx$$

  where F is the load applied, x is the extension, and k is called the **spring constant** and is a measure of the stiffness of the spring. The SI units for k are $N\,m^{-1}$.

  Point A is called the **elastic limit**. The spring stretches elastically. If the load is removed at any point between O and A, the spring will return to its original length.

- **A–B:** If the spring is stretched some plastic deformation occurs. When the load is removed the load–extension graph follows the dotted line. With the load removed completely, there is a permanent extension of the spring.

Energy considerations

The work done in stretching a spring is the area under the load–extension graph (see Figure 9.3). If the spring obeys Hooke's law and is not stretched beyond the elastic limit, the work done on the spring (called the strain energy) is:

$$E = \tfrac{1}{2} Fx$$

Substituting from $F = kx$ gives

$$E = \tfrac{1}{2} kx^2 \text{ or } E = \tfrac{1}{2} \frac{F^2}{k}$$

As the spring stretches elastically, all the energy stored is recoverable as mechanical energy when the load is removed from the spring.

> **Remember**
>
> Hooke's Law:
>
> $$F = kx$$
>
> k is the spring constant; it is measured in $N\,m^{-1}$.
>
> x is the total **extension** of the spring, **not** the total length of the spring.

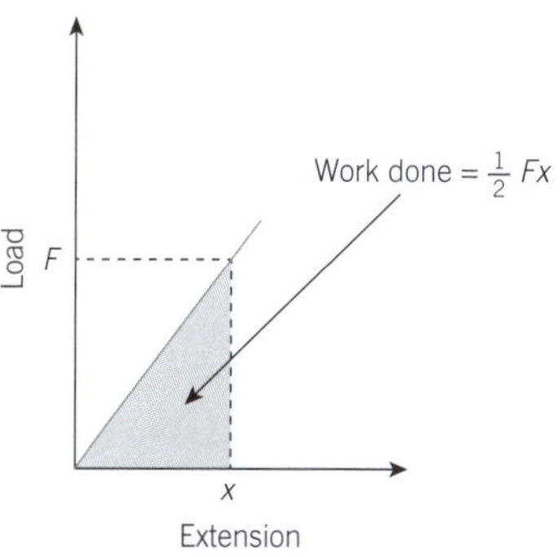

▲ **Figure 9.3** Energy stored in a stretched spring

Worked example

A spring, with unstretched length 12.0 cm, stretches 4.0 cm when supporting a load of 5.0 N.

a) Determine the spring constant k of the spring.

b) Calculate the total length of the spring if it supports a load of 8.0 N.

c) Find the energy stored in the spring when it is stretched by 7.0 cm.

State any assumptions you make.

Answer

a) $k = \dfrac{F}{x} = \dfrac{5.0}{4.0 \times 10^{-2}} = 125\,N\,m^{-1}$

b) $x = \dfrac{F}{k} = \dfrac{8}{125} = 0.064\,m$

   Total length of spring $= 12.0 + 6.4 = 18.4$ cm

c) $E = \dfrac{1}{2} kx^2 = \dfrac{1}{2} \times 125 \times (7.0 \times 10^{-2})^2 = 0.31\,J$

Assumptions made: the spring obeys Hooke's law, and the elastic limit is not exceeded.

> **Exam tip**
>
> When using an equation like the equation for the energy stored in a spring, the units used must be **consistent**.
>
> The extension must be in metres (m), the force applied in newtons (N), and the stiffness k in newtons/metre ($N\,m^{-1}$), so that the final answer is in joules (J).

Stretching materials

The graph (Figure 9.4) shows how a ductile metal (one that can be drawn into a thin wire, such as copper) stretches when supporting different loads.

- **O–A:** The material obeys Hooke's law – the extension of the material is directly proportional to the load applied. If the load is removed, the material returns to its original length. It has behaved elastically. A is the **Hooke's law limit**.

- **A–B:** The material is now past the Hooke's law limit, but still behaves elastically. If the load is removed, the material again returns to its original length – B is the **elastic limit**.

- **B–C:** The material has been stretched beyond its elastic limit. If the material is stretched to point C and the load then removed, the material will not return to its original length, but instead return along the dotted line on the graph. There is now some **permanent deformation** of the material.

If a material is stretched beyond its elastic limit (see Figure 9.5), the work done in stretching the material is the area under the loading curve (solid line). The recoverable mechanical energy is the area under the unloading curve (dotted line).

The difference between the two areas is the energy lost as internal energy (heat) in the material.

Stress, strain, and the Young modulus

When stetching different materials, such as metal wires, **stress** and **strain** are more useful quantities to calculate than just the force applied and the extension, as the stress–strain graph illustrates properties of the material itself rather than one specific length or diameter of the material.

Stress σ

$$\text{stress } \sigma = \frac{\text{force applied}}{\text{cross-sectional area}} = \frac{F}{A}$$

The SI units for stress are $N\,m^{-2}$, or pascal (Pa).

Strain ε

$$\text{strain } \varepsilon = \frac{\text{extension}}{\text{original length}} = \frac{x}{l_0}$$

Strain has no units. It can be expressed as a number or a percentage.

Worked example

A steel wire of length 3.0 m and diameter 1.0 mm stretches 1.9 mm when supporting a mass of 10 kg. [Use $g = 9.81\ m\,s^{-2}$.] Calculate:

a) the stress   **b)** the strain.

Answer

a) $\sigma = \dfrac{F}{A} = \dfrac{10 \times 9.81}{\pi \times (0.5 \times 10^{-3})^2} = 1.23 \times 10^8\ N\,m^{-2}$

b) $\varepsilon = \dfrac{x}{l_0} = \dfrac{1.9 \times 10^{-3}}{3.0} = 6.3 \times 10^{-4}\ (0.063\%)$

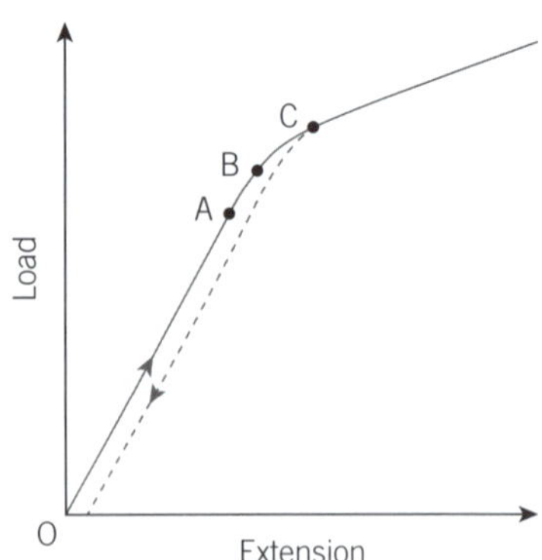

▲ **Figure 9.4** Load against extension graph for ductile metals

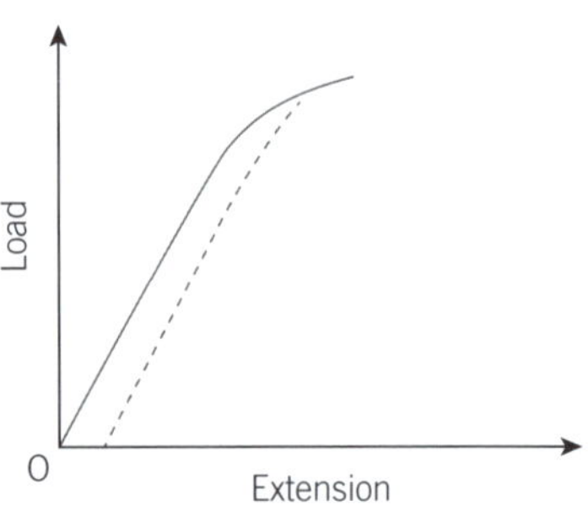

▲ **Figure 9.5** Stretching beyond the elastic limit

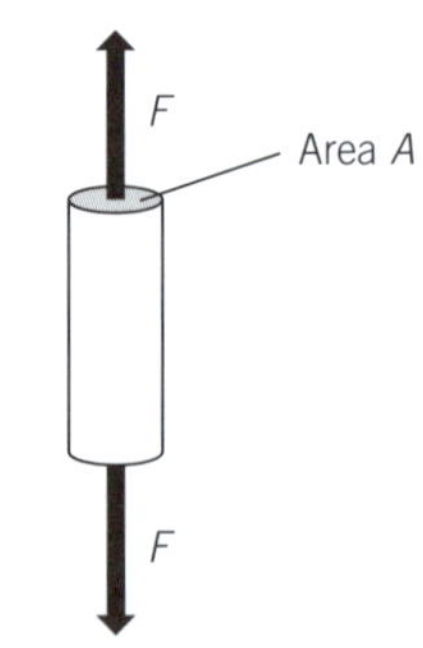

▲ **Figure 9.6** Stress

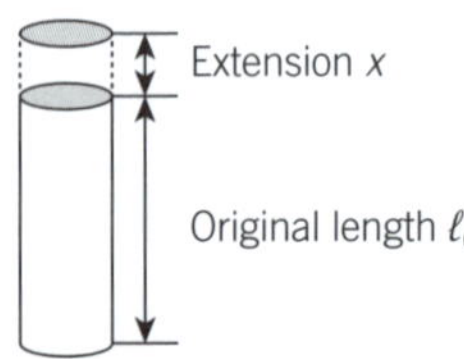

▲ **Figure 9.7** Strain

★ Exam tip

When calculating A (the cross-sectional area) don't forget to halve the diameter to find the **radius** when using $A = \pi r^2$.

Young modulus E

The **Young modulus** of a material is a measure of the stiffness of that material. The larger the value of E, the stiffer the material – the greater the stress needed to produce a particular strain.

$$\text{Young modulus} = \frac{\text{stress}}{\text{strain}}$$

$$E = \frac{\sigma}{\varepsilon}$$

The units for the Young modulus are N m^{-2} or pascal (Pa).

> **Exam tip**
>
> Be careful not to confuse E for the Young modulus with E for energy.

Remember

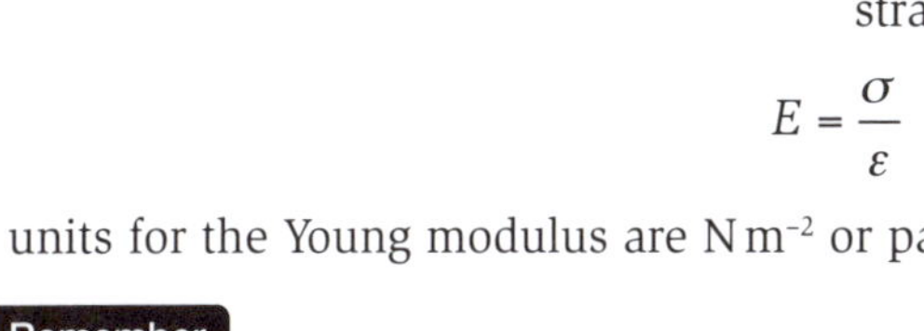

$$\text{stress} = \frac{\text{force applied}}{\text{cross-sectional area}} \quad \text{or} \quad \sigma = \frac{F}{A}$$

$$\text{strain} = \frac{\text{extension}}{\text{original length}} \quad \text{or} \quad \varepsilon = \frac{x}{l_0}$$

$$\text{Young modulus} = \frac{\text{stress}}{\text{strain}} \quad \text{or} \quad E = \frac{\sigma}{\varepsilon}$$

Worked examples

1 A nylon rope of length 3.5 m and diameter 5.0 mm stretches 12 mm when supporting a load of 80 kg. Determine the Young modulus of the nylon.

Answer

$$E = \frac{\sigma}{\varepsilon} = \frac{\left(\dfrac{F}{A}\right)}{\left(\dfrac{x}{l_0}\right)} = \frac{\left(\dfrac{80 \times 9.81}{\pi \times (2.5 \times 10^{-3})^2}\right)}{\left(\dfrac{12 \times 10^{-3}}{3.5}\right)} = 1.2 \times 10^{10}\,\text{Pa}$$

2 A car hoist is supported by four solid steel columns. Each column is 2.0 m high and 2.0 cm in diameter. Determine how far the hoist will descend when supporting a car of mass 2000 kg.
   [Young modulus of steel $= 2.0 \times 10^{11}\,\text{N m}^{-2}$]

Answer

$$x = \frac{Fl_0}{EA} = \frac{(2000 \times 9.81) \times 2.0}{2.0 \times 10^{11} \times 4 \times \pi \times (1.0 \times 10^{-2})^2} = 1.6 \times 10^{-4}\,\text{m}\ (0.16\,\text{mm})$$

Measuring Young modulus E

Figure 9.8 shows a simple way of measuring E using a thin wire.

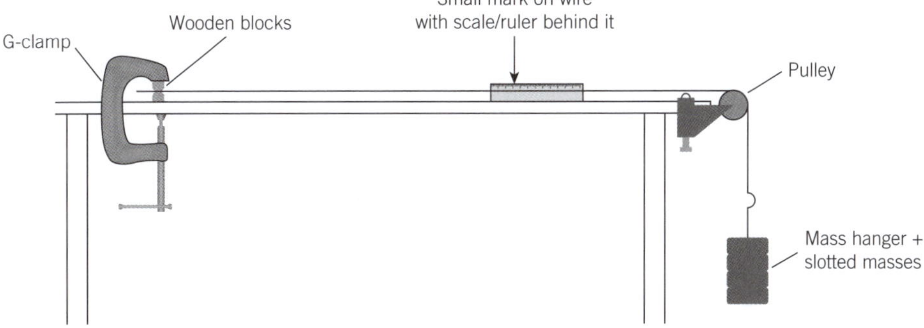

▲ **Figure 9.8** Measuring Young modulus E

This method is useful for testing materials that can be drawn into long, thin strips such as copper or steel wire, nylon or polythene. A mark is made on the wire, and the length of the wire from the wooden blocks recorded. Weights are steadily added, and the extension of the wire recorded for different loads (the weights can be removed periodically to see if the wire returns to its original length; that is, if the material is still elastic).

The length of the wire can be measured with a metre rule and the diameter with vernier calipers (or a micrometer if available). Stress and strain can then be calculated from the load and extension values, and a graph of stress against strain plotted. The value of E can then be found from the graph.

Making improvements

Practical physics papers often ask how a particular experiment or procedure can be improved.

How can this experiment be improved? Good suggestions might include:

- using a longer length of wire (this will reduce the percentage uncertainty in the measurement of the length of the wire)
- measuring the extension with a travelling microscope
- measuring the diameter of the wire with a micrometer (a micrometer is accurate to ±0.01 mm, whereas vernier calipers are only accurate to ±0.1 mm).

Avoid making suggestions for improvements that should be carried out anyway, such as repeating measurements, reading an instrument 'square-on' to avoid parallax errors, or correcting for zero errors in instruments.

Elastic and plastic deformation

The loading and unloading force–extension graphs for different materials provide useful information about their elastic and plastic properties.

In Figure 9.9a, the metal wire is elastic up to the elastic limit. If the load exceeds the elastic limit value, the material will not return to its original length. The area under the loading curve is the work done in stretching the material (the strain energy). Up to the elastic limit this energy is stored as potential energy and can be recovered as mechanical energy.

In Figure 9.9b, the material is elastic but does not obey Hooke's law. The area under the unloading curve is the energy which is **recoverable** as mechanical energy. The area between the two curves is the energy **lost** as internal energy (heat) in the material. This is why car tyres are hot after a long journey.

In Figure 9.9c, the material behaves plastically. When the load is removed there is only a small reduction in length. The area between the two curves is the energy lost as internal energy (heat) in the material.

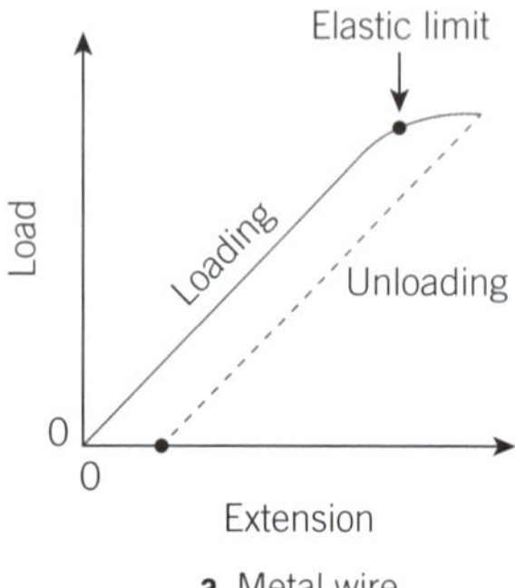

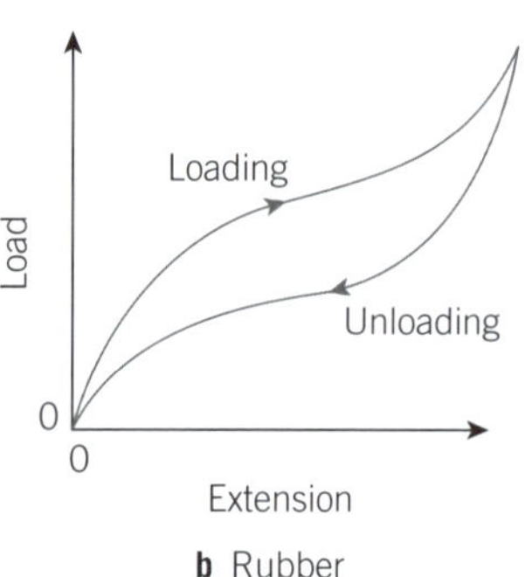

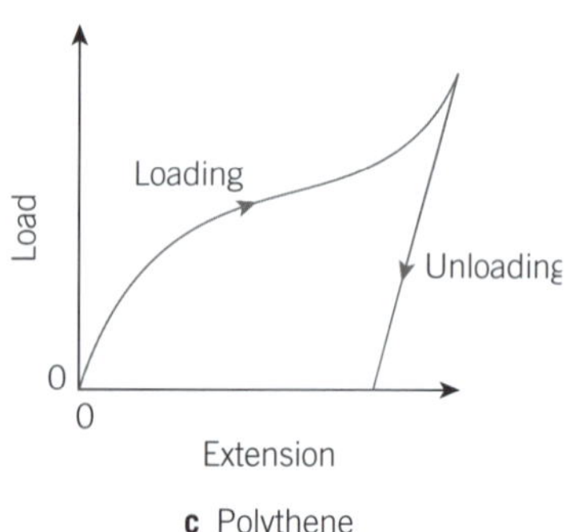

▲ **Figure 9.9** Stretching different materials

 **Raise your grade**

(a) Define, for a metal wire,

(i) stress

Stress is the force on the wire divided by the cross-sectional area of the wire. ✔

A correct definition.

(ii) Young modulus

Young modulus equals $\dfrac{\text{tensile stress}}{\text{tensile strain}}$ ✔

A correct definition. [2]

(b) A steel cable of length 0.50 m and diameter 4.0 mm is suspended from the ceiling and supports a chandelier (a light fitting). The mass of the chandelier is 25.0 kg. The Young modulus of steel is 2.1×10^{11} Pa.

Calculate:

(i) the weight of the chandelier

$W = mg = 25.0 \times 9.81 = 245\,\text{N}$ ✔

weight = ...245...... N [1]

(ii) the extension of the cable caused by the weight of the chandelier.

$$x = \frac{FL}{EA} = \frac{245 \times 0.50}{2.1 \times 10^{11} \times \pi \times \left(2 \times 10^{-3}\right)^2} = 4.6 \times 10^{-5}$$ ✔✔✘

Correct method.

Correct substitutions into equation. The candidate has forgotten to convert the answer from metres to millimetres.

extension = 4.6×10^{-5} mm .... [3]

(c) The steel cable consists of a large number of very thin strands of steel bound together, rather than a single wire. Suggest a reason for this. [1]

A lot of thin strands are stronger than a single thick cable. ✘

If the combined cross-sectional area of all the thin strands is the same as the cross-sectional area of the thick cable, the strengths of the two cables are exactly the same. A large number of thin cables is much easier to bend (more flexible) than a single thick cable.

1 A spring stretches 4 cm when supporting a load of 20 N.

Three identical springs to the one described are connected as shown in the diagram.

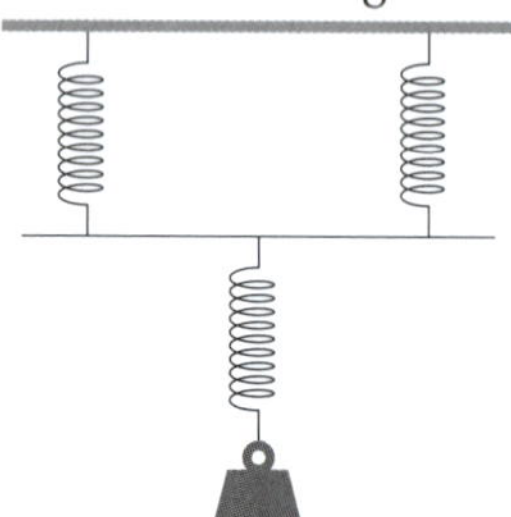

What is the extension of the three springs when supporting a load of 30 N?

A 3 cm

B 6 cm

C 9 cm

D 12 cm [1]

2 The graph shows the extension of a spring for different loads.

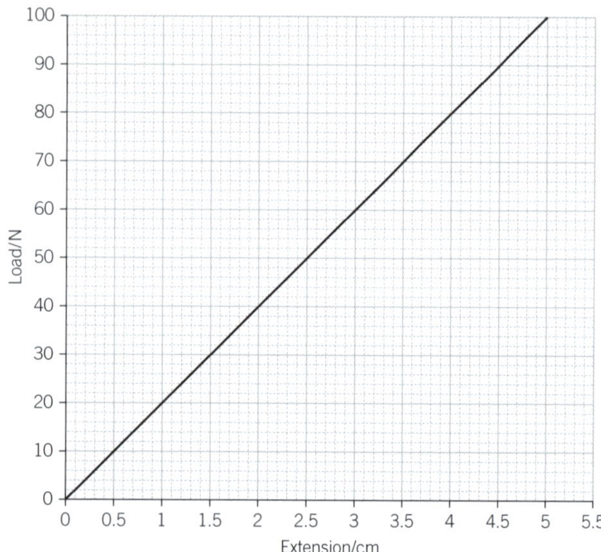

What is the energy stored in the spring when a load of 80 N is applied to the spring?

A 1.6 J

B 3.2 J

C 160 J

D 320 J [1]

3 What are the SI base units of stress?

A $kg\,m^{-1}\,s^{-2}$

B $kg\,m\,s^{-2}$

C $kg\,m^{-1}\,s^{2}$

D $kg\,m\,s^{2}$ [1]

4 A tensile force is applied to a thin wire causing it to stretch by an amount x. The same force is now applied to a wire of the same material, but with three times the length and twice the diameter. How much will this wire stretch?

A $\dfrac{2x}{3}$ **B** $\dfrac{3x}{4}$ **C** $\dfrac{4x}{3}$ **D** $\dfrac{3x}{2}$ [1]

5 A lift in a building is supported by six steel cables, each 20 m long. Each cable consists of 20 strands of steel wire of diameter 2.0 mm. The Young modulus of steel is 2.1×10^{11} Pa.

A man of mass 80 kg steps into the lift. How far will the lift descend?

A 0.01 mm

B 0.05 mm

C 0.10 mm

D 0.20 mm [1]

6 A metal wire of length 1.200 m and diameter 0.61 mm stretches 0.43 mm when a load of 1.30 kg is hung from one end.

(a) Calculate:

    (i) the stress on the wire

    (ii) the strain of the wire

    (iii) the Young modulus of the metal. [4]

(b) Justify the number of significant figures you have given for your answer to (a)(iii). [1]

7 **(a)** Define the Young modulus. [1]

(b) The upper leg bone (femur) in an adult human has a length of 50 cm and a minimum diameter of 2.8 cm.

    The Young modulus of bone is 8.5×10^{9} Pa.

    Estimate:

    (i) the mass of a man

    (ii) the maximum stress on the bone when the man stands on one leg

    (iii) the compression of the bone when the man stands on one leg. [4]

(c) State, with a reason, whether your answer to (b)(iii) is likely to be an overestimate or an underestimate. [1]

8 You have been asked to investigate the mechanical properties of nylon. You are provided with a reel of nylon thread.

Design a laboratory experiment to determine the mechanical properties of the material. Draw a diagram showing the arrangement of your equipment. In your account you should pay particular attention to:

(a) the procedure to be followed

(b) the measurements to be taken

(c) the control of variables

(d) the analysis of the data

(e) the safety precautions to be taken. [15]

Key points

- ☐ Recall and solve problems using the equation of state for an ideal gas $pV = nRT$.
- ☐ Infer from a Brownian motion experiment the evidence for the movement of molecules.
- ☐ State the basic assumptions of the kinetic theory of gases.
  Explain how molecular movement causes the pressure exerted by a gas, and hence deduce the relationship $pV = \frac{1}{3} Nm <c^2>$, where N is the number of molecules.
- ☐ Recall that the Boltzmann constant k is given by the expression $k = \dfrac{R}{N_A}$.
- ☐ Compare $pV = \frac{1}{3} Nm <c^2>$ with $pV = NkT$, and hence deduce that the average translational kinetic energy of a molecule is proportional to T.

Equation of state for an ideal gas

An **ideal gas** is one in which all the collisions between atoms or molecules are perfectly elastic and in which there are no intermolecular forces. Such a gas obeys the equation:

$$pV = nRT$$

where p is the pressure of the gas, V its volume, and T its absolute temperature measured in kelvin (K). n is the number of moles of the gas and R is the molar gas constant. The equation is called the **equation of state for an ideal gas**.

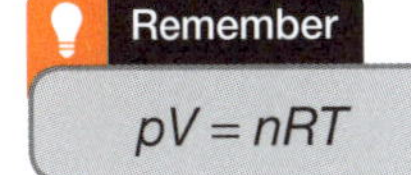

Real gases normally obey this equation except under extremely low values of temperature or very high pressures. For gases at moderate values of temperature and pressure it can be shown by experiment that:

- $p \propto \dfrac{1}{V}$ at constant temperature – Boyle's law (Figure 10.1a)

- $V \propto T$ at constant pressure – Charles' law (Figure 10.1b)

- $p \propto T$ at constant temperature – the pressure law (Figure 10.1c).

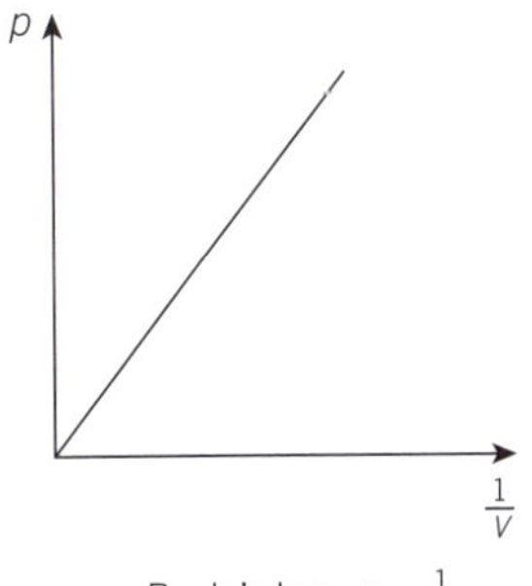
a Boyle's law: $p \propto \frac{1}{V}$
at constant temperature

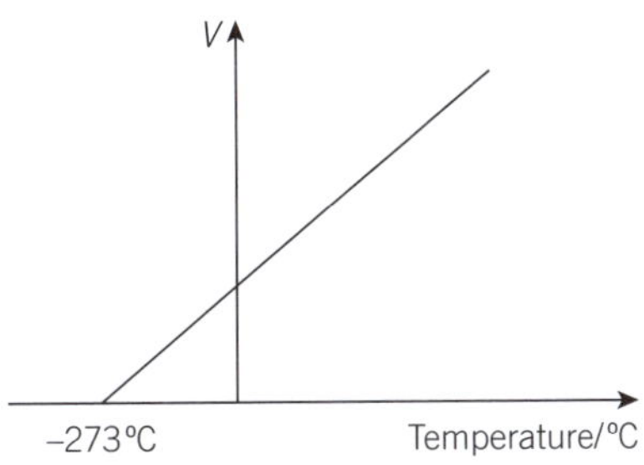
b Charles' law: $V \propto T$ (in **kelvin**)
at constant pressure

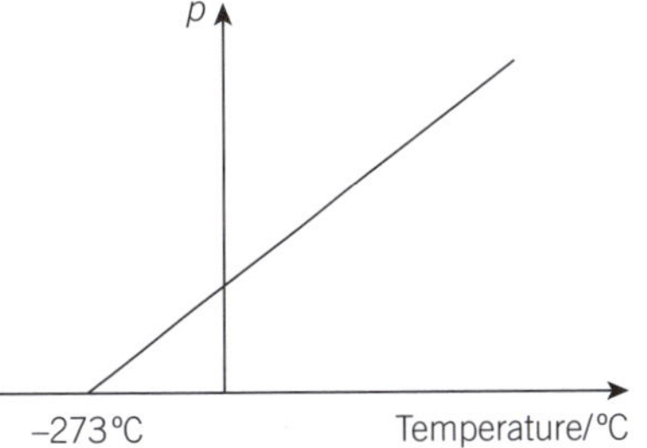
c pressure law: $p \propto T$ (in **kelvin**)
at constant temperature

▲ **Figure 10.1** Gas laws

The three gas laws can be combined:

$$\frac{pV}{T} = \text{constant}$$

The size of the constant depends on the number of gas molecules (e.g., doubling the number of gas molecules in a sample doubles the number of collisions of the molecules with the walls of the container, and so doubles the pressure).

R is the value of the constant for one mole of gas molecules. A **mole** is just a number, **Avogadro's number** N_A (where $N_A = 6.02 \times 10^{23}\,\text{mol}^{-1}$), and so the constant for a different amount of gas is nR where n is the number of moles of gas present.

Exam tip

Don't forget that T in all the gas equations is always the absolute temperature, in kelvin:

$$T(\text{K}) = T(°\text{C}) + 273°$$

The values of R and N_A are provided in Exam Papers 1, 2, and 4.

Worked example

Before setting off on a journey, the air in the tyres of a car was at a temperature of 18 °C and a pressure of 1.92×10^5 Pa. Immediately after completing the journey the pressure in the tyres was 2.25×10^5 Pa. What was the temperature of the air in the tyres?

Answer

The volume of the air in the tyres V is constant, and so from the equation of state for an ideal gas:

$$\frac{p_1 \cancel{V}}{T_1} = \frac{p_2 \cancel{V}}{T_2}$$

so

$$T_2 = \frac{T_1 p_2}{p_1} = \frac{(273 + 18) \times (2.25 \times 10^5)}{1.92 \times 10^5} = 341\,\text{K}$$

Temperature of the air in the tyres = 341 − 273 = 68 °C.

Link

Avogadro's constant N_A is equal to the number of atoms in 12 g of ^{12}C.

$$N_A = 6.02 \times 10^{23}\,\text{mol}^{-1}$$

A *mole* (mol) of anything is N_A of it.

See Unit 1 *Physical quantities and units*.

Kinetic theory of gases

Brownian motion

Particles of smoke, illuminated with a bright light and observed under a microscope, 'jiggle' about randomly, continually changing speed and direction (see Figure 10.2). The effect is explained by air molecules, which are too small to see under the microscope, bombarding each smoke particle from different directions. **Brownian motion** is clear evidence that molecules in gases are in continuous, random motion and is the experimental basis for the kinetic theory of gases.

Brownian motion is named after Robert Brown, a botanist. In 1827 he observed pollen grains in water and noticed tiny particles, ejected by the pollen grains, moving about randomly in the water.

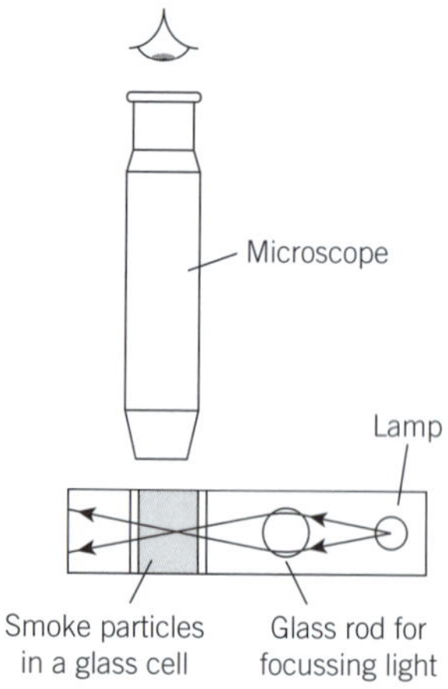

a Observing Brownian motion

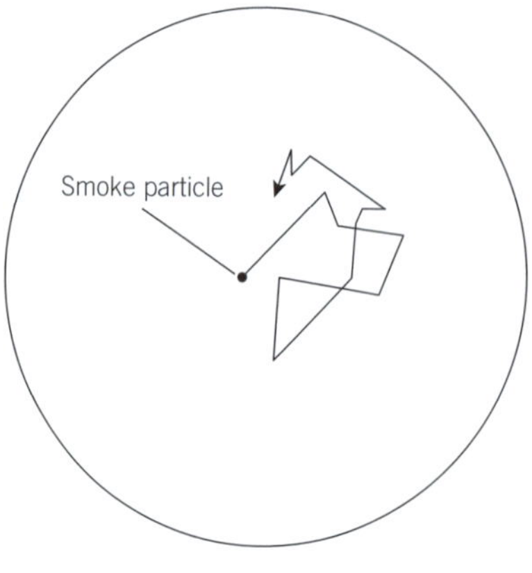

b Motion of smoke particles

▲ **Figure 10.2** Brownian motion

Internal energy of a gas

The molecules of a gas have both kinetic energy (because they are moving) and potential energy (because there are attractive forces between them). The sum total of all the molecules' potential and kinetic energies is the internal energy of the gas.

For an **ideal gas**, there are no intermolecular forces so the molecules have no potential energy – all the internal energy of an ideal gas is kinetic energy.

Kinetic theory of gases

The kinetic theory of gases is based on the model of a gas that has many identical particles moving about randomly, colliding with each other and the walls of their container, constantly changing speed and direction.

The theory makes a number of simplifying assumptions about gases:

* the gas molecules have negligible volume compared to the volume of the gas as a whole

* the forces between gas molecules are negligible, except when colliding with each other

* collisions between molecules, or between molecules and the walls of their container, are perfectly elastic

* for an individual gas molecule, the time for a collision is negligible compared to the time between collisions.

Imagine a single molecule of mass m, inside a box of side L, travelling at speed c towards the right-hand side of the box (Figure 10.3). When it collides with the wall it has a perfectly elastic collision and rebounds with the same speed in the opposite direction:

$$\text{change in momentum of the molecule} = mc - (-mc) = 2mc$$

The molecule moves towards the left-hand wall, rebounds, and returns to hit the right-hand wall again. The molecule will have travelled a total distance $2L$ at speed c between collisions with the right-hand wall:

$$\text{time taken between collisions with the right-hand wall} = \frac{2L}{c}$$

Using Newton's second law (force = rate of change of momentum):

$$\text{force on right-hand wall} = \text{rate of change in momentum}$$

$$= \frac{2mc}{\left(\frac{2L}{c}\right)} = \frac{mc^2}{L}$$

$$\text{pressure on right-hand wall} = \frac{\text{force}}{\text{area}} = \frac{\left(\frac{mc^2}{L}\right)}{L^2} = \frac{mc^2}{L^3}$$

For N molecules in the gas, all moving with different speeds, the pressure p is:

$$p = N\frac{m<c^2>}{L^3} = N\frac{m<c^2>}{V}$$

where $<c^2>$ is the mean of the squares of the velocities and V is the volume of the box. This assumes all the molecules are travelling in the same direction and colliding with the same two opposite faces of the box.

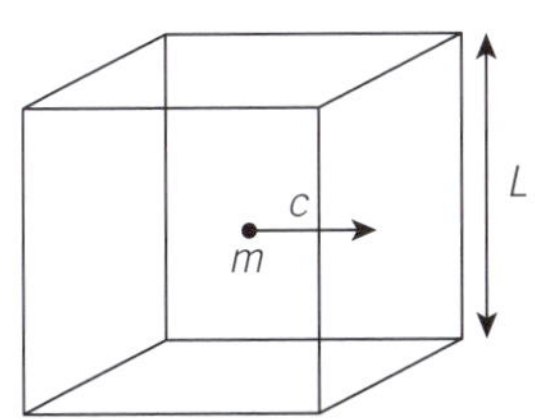

▲ **Figure 10.3** Molecule in a box

At any one time, as many molecules are moving up and down, or forward and backward, as left and right, and so we must divide by three to calculate the pressure in any direction:

$$p = \frac{1}{3} N \frac{m<c^2>}{V}$$

This equation is usually written:

$$pV = \frac{1}{3} Nm <c^2>$$

Re-arranging this equation gives:

$$p = \frac{1}{3}\left(\frac{Nm}{V}\right)<c^2> = \frac{1}{3}\rho<c^2>$$

where ρ is the density of the gas.

Worked example

Estimate the r.m.s. speed of air molecules at room temperature.

Answer

Estimates: atmospheric pressure $\approx 1.0 \times 10^5\,\mathrm{Pa}$

density of air at room temperature $\approx 1.29\,\mathrm{kg\,m^{-3}}$

Rearranging $pV = \frac{1}{3}Nm<c^2>$ gives:

$$p = \frac{1}{3}\left(\frac{Nm}{V}\right)<c^2> = \frac{1}{3}\rho<c^2>$$

so $$<c^2> = \frac{3p}{\rho} = \frac{3 \times 1.0 \times 10^5}{1.29} = 2.3 \times 10^5$$

The root-mean-square speed $(c_{\mathrm{r.m.s.}}) = \sqrt{<c^2>} = 480\,\mathrm{m\,s^{-1}}$.

The r.m.s. or root-mean-square' speed $(c_{\mathrm{r.m.s.}})$ is the square root of the mean value of the squares of the speeds.

Kinetic energy of a gas molecule

Comparing $pV = \frac{1}{3} Nm <c^2>$ with the equation of state for an ideal gas:

$$\frac{1}{3} Nm <c^2> = nRT$$

$$\frac{1}{2} m <c^2> = \frac{3}{2}\left(\frac{n}{N}\right)RT$$

kinetic energy of a gas molecule $= \frac{1}{2}m <c^2> = \frac{3}{2}kT$

where k is the Boltzmann constant $(k = 1.38 \times 10^{23}\,\mathrm{J\,K^{-1}})$.

So the average kinetic energy of a gas molecule is directly proportional to the absolute temperature of the gas.

N = number of molecules

n = number of moles

$\dfrac{N}{n} = N_A$ (Avogadro's constant)

$k = \dfrac{R}{N_A} = \dfrac{8.31}{6.02 \times 10^{23}}$

$= 1.38 \times 10^{23}\,\mathrm{J\,K^{-1}}$

The value of k is provided in Exam Papers 1, 2, and 4.

You may be asked to deduce the relationship between average kinetic energy and absolute temperature.

⬆ Raise your grade

(a) State two assumptions made in the kinetic theory of gases.

The atoms have elastic collisions with each other and the walls of the container.
The volume of individual atoms is negligible compared to the volume of the container. ✔✔ [2]

A good answer. Other answers include: no inter-molecular forces/molecules in random motion/time of collision negligible compared to time between collisions/large number of molecules.

(b) Using the kinetic theory, it can be shown that the pressure p and the volume V of a gas are related by the equation: $pV = \dfrac{1}{3}Nm<c^2>$

State the meaning of:

(i) Nm [2]

The mass of the gas ✔✘

The answer is correct, but note that there are 2 marks available. The second mark is for identifying m as the mass of an individual atom/molecule and/or N as the number of atoms/ molecules.

(ii) $<c^2>$ [1]

The speed of the gas molecules squared ✘

The symbol $< >$ denotes a mean (average) value. $<c^2>$ is the mean of the squares of the velocities of all the gas atoms/molecules.

(c) Use this equation to show that the density ρ of a gas is given by $\rho = \dfrac{3p}{<c^2>}$. [1]

$pV = \dfrac{1}{3}Nm<c^2> \Rightarrow p = \dfrac{1}{3}\dfrac{Nm}{V}<c^2>$ but $\dfrac{Nm}{V} = \rho$ (the density)

so $\rho = \dfrac{3p}{<c^2>}$ ✔

$\dfrac{Nm}{V}$ correctly identified as the density.

(d) (i) An ideal gas obeys the equation $pV = NkT$ where k is the Boltzmann constant. Combine this equation with the equation given in part (b) to show that the kinetic energy of a gas molecule is proportional to T. [1]

$pv = \dfrac{1}{3}Nm<c^2> = NkT$ ✔

N cancels and equation re-arranged to find $\dfrac{1}{2}m<c^2>$.

so $\dfrac{1}{2}m<c^2> = \dfrac{3}{2}kT \rightarrow$ the kinetic energy of a gas molecule is proportional to T.

(ii) A gas, initially at a temperature of $50\,°C$, is heated to a temperature of $150\,°C$. Compare the kinetic energies of a gas molecule at these two temperatures. [2]

Kinetic energy of a gas molecule is proportional to T so kinetic energy at $150\,°C$ is three times kinetic energy at $50\,°C$. ✘✘

A frequent error. The temperature T is the **absolute** temperature, measured in kelvin. The answer is:

$$\frac{\text{kinetic energy at }150°C}{\text{kinetic energy at }50°C} = \frac{(273+150)}{273+50} = 1.31$$

1 **(a)** State what is meant by an *ideal* gas. [1]

  (b) An oxygen cylinder contains oxygen at a pressure of 1.37×10^7 Pa and a temperature of $20\,°C$. Calculate the pressure if the temperature rises to $70\,°C$. [2]

2 **(a)** What is the equation of state for an ideal gas? [1]

  (b) A weather balloon is released from ground level, where the temperature is $20\,°C$. The pressure inside the balloon is 1.2×10^5 Pa.

  The temperature of the atmosphere decreases by $5\,°C$ for each kilometre the balloon rises.

  Calculate:

  (i) the temperature of the atmosphere at a height 3 km above ground level

  (ii) the pressure of the gas inside the balloon at this height. [Assume the volume of the balloon remains constant.] [3]

3 **(a)** Define *Avogadro's constant*. [1]

  (b) A gas cylinder contains 9.0×10^{-3} m³ of argon (^{40}Ar) at a pressure of 1.37×10^7 Pa and a temperature of $20\,°C$.

  Calculate:

  (i) the number of moles of argon

  (ii) the mass of argon in the cylinder. [3]

  (c) For an ideal gas, $pV = \dfrac{1}{3} Nm <c^2>$.

  (i) Explain what is meant by $<c^2>$.

  (ii) Calculate the root-mean-square speed of the argon atoms. [3]

4 **(a)** State what is meant by a *mole* of gas. [1]

  (b) An ideal gas at a temperature of $25\,°C$ exerts a pressure of 2.0×10^5 Pa. The volume of the gas is 4.0×10^{-3} m³. Calculate the number of moles of gas present. [2]

  (c) The gas is heated at constant volume so that the pressure increases to 2.5×10^5 Pa. What is the new temperature of the gas? [1]

5 **(a) (i)** Describe what is meant by the *internal energy* of a gas.

  (ii) Explain how this differs from the internal energy of an ideal gas. [2]

  (b) Calculate the internal energy of:

  (i) 30 g of helium (^{4}He) at $50\,°C$,

  (ii) 0.5 g of krypton (^{84}Kr) at $200\,°C$. [3]

  [Both gases can be treated as ideal gases.]

6 The equation of state for an ideal gas with pressure p and volume V is:

$$pV = nRT$$

  (a) State the meaning of the symbol:

  (i) n **(ii)** R **(iii)** T. [3]

  (b) A gas cylinder contains 5.0×10^{-4} m³ of xenon at a pressure of 20.0×10^5 Pa and a temperature of $27\,°C$. Determine the number of atoms of xenon that are in the cylinder. [3]

7 A piston of mass 0.400 kg can slide freely up and down a cylinder of inner diameter 12.0 cm. The temperature of the air trapped inside the cylinder is $20\,°C$. The piston rests at a height of 8.0 cm above the base of the cylinder.

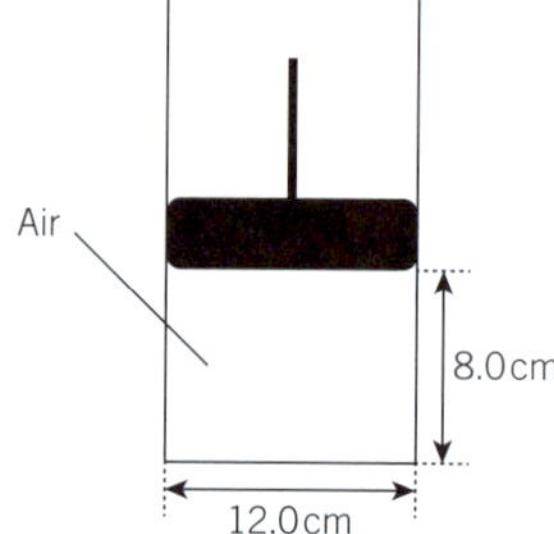

  (a) Calculate the pressure (above atmospheric pressure) of the air trapped in the cylinder. [1]

  (b) The temperature of the air is increased to $80\,°C$. Determine the new height of the piston in the cylinder. [2]

8 A canister of volume 2.50×10^{-4} m³ contains air at a pressure of 1.01×10^5 Pa. The temperature of the air is $20\,°C$.
[Specific heat capacity of air at constant volume is 0.716 kJ kg^{-1} K^{-1}. Molar mass of air is 29 g mol^{-1}.]

  (a) Determine how many moles of air are in the canister. [2]

  (b) The air in the canister is heated to $80\,°C$.

  Calculate:

  (i) the pressure of the air at the new temperature

  (ii) the change in internal energy of the air. [3]

Key points

- ☐ Understand that heat (thermal) energy is transferred from a region of higher temperature to a region of lower temperature.
- ☐ Understand that regions of equal temperature are in thermal equilibrium.
- ☐ Understand that a physical property that varies with temperature may be used for the measurement of temperature and state examples of such properties.
- ☐ Understand that there is an absolute scale of temperature that does not depend on the property of any particular substance (i.e., the thermodynamic scale and the concept of absolute zero).
- ☐ Convert temperatures measured in kelvin to degrees Celsius and recall that $T/K = T/°C + 273.15$.
- ☐ Compare the relative advantages and disadvantages of thermistor and thermocouple thermometers.

Heat and temperature

Heat is the movement of energy caused by a temperature difference. If one body (A) is in thermal contact with another body (B) at a lower temperature, then heat energy will flow from A to B (see Figure 11.1).

The extra energy that B gains can either increase the average energy of the molecules of B or cause B to do physical work. For example, if B is a gas, the extra energy can cause the gas to expand (doing external work) or the internal energy of the gas to increase (or both).

If A and B are at the same temperature, no net heat flow takes place. The two bodies are in thermal equilibrium (see Figure 11.2).

Measuring temperature

A wide range of **thermometric properties** can be used to measure temperature, including:

- expansion of a liquid (e.g., mercury or alcohol)
- change in electrical resistance of a metal wire (e.g., platinum)
- change of electrical resistance of a thermistor
- change in the output p.d. of a thermocouple
- change in pressure of a fixed volume of gas.

Some of these physical properties vary linearly with a change in temperature over a large range of temperatures (e.g., the resistance of a metal wire); others do not vary linearly, or do so only over a small range of temperatures (e.g., the resistance of a thermistor).

If $T_1 > T_2$ heat energy flows from A to B

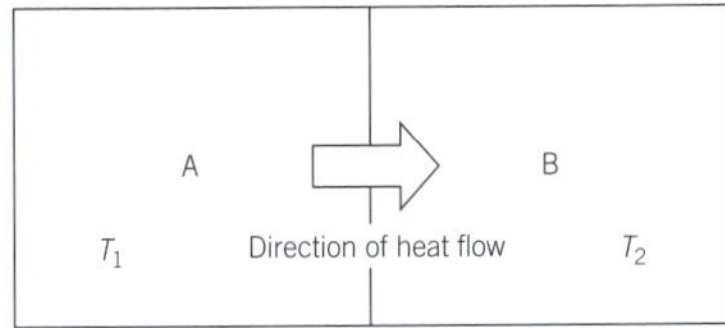

▲ **Figure 11.1** Heat flow when $T_1 > T_2$

Thermal equilibrium: $T_1 = T_2$ no **net** heat energy flows from A to B

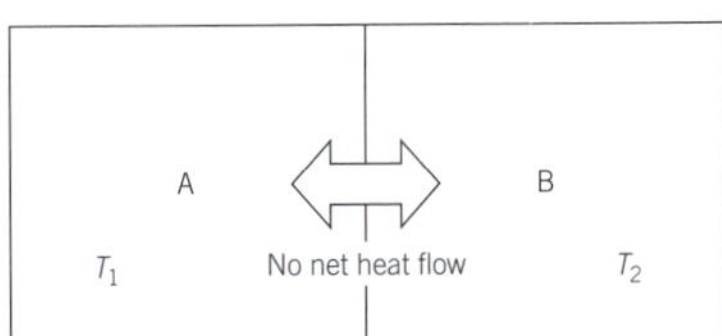

▲ **Figure 11.2** Thermal equilibrium when $T_1 = T_2$

> 💡 **Remember**
>
> A **thermometric property** is one which changes with temperature; e.g., electrical resistance.

Liquid-in-glass thermometers

Liquids such as mercury and alcohol expand as their temperature rises. The volume of liquid is small, allowing the thermometer to respond quickly to changes in temperature, and is contained in a thin-walled glass bulb to aid the conduction of heat energy (see Figure 11.3). Making the capillary tube finer increases the sensitivity of the thermometer.

Liquid-in-glass thermometers are relatively cheap and portable, but cannot be used to read temperatures remotely or electronically. They have a restricted range, limited by the freezing and boiling points of the liquids. They are also fragile and mercury is poisonous.

They are usually calibrated by first placing the uncalibrated thermometer in melting ice (Figure 11.4a) and then directly above boiling water (Figure 11.4b) to establish two 'fixed points' on the temperature scale, as shown in Figure 11.3. The distance between the two fixed point marks can then be subdivided into 100 equal degrees ('centigrade').

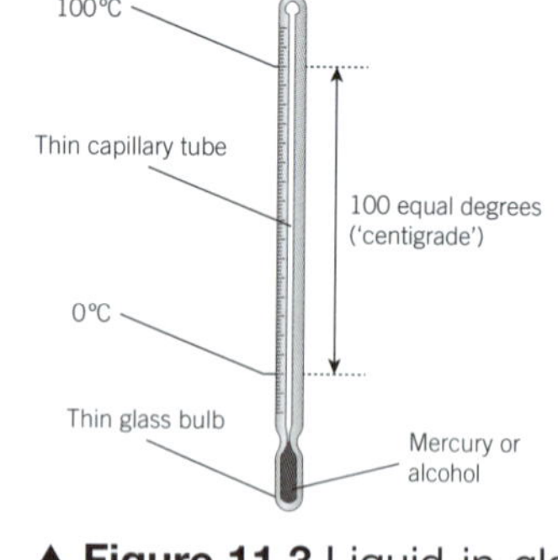

▲ **Figure 11.3** Liquid-in-glass thermometer

Metal resistance thermometers

Metal resistance thermometers (sometimes called RTDs – resistance temperature detectors) consist of a length of fine wire wrapped around a ceramic (or glass) strip or rod, as shown in Figure 11.5. The wire is usually platinum, nickel or copper. Platinum is particularly good as the relationship between resistance and temperature is extremely linear.

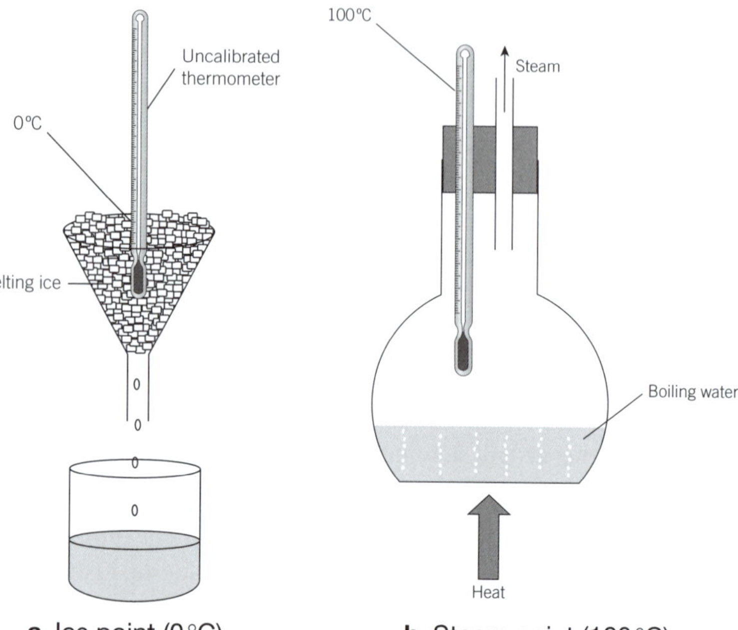

▲ **Figure 11.4** Calibrating a thermometer

Platinum resistance thermometers have a wide range of operating temperatures (typically −200 °C to 1200 °C) and are very accurate, but are unsuitable for rapidly changing temperatures because of the relatively high heat capacity of the wire.

Thermistors

Thermistors are semiconductors consisting of a mixture of metals and metal oxides (see Figure 11.6). The electrical resistance of most thermistors decreases as the temperature increases (NTC – negative temperature coefficient) though the resistance of some thermistors increases with increasing temperature (PTC – positive temperature coefficient). In both cases the relationship between temperature and resistance is non-linear so a **calibration curve** is also needed.

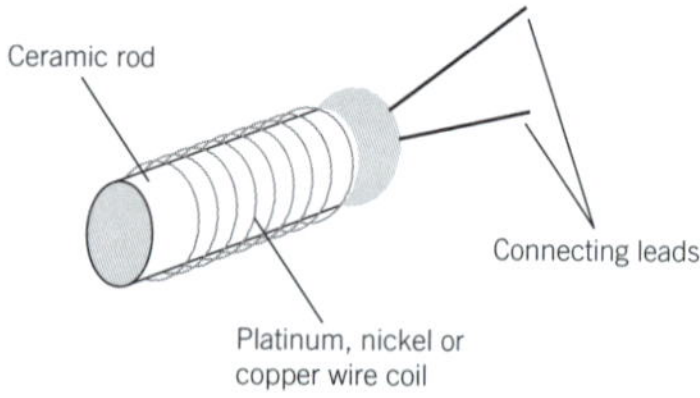

▲ **Figure 11.5** Metal resistance thermometer

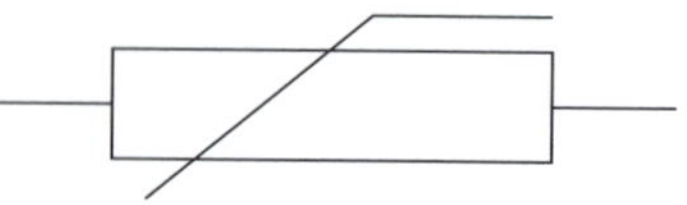

▲ **Figure 11.6** Symbol for a thermistor

Thermocouples

A **thermocouple** consists of two different metal wires, such as iron and copper, or platinum and constantan (an alloy of copper and nickel) joined together to form two junctions as shown in Figure 11.7.

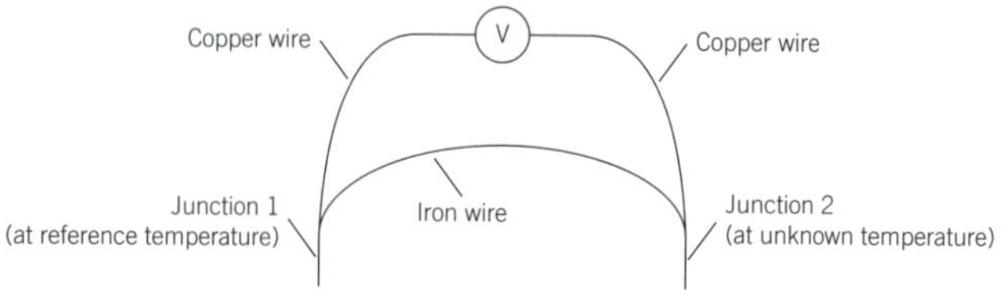

▲ **Figure 11.7** Thermocouple

When the two junctions are at different temperatures an e.m.f. is produced. The larger the difference in temperature, the greater the e.m.f., but the relationship is not linear, and so a calibration graph of p.d. against T is also needed.

The thermocouple junctions require little heat energy to warm up so they respond very quickly to changes in temperature. They can also be made quite small so that the temperature at precise locations can be found. As the output from the thermocouple is electrical it can be recorded and used as part of a control system to monitor temperature and give warnings if a temperature becomes too high or too low. Temperatures up to 2000 °C can be measured.

Link

For more about calibration curves, see Unit 2 *Measurement techniques*

Constant-volume gas thermometer

If the volume of a fixed mass of gas is kept constant when heated, the change in pressure is proportional to the change in temperature. Constant-volume gas thermometers have a very wide range of temperatures, but are bulky and slow to respond (see Figure 11.8). They are used mainly as a standard to help calibrate other, more practical thermometers.

In Figure 11.8, as the gas is heated by the surrounding liquid, the increased pressure pushes the mercury column down on one side of the tube and up on the other. The difference in height h is proportional to the change in temperature.

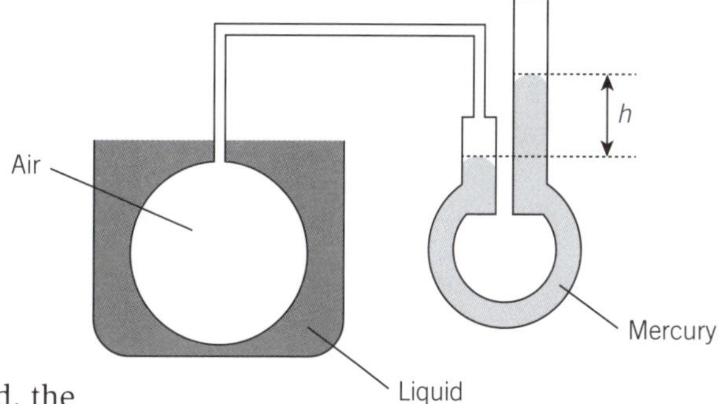

▲ **Figure 11.8** Constant-volume gas thermometer

Comparing different types of thermometer

▼ **Table 11.1** Advantages and disadvantages of different types of thermometer

Type of thermometer	Thermometric property	Advantages	Disadvantages
Liquid-in-glass (e.g., mercury, alcohol)	Liquid expands as temperature rises	Easy to use, portable, high precision	Fragile, restricted range, must be read directly, cannot measure the temperature of small objects
Platinum resistance	Resistance increases (linearly) as temperature rises	Very accurate, wide range of temperatures, can be used remotely	Slow response – not suitable for rapidly changing temperatures, or measuring the temperature of small objects
Thermistor	Resistance decreases (NTC) or increases (PTC) as temperature rises	Low cost, small size	Non-linear and relatively small range of operating temperatures (–100 °C to +300 °C)
Thermocouple	E.m.f. produced across thermocouple which depends on the difference in temperature between the two thermocouple junctions	Fast response, wide range, remote readings possible with long leads. Can measure temperature at a specific point	Metals can corrode. One junction needs to be maintained at a constant temperature. E.m.f. produced is small (a few microvolts/°C) so sensitive instrument needed to measure e.m.f.
Constant-volume gas thermometer	Pressure of gas increases with increasing temperature	Wide range of temperatures, very accurate. Used as a 'standard' thermometer to calibrate other types of thermometer	Bulky and slow to respond

Worked example

A platinum resistance wire has a resistance of $760\,\Omega$ when the temperature is $30\,°C$ and a resistance of $830\,\Omega$ when the temperature is $70\,°C$.

a) Determine the temperature of the wire when its resistance is $795\,\Omega$.

b) Calculate the resistance of the wire when the temperature is $55\,°C$

c) State any assumptions you make in deriving your answers.

Answer

a) $\dfrac{\text{Change in temperature}}{\text{change in resistance}} = \dfrac{70-30}{830-760} = 0.571°C\,\Omega^{-1}$, so when $R=795\,\Omega$,

temperature $= 30 + (795 - 760) \times 0.571 = 50\,°C$

b) $30 + (R - 760) \times 0.571 = 55\,°C,\ R - 760 = \dfrac{55-30}{0.571} = 43.8$, so $R = 804\,\Omega$

c) The resistance of the wire varies linearly with temperature for the range of temperatures in the question.

Temperature scales

Most temperature scales are thermometric scales in that they rely on the properties of a particular substance to establish the fixed points of the scale. The **Celsius scale**, for example, uses the melting point of ice ($0\,°C$) and the boiling point of water ($100\,°C$) as its fixed points.

The **absolute (thermodynamic) scale** of temperature does not depend on physical properties. Instead the scale relies on two fixed points:

Remember

To convert temperature from Celsius scale to absolute scale:

$$T/\mathrm{K} = T/°C + 273.15$$

- **absolute zero**: the temperature at which a substance has minimum internal energy (the atoms or molecules of the substance have no random kinetic energy, but may still have some potential energy),

- **triple point of water**: the temperature and pressure at which water exists in equilibrium as a solid, liquid and vapour ($0.01\,°C$ and $611.2\,Pa$).

The size of the unit of temperature on the thermodynamic scale of temperature is chosen to be the same as the size of a degree on the Celsius scale – a $1\,°C$ change in temperature is the same as a change of 1 kelvin ($1\,K$).

 # Raise your grade

A student is investigating how the resistance $R(\Omega)$ of a thin copper wire changes with temperature $T\,(°C)$.

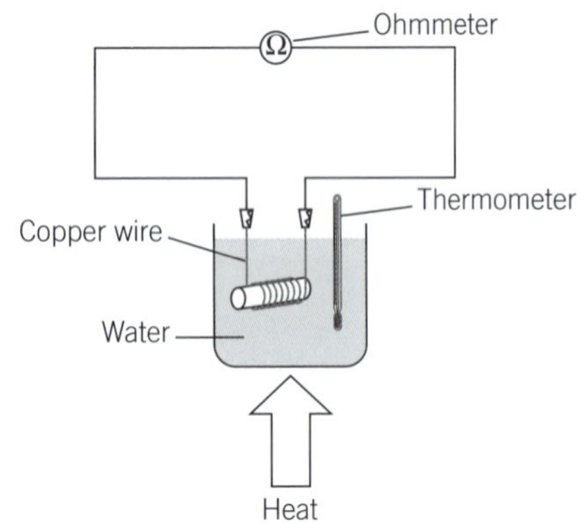

$T/°C$	R/Ω
25	4.4 ± 0.05
38	4.6 ± 0.05
51	4.8 ± 0.05
65	5.0 ± 0.05
69	5.1 ± 0.05
89	5.4 ± 0.05

(a) (i) Plot a graph of R/Ω against $T/°C$. Include error bars for R. [2]

(ii) Draw the straight line of best fit and a worst acceptable straight line on your graph. Both lines should be clearly labelled. [2]

(iii) Determine the gradient and intercept of the line of best fit.

$$\text{Gradient} = \frac{5.16 - 4.44}{74.0 - 28.0} = 0.0157\,\Omega^{-1}\ ✔$$

Correct read-offs and substitution into $\dfrac{\Delta y}{\Delta x}$ to find the gradient.

Using the point (84.0, 5.32) in $y = mx + c$:
$c = y - mx = 5.32 - 0.0157 \times 84.0 = 4.00\,\Omega$ ✔ [2]

gradient = $0.0157\,\Omega\,°C^{-1}$

Correct read-off and substitution into $y = mx + c$.

intercept = $4.00\,\Omega$

'Worst acceptable' line drawn – 'bottom-right to top-left'.

Points plotted accurately.

Error bars drawn correctly.

'Best fit' line drawn accurately.

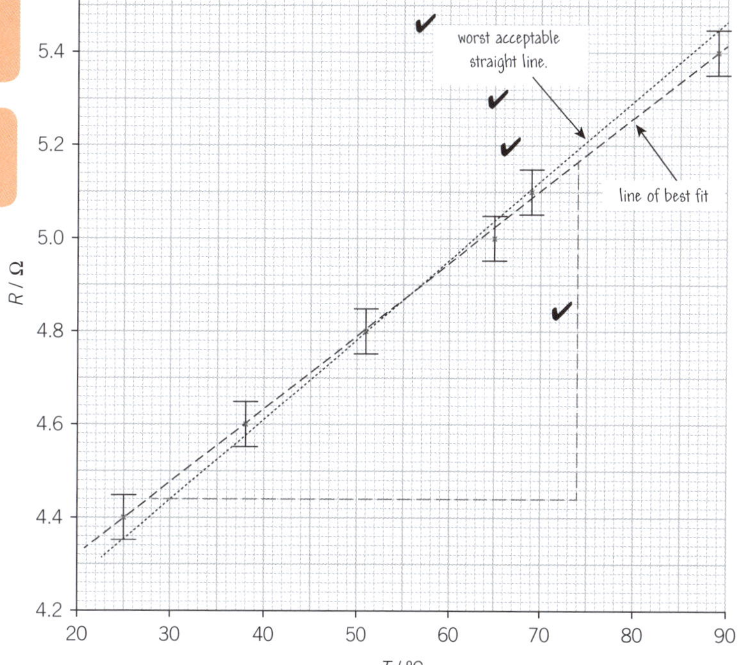

(b) R and T are related by the equation:

$$R = a(1 + bT) \text{ where } a \text{ and } b \text{ are constants.}$$

Use your values from (a)(iii) to determine the values of a and b. Give appropriate units.

$a = y\text{-intercept} = 4.00$

$ab = \text{gradient} = 0.0157$

$b = \dfrac{\text{gradient}}{a} = \dfrac{0.0157}{4.00} = 3.93 \times 10^{-3}$ ✔✗

The candidate has matched a with the y-intercept and ab with the gradient, but has omitted the units for a and b.

The units for a must be the same as the units for R/Ω.

bT must be dimensionless, so b has units of $°C^{-1}$.

$a = $ _____4.00_____ [2]

$b = $ _____3.93×10^{-3}_____

1 **(a)** Convert into kelvin, and to an appropriate number of decimal places:

   (i) 100.00 °C

   (ii) 0 °C

   (iii) − 80.7 °C [3]

   (b) Convert into °C, and to an appropriate number of decimal places:

   (i) 273 K

   (ii) 376.2 K

   (iii) 0.0 K [3]

   (c) State and explain what is meant by *thermal equilibrium*. [2]

2 Describe a method of calibrating a liquid-in-glass thermometer between 0 °C and 100 °C. [3]

3 Describe one advantage and one disadvantage of a thermistor used as a thermometer compared to a thermocouple. [2]

4 **(a) (i)** State the two *fixed points* used for the absolute temperature scale.

   (ii) Describe the difference between the absolute temperature scale and other temperature scales. [3]

   (b) State what is meant by the absolute zero of temperature. [1]

   (c) A heater raises the temperature inside an incubator from 19.7 °C to 37.5 °C.

   Determine, in kelvin and to an appropriate number of decimal places:

   (i) the rise in temperature of the incubator

   (ii) the final temperature of the incubator. [2]

5 A thermocouple produces an e.m.f. of 5.5 µV for each 1.0 °C temperature difference between the two junctions of the thermocouple. The cold junction is maintained at a constant temperature of 20.0 °C.

   (a) Calculate:

   (i) the temperature of the hot junction when the e.m.f. is 1.21 mV

   (ii) the e.m.f. when the temperature of the hot junction is 1200 °C. [2]

   (b) State any assumptions you make. [1]

6 A mercury-in-glass thermometer and a platinum resistance thermometer are both used to measure the temperature of a water bath.

   (a) The length of the mercury column at the ice point is 16.3 mm; at the steam point the length of the column is 53.1 mm. When placed in the water bath, the length of the mercury column is 32.8 mm.

   Determine the temperature of the water bath on the centigrade scale, as measured by the mercury-in-glass thermometer. [2]

   (b) The resistance of the platinum resistance thermometer at the ice point is 2.875 Ω; at the steam point it is 4.621 Ω. When placed in the water bath its resistance is 3.663 Ω.

   Determine the temperature of the water bath on the centigrade scale, as measured by the platinum resistance thermometer. [2]

   (c) Suggest a reason why your answers to (a) and (b) are not the same. [1]

7 State, with reasons, which type of thermometer you would use to measure the following:

   (a) the melting point of wax [1]

   (b) the temperature of a Bunsen flame [1]

   (c) the air temperature in the Antarctic. [1]

8 A student is investigating the relation between the resistance R of a platinum wire and its temperature T.

   It is suggested that the relationship is:

   $$R = R_0(1 + \alpha T)$$

   where R_0 and α are constants.

   Design a laboratory experiment to test the relationship between R and T. Explain how your results could be used to determine values for R_0 and α. You should draw a diagram showing the arrangement of your equipment. In your account you should pay particular attention to:

   • the procedure to be followed

   • the measurements to be taken

   • the control of variables

   • the analysis of the data

   • any safety precautions to be taken. [15]

Key points

- Explain, using a simple kinetic model for matter:
  - the structure of solids, liquids, and gases
  - why melting and boiling take place without a change in temperature
  - why the specific latent heat of vaporisation is higher than the specific latent heat of fusion for the same substance
  - why a cooling effect accompanies evaporation.
- Define and use the concept of specific heat capacity, and identify the main principles of its determination by electrical methods.
- Define and use the concept of specific latent heat, and identify the main principles of its determination by electrical methods.
- Understand that internal energy is determined by the state of the system, and that it can be expressed as the sum of a random distribution of kinetic and potential energies associated with the molecules of a system.
- Relate a rise in temperature of a body to an increase in its internal energy.
- Recall and use the first law of thermodynamics $\Delta U = q + w$ expressed in terms of the increase in internal energy, the heating of the system (energy transferred to the system by heating) and the work done on the system.

Specific heat capacity and specific latent heat

Solids, liquids and gases

Solids

The atoms and molecules in a solid substance are close together, held in place by strong forces of attraction between them (see Figure 12.1a). Heating a solid causes the molecules to vibrate more (the temperature of the solid increases). If enough heat energy is supplied, the molecules can break free of each other, and the substance starts to melt.

Liquids

In liquids, the atoms and molecules are free to slip past each other and move about at random (see Figure 12.1b). The forces of attraction between molecules are much smaller, allowing the liquid to flow. Heating a liquid cause the temperature to rise as the molecules gain more kinetic energy. The fastest molecules, near the surface of the liquid, may have enough energy to escape completely (evaporate). If enough heat energy is supplied, all the molecules have enough energy to break free and the liquid starts to boil.

Gases

The atoms or molecules that make up a gas are much further apart and moving at high speeds, continually having elastic collisions with the walls of their container and each other (see Figure 12.1c). Heating a gas causes the average molecular speed to increase and the temperature of the gas to rise.

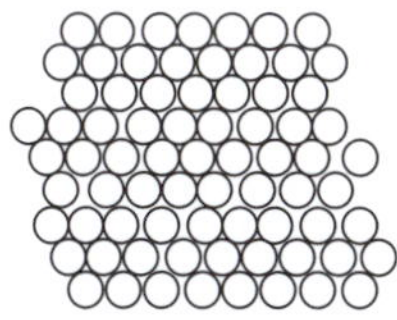
a Solids

b Liquids

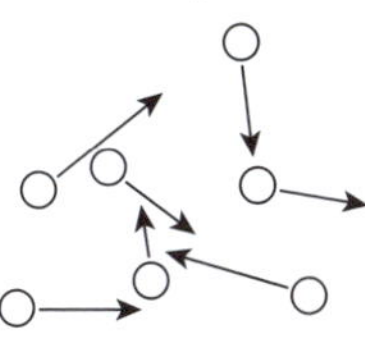
c Gases

▲ **Figure 12.1** Solids, liquids, and gases

Melting and boiling

If a solid such as ice is heated at a uniform rate, its temperature rises until it reaches its **melting point**. Once completely melted, its temperature continues to rise until it reaches its **boiling point**. Once boiled, the temperature of the gas continues to rise. Figure 12.2 illustrates this process.

- As a solid is heated from A to B, its temperature rises. The energy needed to raise the temperature of 1 kg of a substance by 1 °C is called the **specific heat capacity** c of the substance.

- From B to C the substance is melting. All the heat energy supplied is being used to weaken or break the bonds between molecules, so the molecules do not gain any extra kinetic energy during this stage – the temperature of the substance remains constant.

- The energy needed to melt 1 kg of the substance at its melting point is called the **specific latent heat of fusion** L_f. (When a liquid solidifies energy is released.) The amount of energy needed to melt m kg of a substance at its melting point is mL_f.

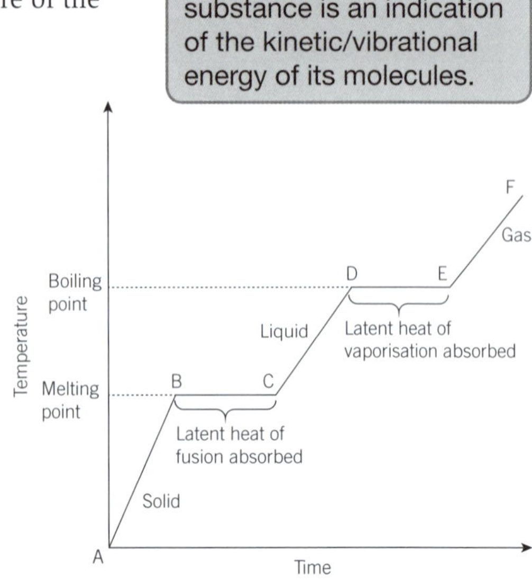

▲ **Figure 12.2** Melting and boiling

- From C to D the temperature of the liquid rises until it reaches its boiling point. If the specific heat capacity of the liquid is c, then the heat energy needed to raise the temperature of a mass m of the liquid by an amount $\Delta\theta$ is $mc\,\Delta\theta$.

- From D to E all the heat energy supplied is being used to break the bonds between molecules completely, and the temperature of the liquid/gas remains constant until this process is complete. The energy needed to change 1 kg of a liquid at its boiling point into gas is called the **specific latent heat of vaporisation** L_v. This is much larger than the specific latent heat of fusion because the molecules have to break away from each other completely during vaporisation. When a gas condenses back into a liquid this energy is released as heat into the surroundings. The amount of energy needed to vaporise m kg of a substance at its boiling point is mL_v.

- From E to F the molecules gain more kinetic energy and the temperature of the gas rises.

Cooling by evaporation

At any one instant the molecules of a liquid have a range of kinetic energies (see Figure 12.3). The more energetic particles which also happen to be at the surface of the liquid may have enough energy to 'escape' completely (become vaporised). If the fastest molecules evaporate, the average energy of the molecules left behind decreases; its temperature falls slightly compared to its surroundings.

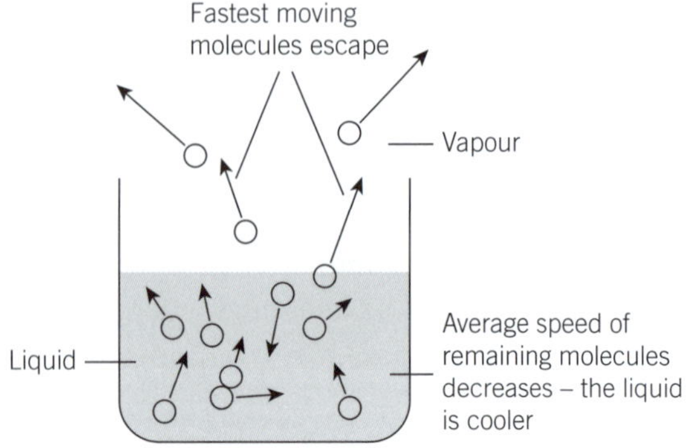

▲ **Figure 12.3** Cooling by evaporation

Worked example

a) State two differences between evaporation and boiling.

b) Explain why the specific latent heat of vaporisation of a substance is much greater than the specific latent heat of fusion.

Answer

a) Evaporation can occur at any temperature but only at the surface of a liquid. Boiling only occurs at the boiling point of the substance but takes place throughout the liquid. Boiling usually requires a heat source but is a relatively rapid process; evaporation draws heat energy from the surroundings but is a relatively slow process.

b) The energy required to completely separate molecules when changing from liquid to gas is much greater than that needed to 'loosen' the bonds between molecules when changing from solid to liquid.

> ★ **Exam tip**
>
> **Specific** in a definition means 'for 1 kg'. The units of specific heat capacity are $J\,kg^{-1}\,K^{-1}$. The units of specific latent heat are $J\,kg^{-1}$.

> 💡 **Remember**
>
> The *specific latent heat* of a substance is the energy needed to change the state of 1 kg of the substance **at constant temperature**.

Finding the specific heat capacity of a metal

A metal of known mass m is placed in an insulated container and heated with an electrical heater which fits into a hole drilled into the metal, as shown in Figure 12.4. A thermometer, fitted into a second drilled hole, measures the temperature change.

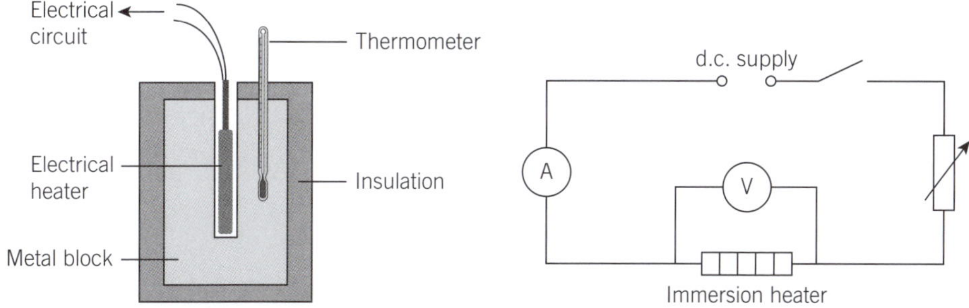

▲ **Figure 12.4** Measuring specific heat capacity

The electrical power of the heater is IV, where I is the current through the heater and V the potential difference across it. A stopwatch records the time t the heater is switched on and the thermometer measures the change in temperature $\Delta\theta$.

If no heat is lost to the surroundings: $mc\Delta\theta = IVt$ so $c = \dfrac{IVt}{m\Delta\theta}$

The value is likely to be an overestimate of the true value as some energy will always be lost to the surroundings.

Worked example

A cylinder of aluminium, of mass 1.0 kg, is heated by a 50 W electric immersion heater. After 4 minutes the temperature of the cylinder has risen 11 °C. Determine the specific heat capacity of aluminium.

Answer

energy supplied to aluminium $= 50 \times (4 \times 60) = 12\,000\,J$

specific heat capacity of aluminium $= \dfrac{12\,000}{1 \times 11} = 1.09 \times 10^3\,J\,kg^{-1}\,K^{-1}$

> The correct value is $920\,J\,kg^{-1}\,K^{-1}$.

Finding the specific heat capacity of a liquid using a calorimeter

A simple calorimeter is a metal canister (made from copper or aluminium) of known mass and specific heat capacity, with a tight-fitting lid (see Figure 12.5). The calorimeter is insulated to reduce heat lost to the surroundings as much as possible. A known mass of the liquid is heated using a small electrical heater and $\Delta\theta$, the rise in temperature, is recorded.

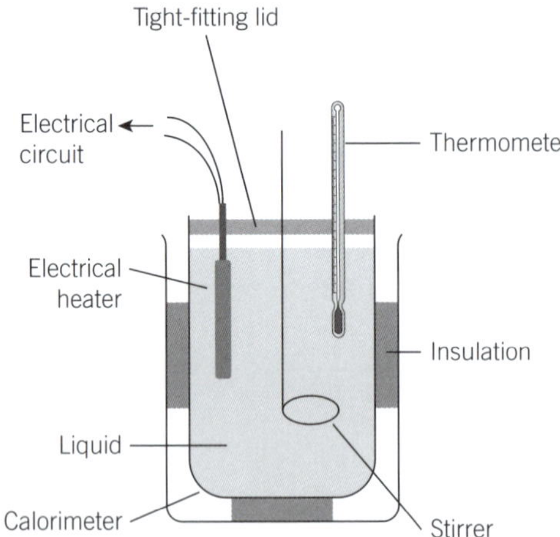

▲ **Figure 12.5** Measuring the specific heat capacity of a liquid

The same method for measuring the electrical energy supplied to find the specific heat capacity of a solid is used, but the heat energy absorbed by the calorimeter also has to be taken into account.

Worked example

500 cm³ of water are heated from 18 °C to 26 °C in a copper calorimeter of mass 200 g. The water is heated by a 50 W electrical immersion heater for 6.0 minutes. Calculate the specific heat capacity of water c_w.

[Specific heat capacity of copper is 390 J kg⁻¹ K⁻¹; density of water is 1.0 g cm⁻³.]

Answer

Mass of 500 cm³ water = 500 g

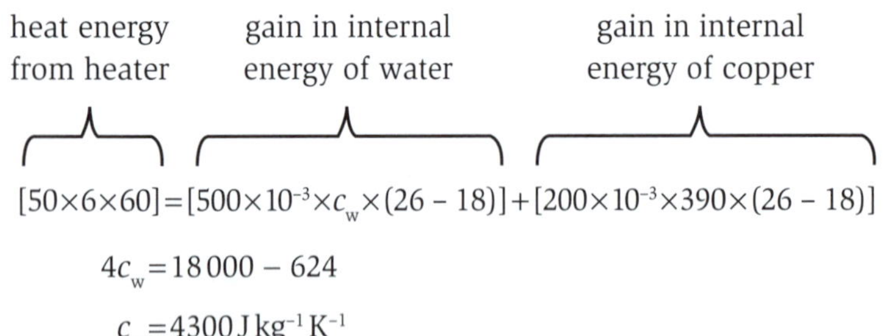

$$[50 \times 6 \times 60] = [500 \times 10^{-3} \times c_w \times (26 - 18)] + [200 \times 10^{-3} \times 390 \times (26 - 18)]$$

$$4c_w = 18\,000 - 624$$

$$c_w = 4300 \text{ J kg}^{-1} \text{ K}^{-1}$$

Finding the specific latent fusion of a solid

Figure 12.6 shows how the specific latent heat of fusion of ice can be found. Ice at its melting point is heated by an electrical immersion heater. The amount of ice melted in a fixed time can be found by weighing the beaker before and after heating. The heat energy needed to melt an amount m of ice at its melting point is mL_f.

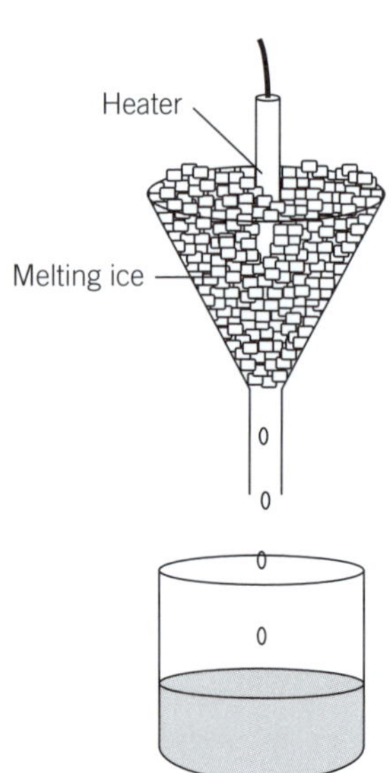

▲ **Figure 12.6** Measuring the specific latent heat of fusion

Worked example

Ice at its melting point is heated by a 50 W immersion heater. After 10 minutes it is found that 100 g of ice has melted.

a) Calculate the specific latent heat of ice.

b) State, with a reason, why your answer is an overestimate or an underestimate of the true value.

Answer

a) $(100 \times 10^{-3}) \times L_F = 50 \times (10 \times 60)$

$$L_F = 3.0 \times 10^5 \, \text{J kg}^{-1}$$

> The true value is $3.3 \times 10^5 \, \text{J kg}^{-1}$.

b) This is an underestimate because some of the heat energy to melt the ice comes from the surroundings. This can be compensated for by repeating the experiment with the heater removed or switched off. The amount of water that collects in the beaker can be subtracted from the first value to find a more accurate value for L_F.

Finding the specific latent heat of vaporisation of a liquid

The liquid is first heated to its boiling point, as shown in Figure 12.7.

As the liquid boils the vapour passes through a condenser and collects in a beaker. The liquid is boiled for a fixed time t, and the current I in the heater and potential difference V across the heater are recorded. The amount of condensed vapour collected m can be found by weighing the beaker before and after the liquid has boiled. The amount of energy needed to boil m kg at its boiling point is mL_V.

The experiment can be repeated with different values of I and V but boiling the liquid for the same fixed time t. The results can be used to obtain an accurate value for L_V, the specific latent heat of vaporisation of the liquid, by eliminating the heat energy lost to the surroundings during the experiment.

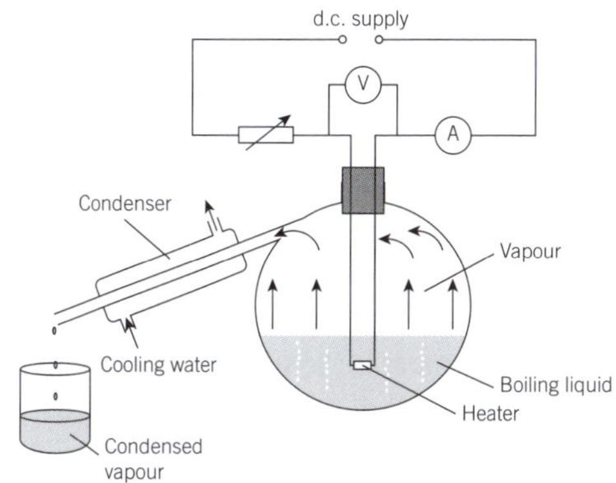

▲ **Figure 12.7** Measuring the specific latent heat of vaporisation

Worked example

A liquid is boiled for 5 minutes, using the apparatus shown in Figure 12.7. The mass m of condensed liquid collected is found by weighing the beaker before and after boiling the liquid. The experiment is repeated with a different set of values for the current I in the heater and p.d. V across the heater. The results are shown in Table 12.1.

▼ **Table 12.1**

I/A	V/V	m/g
3.5	18	43
2.5	11	16

Determine the specific latent heat capacity L_V of the liquid.

Answer

energy supplied by heater (IVt) = energy used to vaporise liquid (mL_V) + energy lost to surroundings (H)

1st experiment: $3.5 \times 18 \times (5 \times 60) = (43 \times 10^{-3}) \times L_V + H$ (eqn 1)

2nd experiment: $2.5 \times 11 \times (5 \times 60) = (16 \times 10^{-3}) \times L_V + H$ (eqn 2)

(eqn 1) − (eqn 2): $27 \times 10^{-3} L_V = 1.065 \times 10^4$

$$L_V = 3.94 \times 10^5 \, \text{J kg}^{-1}$$

Internal energy and the first law of thermodynamics

The **internal energy of a system** is defined as:

internal energy of a system	=	sum of the random distribution of kinetic and potential energies of its molecules

The internal energy of an object or system can be increased by:

* heating the object

* doing mechanical work on the object.

If a gas is heated, its molecules move faster, and so have more kinetic energy. If the gas is compressed, the 'squashing' imparts kinetic energy to the molecules – when a bicycle pump is compressed quickly the end becomes hotter. The change in internal energy, the external work done, and the heat energy supplied to an object are linked together by the **first law of thermodynamics**.

The first law is really a statement of the **conservation of energy**. It states that the change in internal energy of a system (ΔU) is equal to the sum of the energy entering the system by heating (q) and the energy entering the system by work being done on it (w).

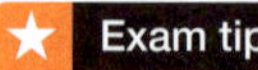

Make sure you understand the sign convention used for the 1st law:

* $+q$ means heat energy is supplied *to* the system

* $-q$ means heat energy is supplied *by* the system

* $+w$ means work is done *on* the system

* $-w$ means work is done *by* the system.

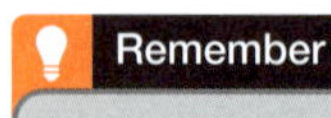

$$\Delta U = q + w$$

Worked examples

1. A litre of water at $100\,°C$ is left to evaporate. When it has all evaporated, the water vapour occupies a volume of $1.7\,m^3$. Calculate the change in internal energy of the water.

   [atmospheric pressure $= 1.01 \times 10^5\,Pa$; specific latent heat of vaporisation of water $= 2.26 \times 10^6\,J\,kg^{-1}$; density of water $= 1.0 \times 10^3\,kg\,m^{-3}$]

Answer

heat energy supplied to the water to evaporate it, q

$$q = (1.0 \times 10^{-3}) \times (1.0 \times 10^3) \times 2.26 \times 10^6 = 2.26 \times 10^6\,J$$

work done on atmosphere, w

$$w = -1.01 \times 10^5 \times (1.7 - 1.0 \times 10^{-3}) = -1.72 \times 10^5\,J$$

change in internal energy $\Delta U = q + w = 2.26 \times 10^6 - 1.72 \times 10^5$

$$\Delta U = 2.09 \times 10^6\,J$$

> w is negative because work is done on the atmosphere

2. When a rubber band is quickly stretched and released several times, its temperature increases. Describe what happens to the rubber band in relation to the first law of thermodynamics. State whether each of the following is positive, negative, or zero: ΔU, q, and w.

Answer

To stretch the rubber band a force is moved in the direction of the force; that is external work is done to the rubber band (w is positive). No heat energy is supplied from an external source ($q = 0$) so the change in internal energy $\Delta U = w$. As the rubber band becomes warmer than its surroundings it will start to emit heat energy (q is now negative).

Eventually the external work done w will be equal to the heat energy lost q, and the internal energy of the rubber band will remain constant:

$$\Delta U = w - q = 0$$

⬆ Raise your grade

(a) Define *specific latent heat.* [3]

The energy needed to change state ✔✗✗

> A better answer would be 'the energy needed, **per kg**, to change state, **at constant temperature.**' In the case of a liquid, it is the energy needed to change 1 kg from liquid at its boiling point to vapour.

(b) A student carries out an experiment to find the specific latent heat of vaporisation of water.

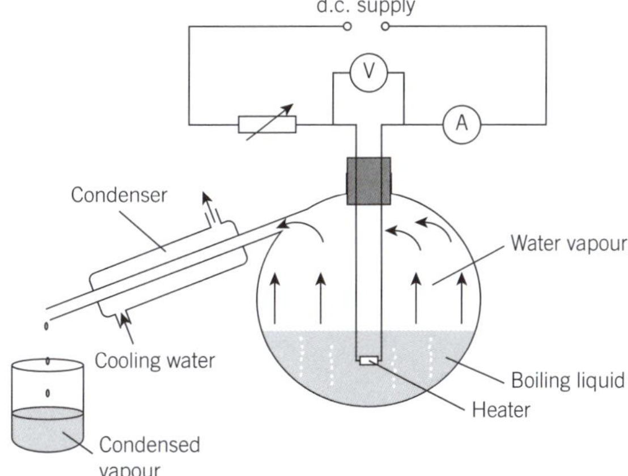

The water is heated up to its boiling point, and then boiled for 6.0 minutes. The mass m of condensed water vapour collected is found by weighing the beaker before and after boiling the water. The experiment is repeated with a different set of values for the current I in the heater and p.d. V across the heater.

The results of the experiment are shown.

I / A	V / V	m / g
3.0	16	47
2.0	12	21

Suggest why the same amount of heat energy H is lost to the surroundings in each experiment.

A good answer. [1]

The apparatus is at the same temperature for the same time in both experiments. ✔

(c) Calculate:

(i) the specific latent heat of vaporisation of the liquid L_v [3]

$$3.0 \times 16 \times (6 \times 60) = (47 \times 10^{-3}) \times L_v + H \quad (1) ✔$$

$$2.0 \times 12 \times (6 \times 60) = (21 \times 10^{-3}) \times L_v + H \quad (2) ✔$$

$$(1) - (2): 26 \times 10^{-3} \times L_v = 17280 - 8640$$

$$L_v = 3.3 \times 10^5 ✔$$

Correct method.

Correct analysis.

Correct final answer.

$$L_v = 3.3 \times 10^5 \, \text{J kg}^{-1} \quad [3]$$

(ii) the heat energy H lost to the surroundings.

From equation 2:

$$H = 2.0 \times 12 \times (6 \times 60) - (21 \times 10^{-3}) \times 3.3 \times 10^5$$

$$= 1710 \, \text{J} ✔$$

Correct calculation.

$$H = 1710 \, \text{J} \quad [1]$$

1 **(a)** Define *specific heat capacity*.

(b) Calculate: [1]

    (i) the energy needed to raise the temperature of 8.0 kg of aluminium by 5 °C

      [Specific heat capacity of aluminium is 920 J kg⁻¹ K⁻¹.]

    (ii) the specific heat capacity of lead if 0.48 kJ are needed to raise the temperature of 200 g of lead by 18 °C. [2]

2 A 3.0 kW electric heater is used to heat 140 kg of water in a tank from 20 °C to 65 °C. [Specific heat capacity of water is 4200 J kg⁻¹ K⁻¹.]

(a) Determine the time taken to heat the water. [2]

(b) State, with a reason, whether your answer to (a) is an under-estimate or an over-estimate. [2]

3 The height of the Angel Falls waterfall in Venezuela is 980 m.
[Specific heat capacity of water is 4200 J kg⁻¹ K⁻¹.]

(a) Describe the energy changes involved as the water falls from the top to the bottom of the waterfall. [2]

(b) Assuming 70 % of the change in potential energy is converted into internal energy of the water, determine the difference in temperature between the top and the bottom of the waterfall. [1]

4 A continuous flow calorimeter is used to find the specific heat capacity of a liquid.

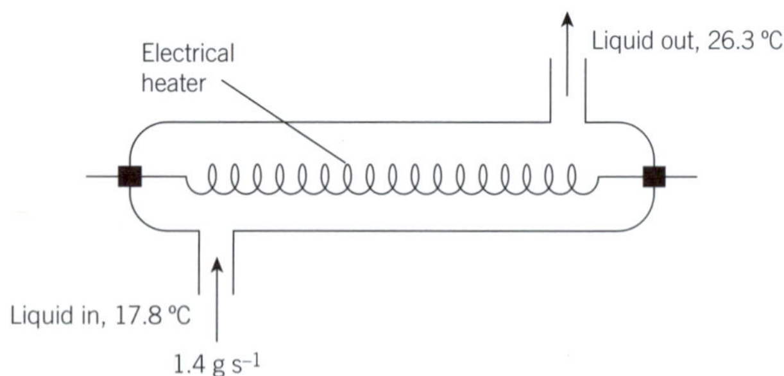

Liquid enters the tube at a constant rate of 1.4 g s⁻¹. The temperature of the liquid at the inlet is 17.8 °C. As it flows through the calorimeter, the liquid is heated by an electrical heater of output power 37.4 W. The liquid leaves the calorimeter at a temperature of 26.3 °C.

The flow rate is now doubled. The output power of the heater is increased until the temperature of the liquid leaving the calorimeter is again 26.3 °C. The output power of the heater is now 66.7 kW.

Calculate:

(a) the specific heat capacity of the liquid [2]

(b) the heat energy lost each second to the surroundings. [2]

5 State what is meant by the *internal energy* of a system. [2]

6 **(a)** State the first law of thermodynamics. Explain the meaning of any symbols you use. [3]

(b) A balloon bursts and all the air escapes rapidly. State and explain, using the first law of thermodynamics, what happens to the internal energy of the air that was inside the balloon. [2]

(c) A block of ice is removed from a freezer and placed in a warm room. The ice starts to melt.

    (i) Explain why work is done *on* the ice as it melts.

    (ii) Compare the work done on the ice with the change in internal energy of the ice as it melts. [3]

7 A cube of aluminium, with sides of length 2.0 cm, is heated so that its temperature increases from 20 °C to 450 °C. The cube expands so that its volume increases by 7 × 10⁻³ %.

[Density of aluminium = 2.7 g cm⁻³; specific heat capacity of aluminium = 900 J kg⁻¹ K⁻¹]

(a) The first law of thermodynamics states that

$$\Delta U = q + w$$

For the aluminium cube described, state whether:

    (i) q is positive or negative

    (ii) w is positive or negative. [2]

(b) Calculate the change in internal energy of the aluminium cube. [3]

13 Oscillations

Key points

- [ ] Describe simple examples of free oscillations and investigate the motion of an oscillator using experimental and graphical methods.
- [ ] Understand and use the terms amplitude, period, frequency, angular frequency, and phase difference, and express the period in terms of both frequency and angular frequency.
- [ ] Recognise and use the simple harmonic motion equation $a = -\omega^2 x$.
- [ ] Recall and use $x = x_0 \sin \omega t$ as a solution to the simple harmonic motion (SHM) equation.
- [ ] Recall and use the equations $v = v_0 \cos \omega t$ and $v = \pm \omega \sqrt{(x_0^2 - x^2)}$.
- [ ] Use graphs to describe the changes in displacement, velocity, and acceleration during SHM.
- [ ] Describe the interchange between kinetic and potential energy during SHM.
- [ ] Describe practical examples of damped oscillations, including the effects of the degree of damping and the importance of critical damping.
- [ ] Describe practical examples of forced oscillations and resonance.
- [ ] Use graphs to show how the amplitude of a forced oscillation changes with frequency near to the natural frequency of the system, and understand qualitatively the factors that determine the frequency response and sharpness of the resonance.
- [ ] Give examples where resonance is useful and where resonance should be avoided.

Oscillations

Any to-and-fro motion about a fixed point, such as a pendulum clock (Figure 13.1) or a child bouncing up and down on a trampoline, is an example of an **oscillation**. Starting from the **equilibrium position**, a complete oscillation is the movement to the **maximum displacement** in one direction, back through the equilibrium position to the maximum displacement in the other direction and back again to the equilibrium position.

Investigating the motion of an oscillator

Using a variable resistor and an oscilloscope

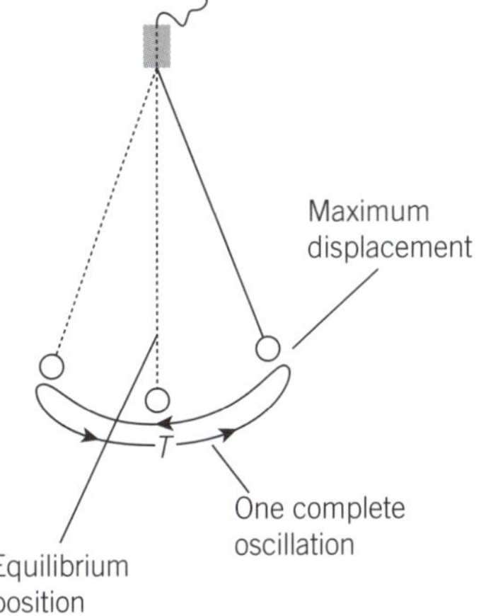

▲ **Figure 13.1** Simple pendulum

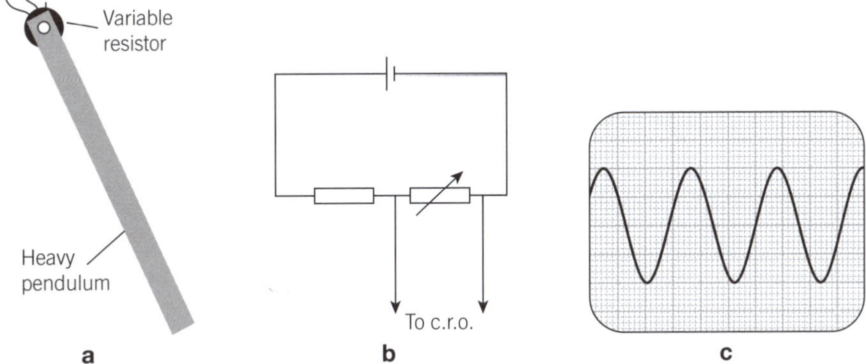

▲ **Figure 13.2** Using a variable resistor to obtain a displacement against time graph

A variable resistor, which is part of a potential divider circuit, is connected to a heavy pendulum. As the pendulum swings, the resistance of the variable resistor changes, changing the potential difference (p.d.) across it. The variation in p.d. can be displayed on a cathode ray oscilloscope (c.r.o).

Using a motion sensor

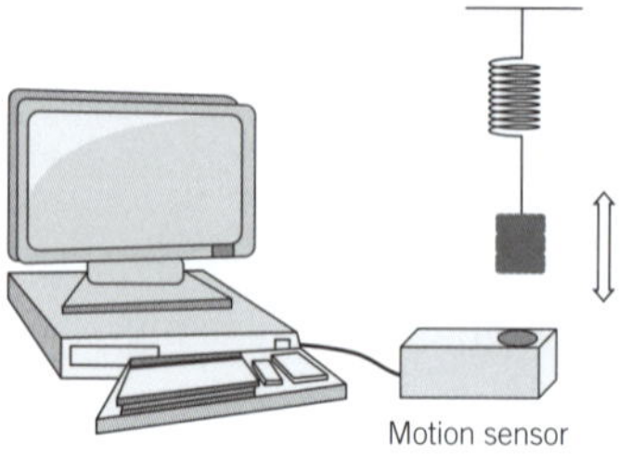
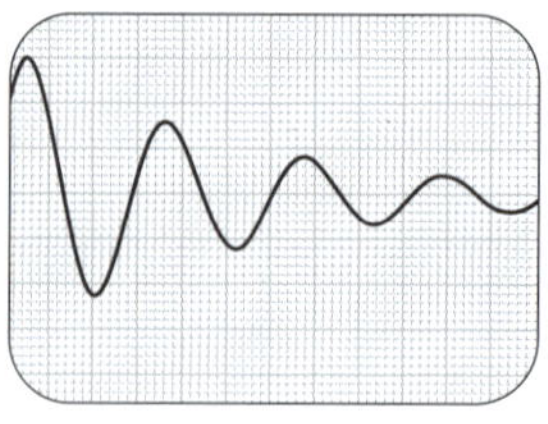

▲ **Figure 13.3** Using a motion sensor to obtain a displacement against time graph

A motion sensor is placed beneath a mass. The sensor emits pulses of ultrasound which are reflected back and detected by the sensor. The time it takes for each pulse to return is used to calculate the position of the mass.

Key terms

Amplitude is the maximum displacement from the equilibrium position. It can be measured in a variety of units (e.g., metres or degrees) depending on the type of oscillation. If the amplitude is constant, no energy is being lost and the oscillations are described as **free oscillations**.

Period T (s) is the time for one complete to-and-fro oscillation.

Frequency f Is the number of oscillations per second, measured in hertz (Hz). 1 Hz is one oscillation, or cycle, per second.

$$f = \frac{1}{T}$$

Angular frequency ω is defined as $\frac{2\pi}{T}$ ($= 2\pi f$), measured in rad s^{-1}.

$$T = \frac{1}{f} = \frac{2\pi}{\omega}$$

Maths skills

kHz = kilohertz = 10^3 Hz

MHz = megahertz = 10^6 Hz

GHz = gigahertz = 10^9 Hz

THz = terahertz = 10^{12} Hz

Maths skills

For more on radians see the *Maths skills* section.

Displacement–time graphs

Worked example

The graph in Figure 13.4 shows the displacement of a mass on a spring against time. Find the period, frequency, angular frequency, and amplitude of oscillation of the mass.

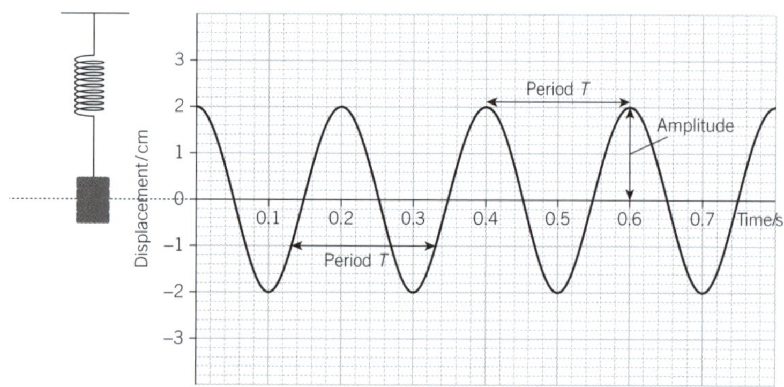

▲ **Figure 13.4** Displacement–time graph for an undamped oscillator

Exam tip

The amplitude is **not** the distance from peak to trough – it is the distance from the maximum displacement to the equilibrium position.

Answer

period T of the oscillation is 0.20 s

frequency $f = \dfrac{1}{T} = 5.0$ Hz

angular frequency $\omega = 2\pi f = 31.4$ rad s^{-1}

amplitude of the oscillation is 2.0 cm

Damped oscillations are oscillations in which the amplitude decreases (due to friction). The greater the amount of damping, the more quickly the amplitude of the oscillations falls to zero.

Phase difference

The **phase difference** between two or more oscillations can be described in terms of fractions of a cycle, in radians or in degrees (see Table 13.1). Two oscillators that are out of step by half a cycle (in **antiphase**) are π radians out of phase; if one oscillator is at a maximum value when another is at zero displacement, the oscillators are one-quarter of a cycle out of phase or $\pi/2$ radians out of phase.

Figure 13.5 shows the displacement against time of two oscillators, A and B. Oscillator B is out of phase with oscillator A (oscillator A reaches its maximum displacement before oscillator B, so **leads** B (B **lags** behind A)). The phase difference is $\dfrac{\Delta t}{T}$ cycles or $2\pi\dfrac{\Delta t}{T}$ radians or $360\,\dfrac{\Delta t}{T}$ degrees. In Figure 13.5 the phase difference is one-quarter of a cycle (or $\pi/2$ rad or 90°).

▼ **Table 13.1** Phase difference

Cycles	Radians	Degrees
0	0	0
¼	$\dfrac{\pi}{2}$	90°
½	π	180°
¾	$\dfrac{3\pi}{2}$	270°
1	2π	360°
n	$2n\pi$	$360n°$

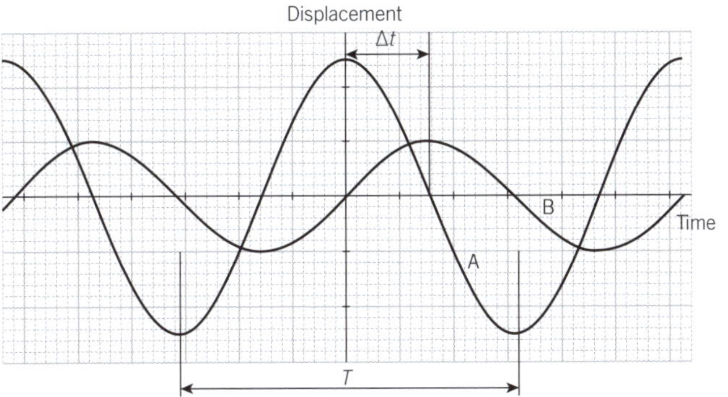

▲ **Figure 13.5** Phase difference

Simple harmonic motion (SHM)

Simple harmonic motion (SHM) is defined as motion in which the acceleration is:

- proportional to the displacement from a fixed point

- in the opposite direction to the displacement.

SHM can be expressed by the equation: $a = -\omega^2 x$

where a is the acceleration, x the displacement, and ω the angular frequency. The minus sign is important as ω^2 must be positive; it means that the acceleration will always have the opposite sign to the displacement. Many oscillations, including that of a mass on a spring and a simple pendulum, approximate to SHM.

A trolley of mass m is attached to a spring of stiffness k. It is pulled to one side and released (see Figure 13.6).

When the trolley is a distance x to the left of the equilibrium position, the spring exerts a force kx to the right. Using $F = ma$:

$$-kx = ma$$

Rearranging this equation:

$$a = -\frac{k}{m}x$$

As this is SHM, $a = -\omega^2 x$, so:

$$\omega^2 = \frac{k}{m}$$

Hence:

$$\omega = \sqrt{\frac{k}{m}} \qquad f = \frac{1}{2\pi}\sqrt{\frac{k}{m}} \qquad T = 2\pi\sqrt{\frac{m}{k}}$$

> For SHM the period T is independent of the amplitude of the oscillation.

> ★ **Exam tip**
>
> The equation $a = -\omega^2 x$ is provided in Exam Papers 1, 2, and 4.

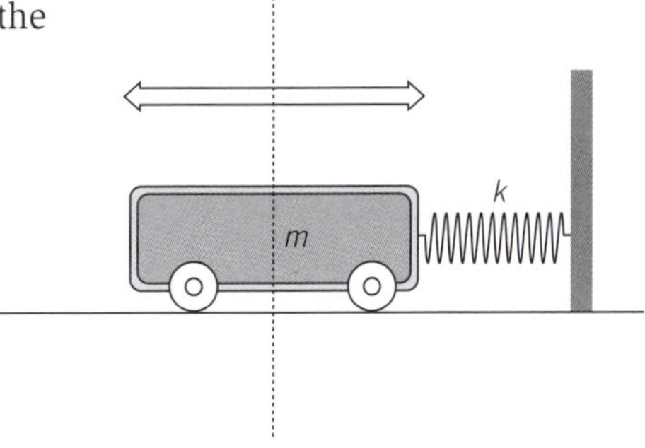

▲ **Figure 13.6** Simple harmonic motion

The SHM equation $a = -\omega^2 x$ can be solved to find the velocity and displacement at time t. If the displacement is x and the velocity is v at time t, then:

$$x = x_0 \sin \omega t$$

and

$$v = v_0 \cos \omega t$$

where x_0 is the amplitude of the oscillation and v_0 is the maximum speed. It can also be shown that:

$$v = \pm \omega \sqrt{(x_0^2 - x^2)}$$

The maximum speed (when $x = 0$) is:

$$v_0 = \omega x_0$$

★ Exam tip

Worked examples

1 A mass on a spring is oscillating with SHM. The amplitude is 0.30 m and the period of oscillation is 1.2 s. Calculate:

    a) the frequency   **b)** the angular frequency

    c) the maximum speed   **d)** the maximum acceleration.

Answer

    a) $f = \dfrac{1}{T} = \dfrac{1}{1.2} = 0.833\,\text{Hz}$

    b) $\omega = 2\pi f = 2\pi \times 0.833 = 5.23\,\text{rad s}^{-1}$

    c) $v_{max} = \omega x_0 = 5.23 \times 0.30 = 1.57\,\text{m s}^{-1}$

    d) $a = -\omega^2 x$, so $a_{max} = \omega^2 x_0 = 5.23^2 \times 0.30 = 8.2\,\text{m s}^{-2}$

2 A trolley, of mass $m = 0.90\,\text{kg}$, is suspended between two springs, each of spring constant $25\,\text{N m}^{-1}$, as shown in Figure 13.7. It is displaced to one side and released.

Calculate:

    a) the angular frequency   **b)** the period of the oscillations.

Answer

    a) When the trolley is displaced a distance x, the restoring force trying to push/pull the trolley back towards its equilibrium position is $2kx$.

    Using $F = ma$: $-2kx = ma$

$$a = -\frac{2k}{m}x$$

$$\omega^2 = \frac{2k}{m} = \frac{2 \times 25}{0.90} \quad \rightarrow \quad \omega = 7.45\,\text{rad s}^{-1}$$

    b) $T = \dfrac{2\pi}{\omega} = \dfrac{2\pi}{7.45} = 0.84\,\text{s}$

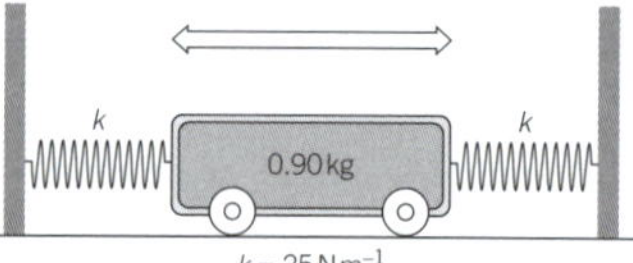

▲ **Figure 13.7** SHM oscillator

Using graphs to analyse SHM

Figure 13.8 shows the relationships between dispacement, velocity and acceleration during simple harmonic motion.

a Displacement against time

b Velocity against time

c Acceleration against time

When the displacement is zero, the velocity is its maximum value

The velocity is the **gradient** of the displacement-time graph

When the displacement is its maximum value (the amplitude), the velocity is zero

When the velocity is zero, the acceleration is its maximum value

The acceleration is the **gradient** of the velocity-time graph

When the velocity is at its maximum value the acceleration is zero

When the acceleration is positive, the displacement is negative (and vice versa)

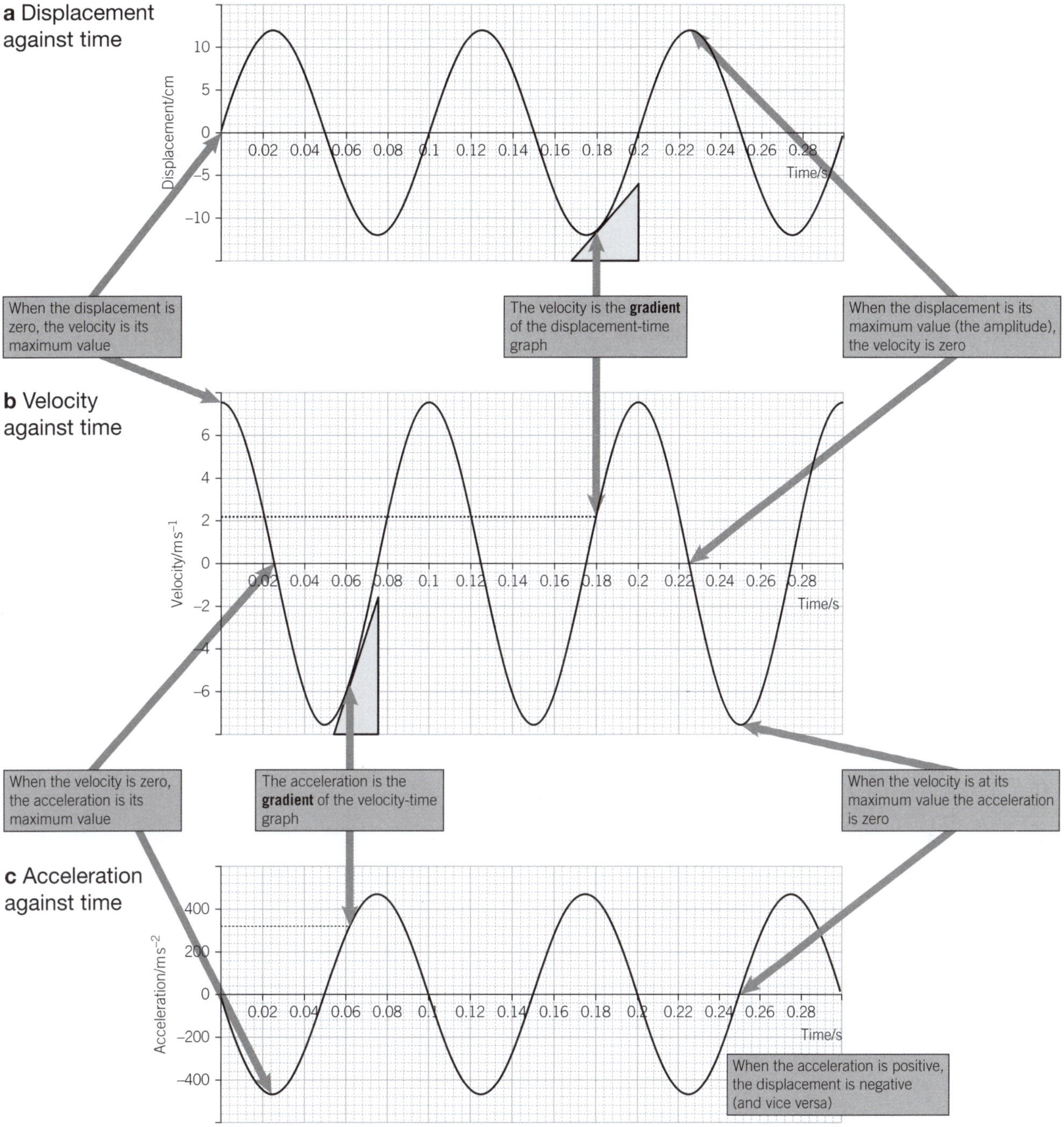

▲ **Figure 13.8** Displacement, velocity, and acceleration against time graphs for SHM.

SHM and energy

The energy of an object oscillating with simple harmonic motion changes from potential energy (P.E.) to kinetic energy (K.E.) and back to potential energy again (see Figure 13.9).

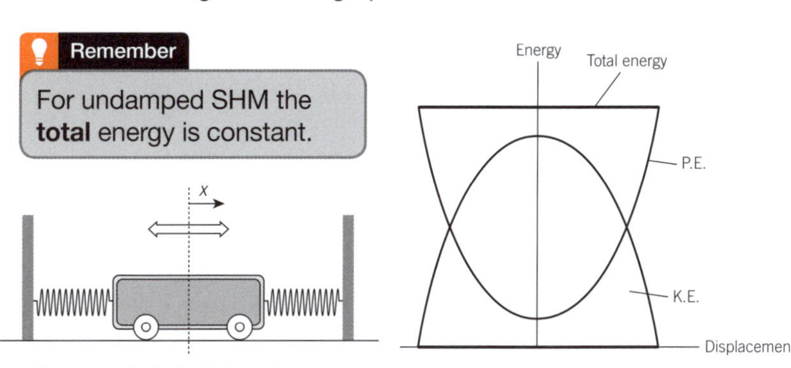

▲ **Figure 13.9** SHM and energy

Worked example

A trolley of mass m is connected to a spring of spring constant k as shown in Figure 13.10. The spring is initially neither stretched nor compressed. The trolley is displaced to the left and released, and oscillates with SHM.

The displacement x of the mass from its equilibrium position at time t is:

$$x = x_0 \sin \omega t$$

where x_0 is the amplitude of the oscillation and ω the angular frequency.

Derive expressions for:

a) the potential energy E_p of the oscillator at time t

b) the kinetic energy E_k of the oscillator at time t

c) the **total** energy of the oscillator at time t.

Answer

a) potential energy $= \dfrac{1}{2}kx^2 = \dfrac{1}{2}kx_0^2 \sin^2 \omega t$

b) kinetic energy $= \dfrac{1}{2}mv^2 = \dfrac{1}{2}m(x_0 \omega \cos \omega t)^2 = \dfrac{1}{2}mx_0^2 \omega^2 \cos^2 \omega t$

c) total energy $= \dfrac{1}{2}kx_0^2 \sin^2 \omega t + \dfrac{1}{2}mx_0^2 \omega^2 \cos^2 \omega t$

but $\omega^2 = \dfrac{k}{m}$ so total energy $= \dfrac{1}{2}kx_0^2(\sin^2 \omega t + \cos^2 \omega t)$

$$= \dfrac{1}{2}kx_0^2$$

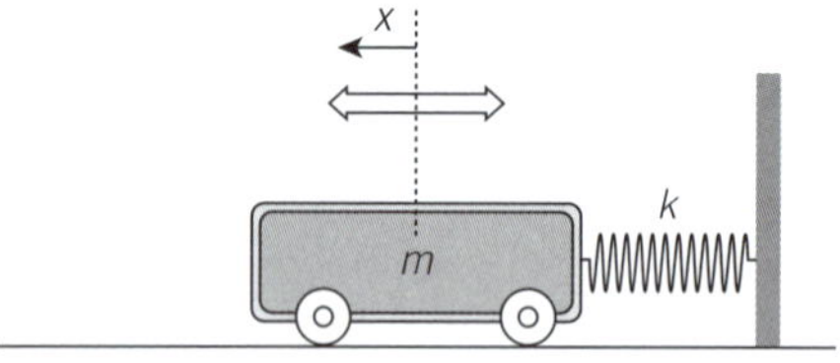

▲ Figure 13.10

$x = x_0 \sin \omega t$

$v = \omega x_0 \cos \omega t$

Link

See Unit 9 *Deformation of solids* for more about the energy stored in a stretched spring.

Maths skills

$\sin^2 \omega t + \cos^2 \omega t = 1$

Remember

- The total energy in SHM is constant.
- The total energy is proportional to the square of the amplitude.

Damped and forced oscillations, and resonance

Damped oscillations

Real oscillators (see Figure 13.11) lose energy over time, mainly as heat caused by friction and air resistance. If a mass on a spring is pulled down and released, it oscillates up and down, the amplitude gradually decreasing (exponentially) over time.

Critical damping

A **critically damped** oscillator returns to its equilibrium position in as short a time as possible without oscillating. The suspension of a car is designed to be critically damped – when the car goes over a bump in the road the suspension returns the car to its equilibrium position as quickly as possible without oscillating.

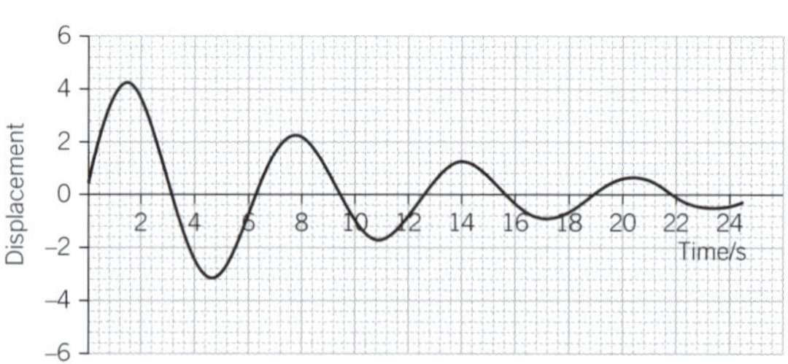

a An **underdamped** system oscillates several times as the amplitude gradually decreases to zero

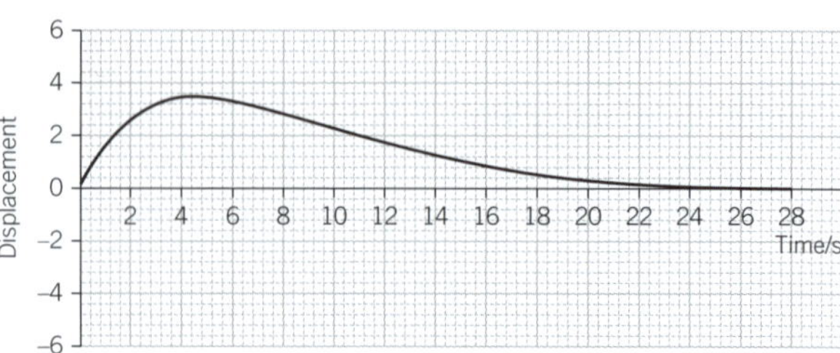

b An **overdamped** system very slowly returns to the equilibrium position without oscillating

▲ Figure 13.11 Damped SHM

Forced oscillations and resonance

Oscillators can be driven by an external system, with energy being transferred from the external system to the oscillator. Figure 13.12 illustrates how a vibration generator can be used to drive the oscillations of a mass on a spring.

At very low driving frequencies, the mass and spring will oscillate with the same amplitude as the vibration generator (see Figure 13.13). As the driving frequency is increased, the amplitude of the oscillation gradually increases as more energy is transferred to the mass on the spring.

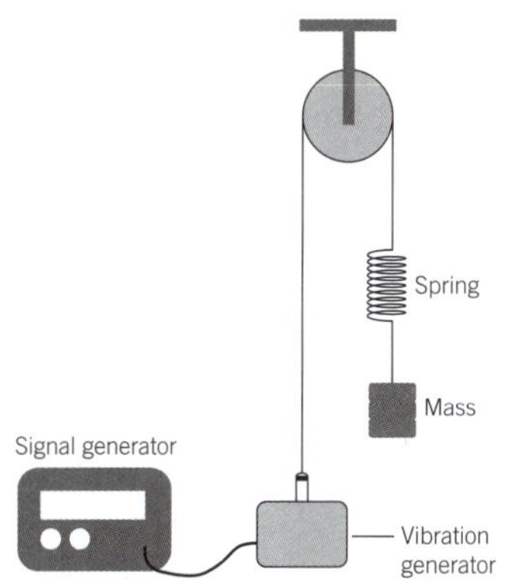

▲ **Figure 13.12** Forced oscillations and resonance

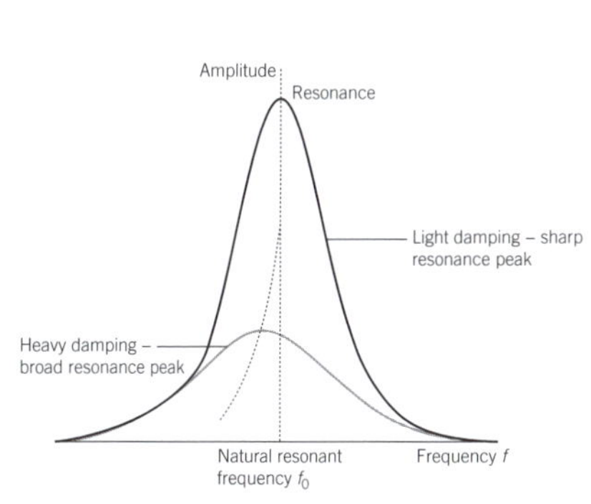

▲ **Figure 13.13** Forced oscillations with damping

When the frequency of the forced oscillations is the same as the natural frequency of the mass and spring (the **natural frequency** f_0), the mass oscillates with its maximum amplitude. This is known as **resonance**. The size of the amplitude at resonance depends on the amount of damping.

> For heavily damped systems the resonant frequency is slightly lower than the natural frequency.

Worked example

A mass is suspended vertically between two stretched springs. The lower spring is attached to a vibration generator connected to a signal generator. When the vibration generator is switched on the mass oscillates vertically, as shown in Figure 13.14.

a) In the context of the apparatus shown, explain what is meant by:

  i) natural frequency   **ii)** forced oscillations   **iii)** resonance.

b) A student measures the amplitude of oscillation for different frequencies. She then plots a graph of amplitude against frequency. Sketch the graph she is likely to obtain. Label this graph A.

c) She replaces the mass with an equal mass in the form of a thin disc. Sketch a second line on your graph showing how the amplitude will vary with frequency for the thin disc. Label this graph B.

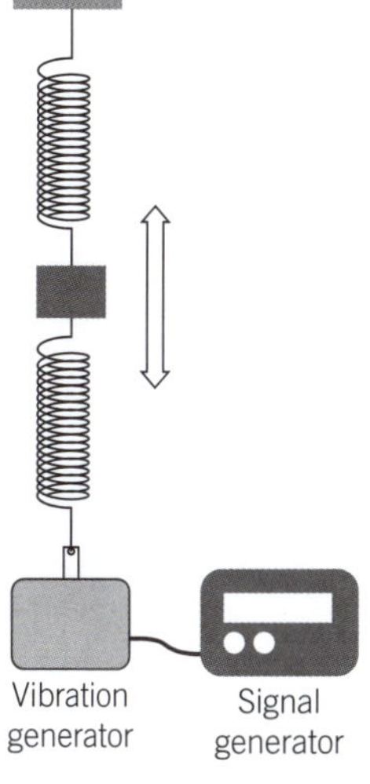

▲ **Figure 13.14**

Answer

a) **i)** If the mass is pulled down and released, the frequency of vibration would be the natural frequency.

  ii) The vibration generator forces the mass and springs to oscillate at the frequency of the generator (the driving frequency).

  iii) As the driving frequency approaches the natural frequency, the amplitude of the oscillations becomes very large, reaching a maximum value when the driving frequency is equal to the natural frequency. The system is then resonating.

(b),(c)

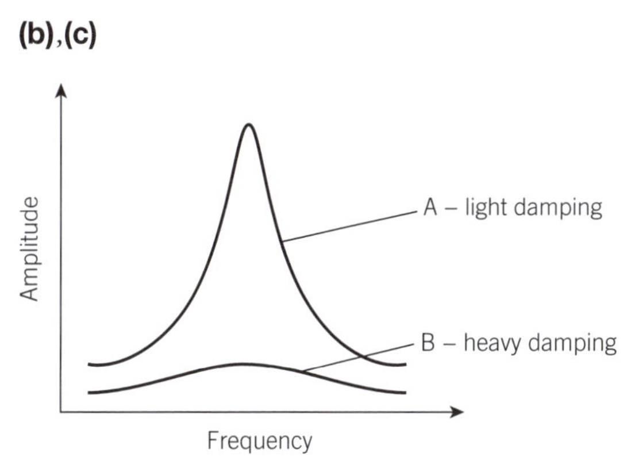

▲ **Figure 13.15**

 **Raise your grade**

1 (a) Define *simple harmonic motion*.

SHM is where the acceleration is proportional to the displacement. ✔ ✘

This is a correct answer but not a complete definition of SHM. It should also state that the acceleration is always in the opposite direction to the displacement.

.... [2]

(b) An aeroplane is moving at a constant speed u, flying at a level height. It experiences some turbulence and is displaced vertically, causing it to oscillate in the vertical direction.

Theory shows that the vertical acceleration a of the aeroplane is given by the equation:

$$a = -\frac{2g^2}{u^2}x$$

where x is the vertical displacement.

(i) Explain how it can be deduced from the equation that the aeroplane oscillates with simple harmonic motion.

g and u are both constants, so the acceleration a is proportional to the displacement. ✔ ✘

This is a correct answer but again not complete. The candidate has ignored the significance of the minus sign – this shows the acceleration is always in the opposite direction to the displacement.

.... [2]

(ii) Calculate the period of oscillation of an aircraft travelling at $300\,km\,h^{-1}$. [Use $g = 9.81\,m\,s^{-2}$.]

$$u = 300\,km\,h^{-1} = \frac{300 \times 1000}{60 \times 60} = 83\,m\,s^{-1}\ ✔$$

The candidate has converted the speed of the aeroplane into SI units correctly, giving an answer to an appropriate number of significant figures.

$$\omega^2 = \frac{2g^2}{u^2} = \frac{2 \times 9.81^2}{83^2} = 0.0279 \rightarrow \omega = \sqrt{0.0279} = 0.167\,rad\,s^{-1}\ ✔$$

The candidate has equated the angular frequency ω to g and u correctly.

$$T = \frac{2\pi}{\omega} = \frac{2\pi}{0.167} = 37.6\,s\ ✔$$

The final calculation is correct.

period of oscillation = 37.6 s [3]

(iii) The oscillations of the aeroplane are *lightly damped*. Explain what *lightly damped* means.

The aeroplane does not complete a full oscillation before the vertical movement ends. ✘

This answer is incorrect – the candidate has described **heavily damped** oscillations. Lightly damped oscillations mean several oscillations will be completed before the amplitude has fallen to zero.

.... [1]

❓ Exam-style questions

1 A ball of mass 200 g is suspended from a spring. When the ball is pulled down 6.0 cm and released, it oscillates with a period of 0.70 s.

Calculate:

(a) the frequency of oscillation [1]

(b) the maximum speed of the ball [1]

(c) the acceleration when the ball is 2.0 cm above the equilibrium position. [2]

2 The graph shows the variation of displacement with time for a mass–spring oscillator.

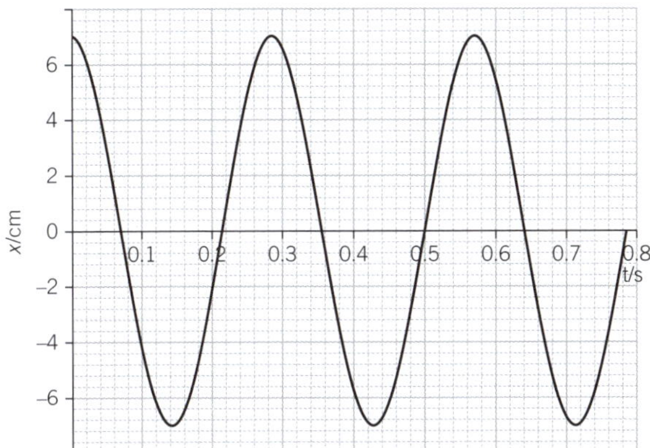

(a) Use the graph to determine:

(i) the period of oscillation

(ii) the frequency

(iii) the angular frequency. [4]

(b) Use the graph to calculate the speed of the mass at time $t = 0.20$ s. [2]

3 The displacement–time graphs of two oscillators, A and B, are shown in the graphs.

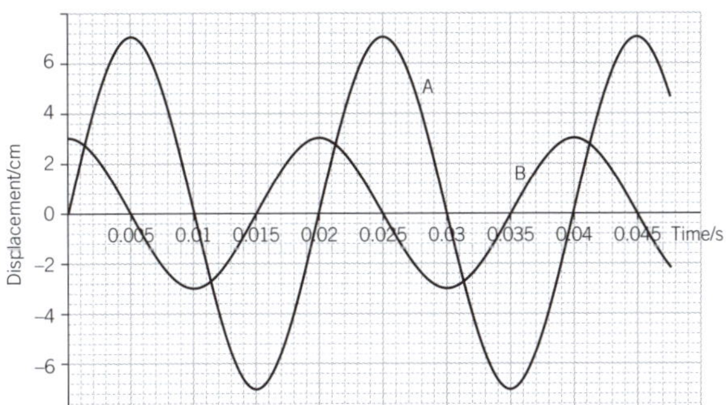

(a) State the phase difference between A and B:

(i) as a fraction of a cycle

(ii) in radians. [2]

(b) Determine the ratio $\dfrac{\text{amplitude of A}}{\text{amplitude of B}}$. [1]

(c) Use the graph to find:

(i) the frequency of A

(ii) the maximum speed of A

(ii) the maximum acceleration of A. [4]

4 **(a)** Define *simple harmonic motion*. [2]

(b) A small mass oscillates vertically according to the equation:

$$y = 15\cos 20\,t$$

where y is the displacement in centimetres of the mass at time t.

Calculate:

(i) the frequency of the oscillation

(ii) the angular frequency

(iii) the maximum speed of the mass. [4]

5 A U-tube contains liquid of density ρ. The tube is briefly tilted to one side and then returned to the vertical position. The liquid oscillates back and forth with simple harmonic motion.

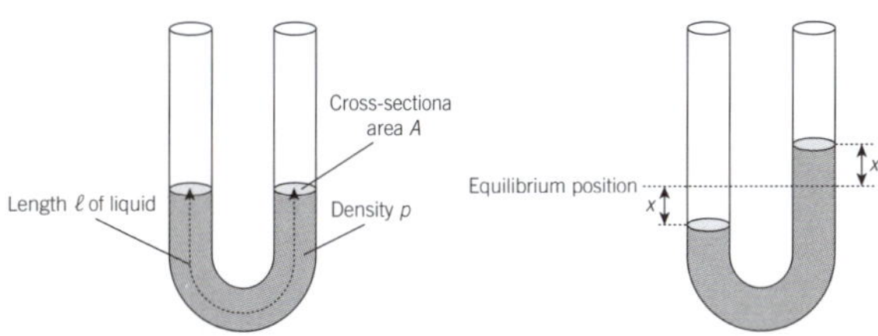

(a) Explain why the liquid oscillator satisfies the conditions for SHM. [2]

(b) Determine:

(i) the mass of liquid in the tube

(ii) the unbalanced force on the liquid at time t

(iii) the period of the oscillations. [4]

6 **(a)** In relation to oscillating systems, explain what is meant by:

(i) *forced vibrations*

(ii) *resonance*. [2]

(b) Describe one situation where resonance can be useful and one situation where resonance should be avoided. [2]

14 Waves

Key points

- ☐ Describe what is meant by wave motion as illustrated by vibration in ropes, springs, and ripple tanks.
- ☐ Understand and use the terms displacement, amplitude, phase difference, period, frequency, wavelength, and speed.
- ☐ Deduce, from the definitions of speed, frequency and wavelength, the wave equation $v = f\lambda$.
- ☐ Recall and use the equation $v = f\lambda$.
- ☐ Understand that energy is transferred by a progressive wave.
- ☐ Recall and use the relationship intensity $\propto$ (amplitude)^2.
- ☐ Compare transverse and longitudinal waves.
- ☐ Analyse and interpret graphical representations of transverse and longitudinal waves.
- ☐ Determine the frequency of sound using a calibrated cathode-ray oscilloscope (c.r.o.).
- ☐ Determine the wavelength of sound using stationary waves.
- ☐ Understand that when a source of waves moves relative to a stationary observer, there is a change in observed frequency (Doppler effect).
- ☐ Use the expression $f_0 = f_s \dfrac{v}{v \pm v_s}$ for the observed frequency when a source of sound waves moves relative to a stationary observer.
- ☐ Appreciate that Doppler shift is observed with all waves, including sound and light.
- ☐ State that all electromagnetic waves travel with the same speed in free space and recall the orders of magnitude of the wavelengths of the principal radiations from radio waves to γ-rays.
- ☐ Explain how ultrasonic waves can be generated and detected using piezo-electric transducers.
- ☐ Explain how ultrasound can be used to obtain diagnostic information about internal structures.
- ☐ Understand specific acoustic impedance and its importance to the intensity reflection coefficient at a boundary.
- ☐ Recall and solve problems by using the equation $I = I_0 e^{-\mu x}$ for the attenuation of ultrasound in matter.

Progressive waves

What are progressive waves?

Progressive waves transfer energy from one point to another.

There are two types of progressive wave.

- **Transverse waves**: Waves on a rope are examples of transverse waves. As a transverse wave passes along the rope, the particles of the rope oscillate in a direction **perpendicular to the direction of energy transfer** (see Figure 14.1).

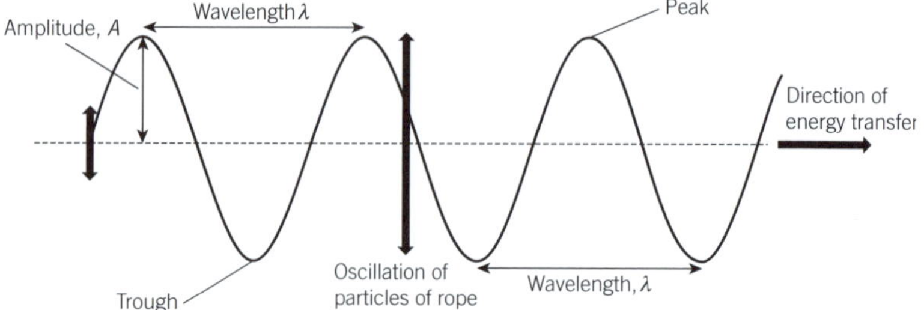

▲ **Figure 14.1** Transverse waves

Water waves, secondary seismic waves and electromagnetic waves are examples of transverse waves.

- **Longitudinal(compression) waves**: Longitudinal, or compression, waves on a slinky (a long spring) can best illustrate the properties of longitudinal waves. The individual rings of the slinky oscillate back and forth **parallel to the direction of energy transfer**.

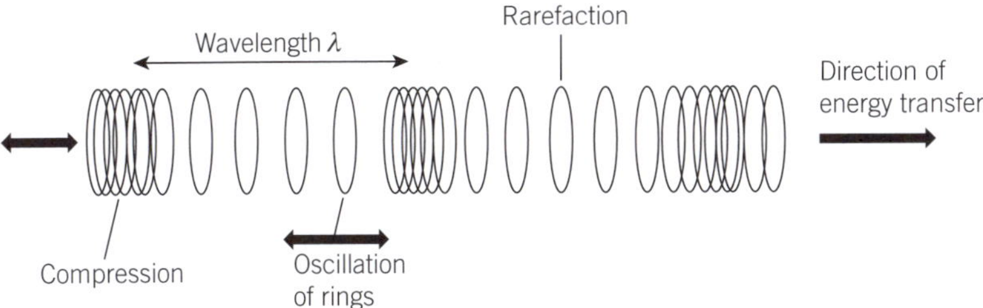

▲ **Figure 14.2** Longitudinal (compression) waves

Sound waves and primary seismic waves are longitudinal waves. As a sound wave travels through air, the air molecules continually move closer together (compression) and then further apart (rarefaction) creating areas of high pressure and low pressure. The greater the amplitude of the sound wave, the greater the pressure difference between the areas of compression and expansion.

Displacement of a particle

The displacement of an individual particle in a transverse or longitudinal wave is shown in Figure 14.3.

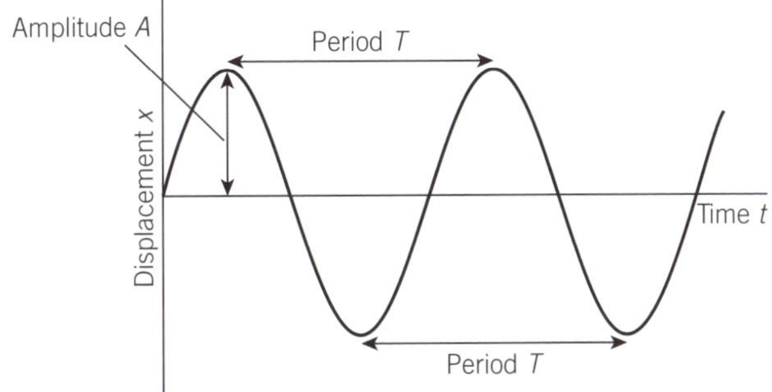

▲ **Figure 14.3** Displacement–time graph for a particle
on the wave

The period T of the oscillation is the time taken for one complete oscillation (and the time for one complete wave to pass any given point). It is related to the frequency of the wave by the equation:

$$f = \frac{1}{T}$$

The frequency f is measured in hertz (Hz). $1\,\text{Hz} = 1$ wave/second.

Wave equation

$$\frac{\text{speed of}}{\text{a wave}} = \frac{\text{length of one wave}}{\text{(the wavelength)}} \times \frac{\text{number of waves passing in}}{\text{one second (the frequency)}}$$

$$v = f\lambda$$

Phase difference

Particles at different points along a wave are out of step with each other – there is a **phase difference** between them.

In Figure 14.4 points P and R are exactly half an oscillation out of step with each other – when P starts to move up, R starts to move down (they are in **antiphase**). Points P and Q are one quarter of an oscillation out of step with each other (when P is at its maximum displacement, Q is at the equilibrium position). Points P and S are exactly one cycle out of phase with each other so are in phase (both are moving up or down at exactly the same time). Phase difference can be expressed in degrees or radians where 360°, or 2π radians, represents one complete cycle (see Table 14.1).

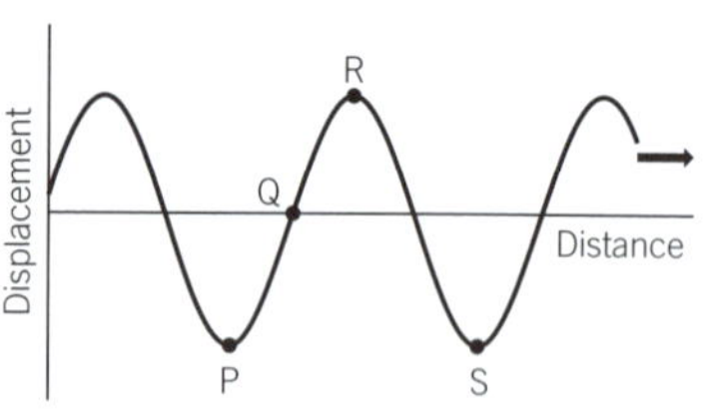

▲ **Figure 14.4** Phase difference

▼ **Table 14.1** Phase difference

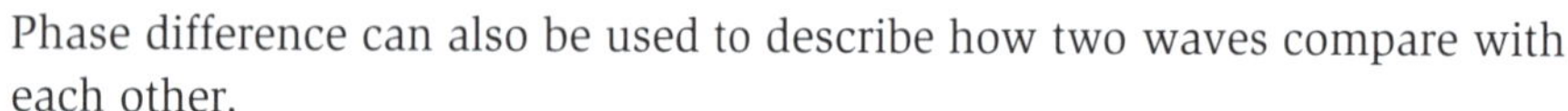

	Phase difference/cycles	Phase difference/°	Phase difference/radians
P → Q	$\frac{1}{4}$	90	$\frac{\pi}{2}$
P → R	$\frac{1}{2}$	180	π
P → S	1	360 (0)	2π (0)

Phase difference can also be used to describe how two waves compare with each other.

In Figure 14.5 the two waves A and B are out of phase with each other. Point P on wave B has reached its maximum displacement and will start to move downwards. The corresponding point Q on wave A is at the equilibrium position and moving upwards; that is, wave A is **lagging** behind wave B by quarter of a cycle. Another way of saying this is wave B **leads** wave A by $\frac{\pi}{2}$ radians or 90°.

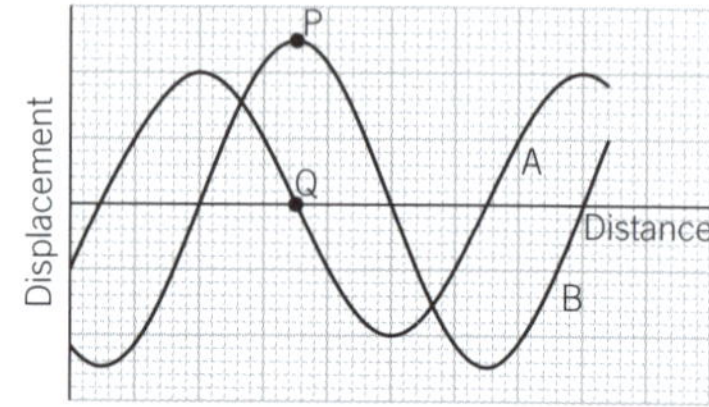

▲ **Figure 14.5** Phase difference

Intensity of a wave

The **intensity** of a wave is a measure of its power. The intensity I of a wave is proportional to the square of the amplitude A of the wave.

$$I \propto A^2$$

Halving the amplitude of a wave reduces its intensity by a factor of four.

Electromagnetic waves

The family of waves which includes visible light is called the **electromagnetic spectrum** (see Table 14.2). All electromagnetic waves:

- are transverse waves
- travel at the speed of light ($3.0 \times 10^8 \, \text{m s}^{-1}$) in a vacuum.

Remember

$I \propto A^2$

★ Exam tip

Try to memorise Table 14.2. You may be asked to recall the order of magnitude of the main parts of the spectrum and from these the corresponding frequencies can be calculated using $c = f\lambda$, where c is the speed of light ($3.0 \times 10^8 \, \text{m s}^{-1}$).

▼ **Table 14.2** Electromagnetic spectrum

Electromagnetic wave	gamma rays	X-rays	ultraviolet	visible	infrared	microwaves	radio waves
Typical wavelengths/m	10^{-12}	10^{-10}	10^{-8}	4×10^{-7}–7×10^{-7} (400–700 nm)	10^{-5}	10^{-2}	10^{-1}–10^5

Measuring the wavelength of sound using stationary waves

Stationary waves in an air column formed inside a pipe or tube can be used to find the wavelength of sound waves.

A loudspeaker is connected to a signal generator and placed at the end of a tube which is closed at one end (see Figure 14.6). As the frequency of the signal generator is slowly increased, the tube resonates (produces a louder sound) at particular frequencies due to stationary waves forming in the tube.

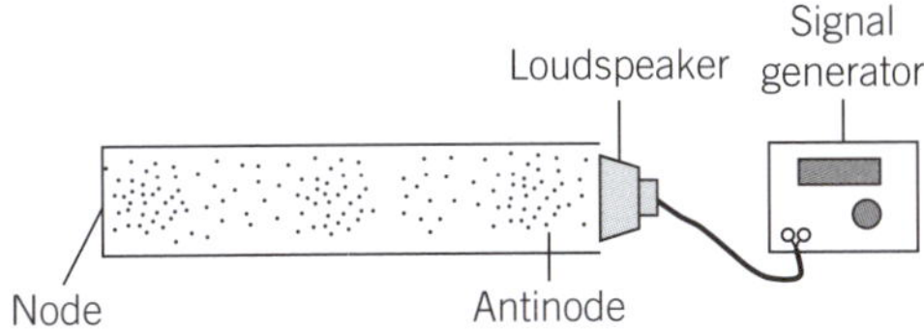

▲ **Figure 14.6** Stationary waves in air columns

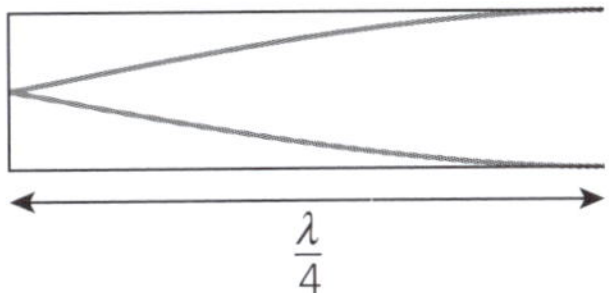

First-order resonance

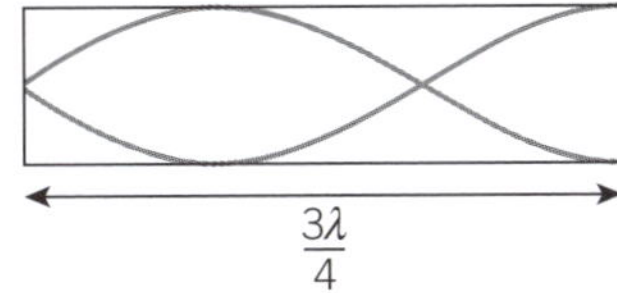

Second-order resonance

▲ **Figure 14.7** First-order and second-order longitudinal stationary waves

Figure 14.7 shows the first two stationary waves in the tube. The closed end is a **displacement node** (the molecules of air cannot oscillate) and a **pressure antinode**. The free end is a **displacement antinode** and a **pressure node**. If the length of the tube is L then the first resonant frequency occurs when $L = \lambda/4$ so $\lambda = 4L$. For the second resonant frequency, $\lambda = 4L/3$. Tubes and pipes open at both ends can also be used in this way.

Doppler effect

When a source of waves travels towards a stationary observer, the wavelength of the waves decreases and the frequency detected by the observer increases.

This effect is known as the **Doppler effect**. If the source of waves is moving away from the observer the opposite effect occurs, with the observer detecting a lower frequency. A familiar example is the sound of a train as it passes you – the frequency (pitch) of the sound decreases as the train passes you and moves away.

For a source of sound waves moving towards a stationary observer:

$$f_o = f_s \frac{v}{v - v_s}$$

where f_s is the frequency of sound of the source, v_s the velocity of the source, v the velocity of sound, and f_o the frequency detected by the observer. For a source of sound waves travelling away from a stationary observer:

$$f_o = f_s \frac{v}{v + v_s}$$

Remember

Sound waves are **longitudinal** waves so the diagrams showing stationary **transverse** waves in a tube are a little misleading. They can be thought of as graphs of the maximum displacement of air molecules at different points inside the tube.

Link

For more about stationary waves in air columns see Unit 15 *Superposition*.

Exam tip

Numerical questions on the Doppler effect will only be concerned with sound waves, travelling towards or away from a stationary observer.

Remember

$$f_o = f_s \frac{v}{v \pm v_s}$$

– for sound waves moving towards a stationary observer

+ for waves moving away from a stationary observer.

> **Worked example**
>
> A train is travelling towards a station at a speed of $35\,\mathrm{m\,s^{-1}}$. An observer standing on a platform in the station hears the train emitting sound of frequency $800\,\mathrm{Hz}$. What frequency will the observer hear when the train has passed through the station? [Speed of sound in air = $330\,\mathrm{m\,s^{-1}}$.]
>
> **Answer**
>
> For the train travelling towards the observer: $800 = f_{\mathrm{s}}\left(\dfrac{330}{330-35}\right)$ (eqn 1)
>
> For the train travelling away from the observer: $f_{\mathrm{o}} = f_{\mathrm{s}}\left(\dfrac{330}{330+35}\right)$ (eqn 2)
>
> Combining eqn 1 and eqn 2: $f_{\mathrm{o}} = \left(\dfrac{330}{330+35}\right) \times \left(\dfrac{330-35}{330}\right) \times 800 = 650\,\mathrm{Hz}$

Doppler shift

The Doppler effect occurs with many types of wave including light waves and microwaves. A radar speed camera emits pulses of microwaves which are reflected back by a vehicle. The faster the vehicle is travelling the greater the change in frequency between the emitted pulses and the reflected pulses.

The wavelengths of light detected from distant stars are longer than the characteristic wavelengths expected of the line spectra of gases such as hydrogen and helium, suggesting that other stars and galaxies are moving away from the Earth. This is often referred to as the **Doppler shift** or red shift (as the wavelengths are longer, moving towards the red end of the visible spectrum).

Ultrasonic waves (ultrasound)

Ultrasonic waves are sound waves with frequencies above human hearing (typically $20\,\mathrm{kHz}$). Unlike high-frequency electromagnetic waves such as X-rays, they are non-ionising, and so do not damage living tissue and are ideal for use in medical imaging. The frequencies used are usually between $1\,\mathrm{MHz}$ and $10\,\mathrm{MHz}$ (frequencies lower than $1\,\mathrm{MHz}$ are diffracted too much and frequencies higher than $10\,\mathrm{MHz}$ are absorbed too much by body tissue).

Ultrasonic waves are produced using a **piezo-electric transducer** in the shape of a disc. An alternating voltage applied between the faces of the disc causes it to vibrate. If the frequency chosen coincides with the natural frequency of the disc, resonance occurs, and the disc emits ultrasonic waves at the resonant frequency (see Figure 14.8).

In medical imaging, an ultrasound probe emits pulses of ultrasound into a body. The ultrasonic waves are partially reflected each time the ultrasonic waves pass from one material (medium) to another. The reflected waves cause the disc of the probe to vibrate, generating a small p.d. across the disc. The probe thus acts as both transmitter and receiver of the ultrasonic waves, the reflected pulses enabling a 'sound picture' to be constructed.

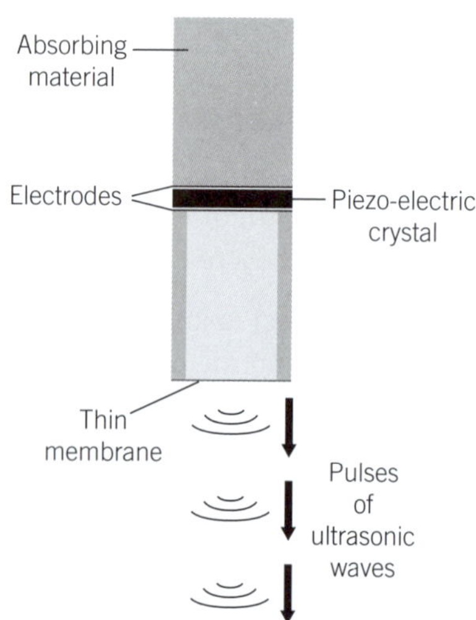

▲ **Figure 14.8** Piezo-electric probe

Transmission and reflection of ultrasound waves

When ultrasonic waves reach a boundary between two different materials, some of the wave energy is reflected and the rest is transmitted (and refracted). The proportion of energy that is reflected is determined by the **acoustic impedance** Z of each of the two materials, where:

$$Z = \rho c$$

ρ is the density of the material and c is the speed of sound in the material. Some typical values are given in Table 14.3.

The **intensity reflection coefficient** is the fraction of ultrasonic wave energy reflected, and is given by the equation:

$$\frac{I_R}{I_0} = \frac{\left(Z_1 - Z_2\right)^2}{\left(Z_1 + Z_2\right)^2}$$

where I_0 is the intensity of the incident waves, I_R the intensity of the reflected waves, and Z_1 and Z_2 the acoustic impedances of the two materials (see Figure 14.9).

▼ Table 14.3 Acoustic impedance

Material	Acoustic impedance $Z/\mathrm{kg\,m^{-2}\,s^{-1}}$
air	430
blood	1.59×10^6
bone	6.80×10^6
muscle	1.70×10^6
soft tissue	1.63×10^6
water	1.50×10^6

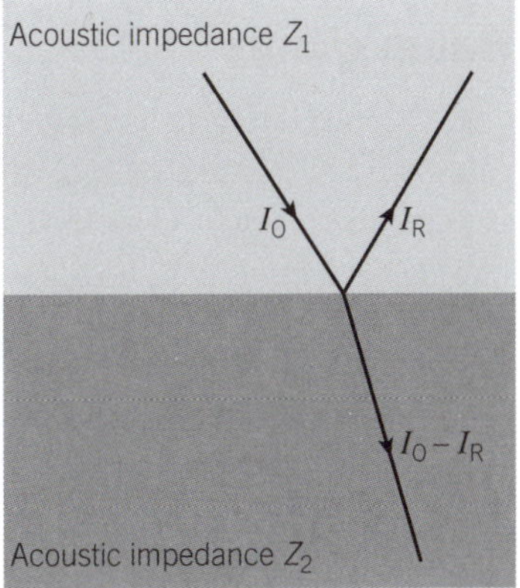

▲ **Figure 14.9** Transmission and reflection

Worked example

Ultrasonic waves from a piezo-electric probe pass from bone into soft tissue. Calculate the fraction of ultrasonic wave energy that is reflected back towards the probe.

Answer

$$\frac{I_R}{I_0} = \frac{\left(Z_1 - Z_2\right)^2}{\left(Z_1 + Z_2\right)^2} = \frac{\left(6.80 - 1.63\right)^2}{\left(6.80 + 1.63\right)^2} = 0.376$$

Approximately a third of the incident wave energy is reflected.

Two key points to remember about the equation for the reflection coefficient:

- if $Z_1 \approx Z_2$ then $\frac{I_R}{I_0} \approx 0$ – almost all the wave energy is transmitted
- If Z_1 is very different from Z_2 (e.g., air and soft tissue) then $\frac{I_R}{I_0} \approx 1$ and almost all the wave energy is reflected.

This explains why a **coupling medium** such as a liquid gel is needed between an ultrasonic probe and soft tissue such as skin (see Figure 14.10). Any air between the probe and the soft tissue would mean virtually all the ultrasonic wave energy is reflected back off the soft tissue as $\frac{I_R}{I_0} \approx 1$. Placing the front of the probe in a gel (which has a similar acoustic impedance to soft tissue) means that almost all the wave energy will be transmitted into the body.

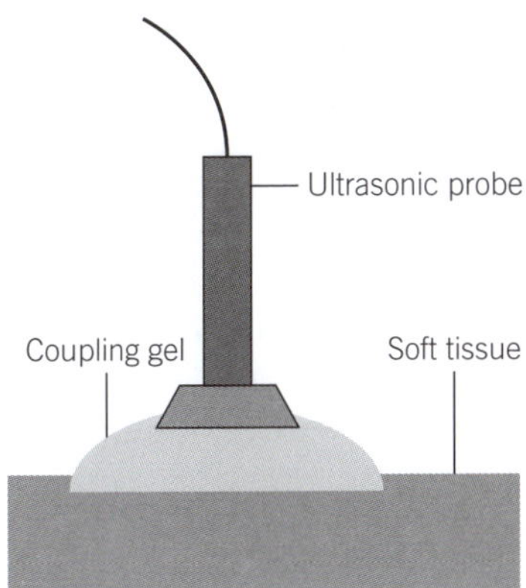

▲ **Figure 14.10** Use of coupling gel to limit reflection from soft tissue

Absorption of ultrasonic waves

When a parallel beam of ultrasonic waves passes through a substance, the intensity of the waves decreases exponentially with distance.

The intensity I of a wave after passing through a distance x of a material is given by:

$$I = I_0 e^{-\mu x}$$

Where I_0 is the incident intensity, and μ the absorption coefficient of the substance (see Figure 14.11).

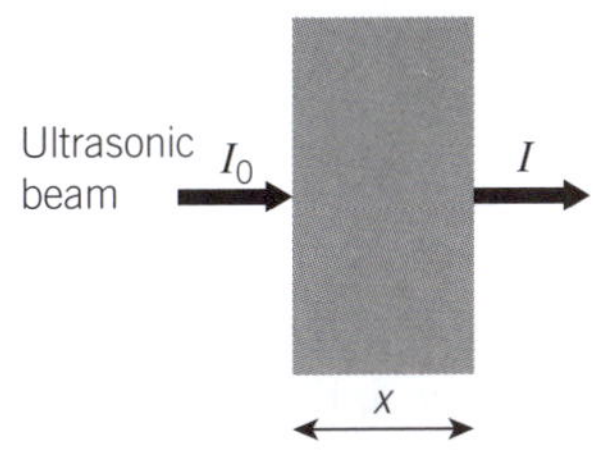

▲ **Figure 14.11** Exponential absorption

 **Raise your grade**

(a) State what is meant by the *specific acoustic impedance* of a medium.

acoustic impedance = density × speed ✓✗

A correct statement, but insufficient for 2 marks. A better answer would be 'specific acoustic impedance is the product of the density of the medium and the speed of ultrasound in the medium'. ... [2]

(b) A vet is using a piezo-electric probe to examine an animal. Pulses of ultrasonic waves pass through the animal and are partially reflected each time the waves reach a boundary between one medium and the next.

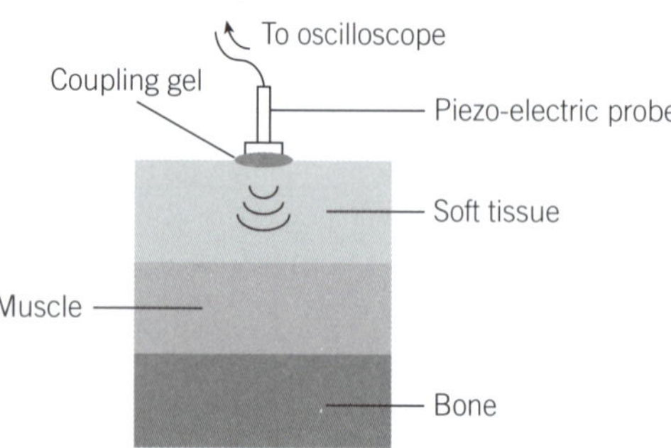

Material	Acoustic impedance Z / $kg\,m^{-2}s^{-1}$
air	430
muscle	1.70×10^6
soft tissue	1.63×10^6
bone	6.80×10^6

(i) Explain why the piezo-electric probe emits pulses of ultrasound.

So that the reflected pulses can be detected between emitting pulses ✓✗

A better answer would include '... which means that the time taken for each pulse to return can be calculated and hence the depth of each reflecting boundary'.

(ii) Explain the purpose of the coupling gel.

So that most of the ultrasound pulse is transmitted through the soft tissue. ✓✗

A better answer would include '... because the coupling gel has a similar acoustic impedance to soft tissue/if there is air between the probe and the soft tissue most of the ultrasound would be reflected'.

... [4]

(c) The intensity reflection coefficient at a boundary between two media is given by the equation:

$$\frac{I_R}{I_0} = \frac{(Z_1 - Z_2)^2}{(Z_1 + Z_2)^2}$$

where I_0 is the intensity of the incident ultrasonic waves, I_R the intensity of the reflected waves, and Z_1 and Z_2 the acoustic impedances of the two media.

Calculate the fraction of the ultrasonic wave energy transmitted when the ultrasonic waves reach the boundary between muscle and bone.

$$\frac{I_R}{I_0} = \frac{(Z_1 - Z_2)^2}{(Z_1 + Z_2)^2} = \frac{(1.63 - 6.80)^2}{(1.63 + 6.80)^2} = 0.376 \ (38\%)$$ ✓✗

Correct substitution of values for Z_1 and Z_2 into the equation and correct calculation for the first mark, but the value obtained is the fraction reflected. The fraction transmitted is $1 - 0.376 = 0.624 \ (62\%)$

fraction transmitted = ...38%... [2]

? Exam-style questions

1 Transverse waves on a rope are travelling at a speed of $10.0\,\mathrm{m\,s^{-1}}$. The frequency of the waves is $5.0\,\mathrm{Hz}$.

What is the phase difference between two points on the rope a distance $1.00\,\mathrm{m}$ apart?

A $\dfrac{\pi}{2}$ **B** π **C** $\dfrac{3\pi}{2}$ **D** 2π [1]

2 Gamma rays, ultraviolet waves and microwaves are all electromagnetic waves. Which option lists these waves in order of increasing frequency?

A gamma rays, ultraviolet, microwaves

B gamma rays, microwaves, ultraviolet

C ultraviolet, microwaves, gamma rays

D microwaves, ultraviolet, gamma rays   [1]

3 An organ pipe of length $50.0\,\mathrm{cm}$ is closed at one end. The speed of sound in air is $330\,\mathrm{m\,s^{-1}}$. What are the two lowest frequencies that can be produced by the pipe?

A $165\,\mathrm{Hz}$ and $330\,\mathrm{Hz}$ **B** $165\,\mathrm{Hz}$ and $495\,\mathrm{Hz}$

C $330\,\mathrm{Hz}$ and $660\,\mathrm{Hz}$ **D** $660\,\mathrm{Hz}$ and $880\,\mathrm{Hz}$ [1]

4 A racing car approaches a stationary observer with a constant speed u. If the speed of sound in air is v, what is the change in frequency heard by the observer as the car passes him?

A $\dfrac{fuv}{\left(v^2 - u^2\right)}$ **B** $\dfrac{2fuv}{\left(v^2 - u^2\right)}$

C $\dfrac{fuv}{\left(v^2 + u^2\right)}$ **D** $\dfrac{2fuv}{\left(v^2 + u^2\right)}$ [1]

5 A special loudspeaker, with a power output of $1.0 \times 10^{-4}\,\mathrm{W}$, emits sound energy equally in all directions.

(a) Calculate the *intensity* of sound (the sound energy per second per $\mathrm{m^2}$) at the following distances from the speaker:

   (i) $3.0\,\mathrm{m}$ **(ii)** $9.0\,\mathrm{m}$ [2]

   [Surface area of a sphere of radius r is $4\pi r^2$.]

(b) Compare the amplitude of vibration of air molecules at the two distances in (a).   [2]

6 **(a)** Define, for a transverse wave:

   (i) the amplitude A **(ii)** the wavelength λ. [2]

(b) A student is investigating waves using a ripple tank. The drawing shows the waves at one particular moment.

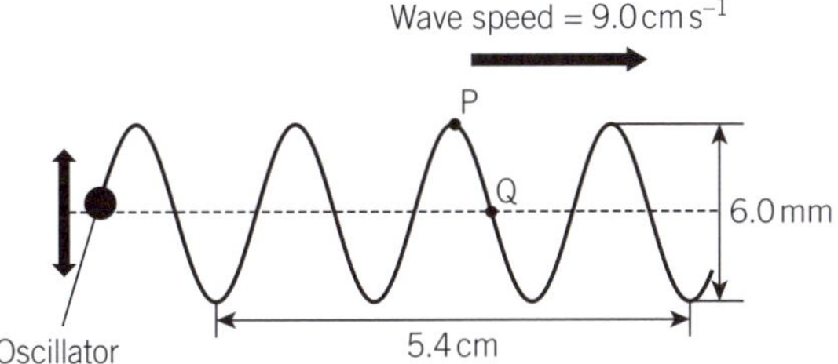

Determine:

   (i) the amplitude of the waves

   (ii) the wavelength of the waves

   (iii) the frequency of the oscillator

   (iv) the phase difference between points P and Q.   [4]

(c) Sketch a graph of the displacement of point P against time, for a period of $0.5\,\mathrm{s}$. Include appropriate scales.   [3]

7 **(a)** Describe the *Doppler effect*.   [2]

(b) An ambulance has a siren which emits a note of frequency $700\,\mathrm{Hz}$. It is travelling towards a stationary observer at a speed of $30\,\mathrm{m\,s^{-1}}$.

Determine:

   (i) the frequency heard by the observer

   (ii) the change in frequency heard by the observer once the ambulance has gone past.   [3]

8 A pregnant woman is having an ultrasound scan using a piezo-electric probe. A gel is spread over the patient's abdomen and the probe placed onto the gel.

(a) Define *specific acoustic impedance*.   [2]

(b) Explain the purpose of the gel.   [2]

(c) Suggest a reason why only high frequency ultrasonic waves are used for this.   [1]

9 Draw a diagram of a piezo-electric probe. Describe and explain how it produces and detects ultrasonic waves.   [5]

Key points

- ☐ Explain and use the principle of superposition in simple applications.
- ☐ Show an understanding of experiments that demonstrate stationary waves using microwaves, stretched strings, and air columns.
- ☐ Explain the formation of a stationary wave using a graphical method, and identify nodes and antinodes.
- ☐ Explain the meaning of the term diffraction.
- ☐ Show an understanding of experiments that demonstrate diffraction, including the diffraction of water waves in a ripple tank with both a wide gap and a narrow gap.
- ☐ Understand the terms interference and coherence.
- ☐ Show an understanding of experiments that demonstrate two-source interference using water ripples, light and microwaves.
- ☐ Understand the conditions required if two-source interference fringes are to be observed.
- ☐ Recall and solve problems using the equation $\lambda = ax/D$ for double-slit interference using light.
- ☐ Recall the equation for diffraction gratings and solve problems using the formula $d \sin \theta = n\lambda$.
- ☐ Describe the use of a diffraction grating to determine the wavelength of light.

Superposition

Principle of superposition

When two waves meet and overlap, the total displacement at any point is the (vector) sum of the individual displacements at that point – this is called the **principle of superposition**.

Interference

When two or more waves combine to produce a new wave, they can interfere constructively or destructively.

- **Constructive interference**: the two waves are **in phase**. They superpose ('add up') to produce a wave that has a larger amplitude than the original waves (see Figure 15.1a).

- **Destructive interference**: two waves are 180° (π radians) out of phase. They superpose so that the amplitude of the resultant wave is smaller. If the two waves have the same amplitude they cancel out completely (see Figure 15.1b).

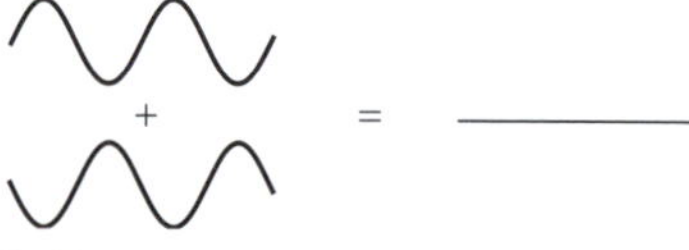

a Constructive interference

b Destructive interference

▲ **Figure 15.1** Interference

Stationary waves

Stationary waves are produced when interference takes place between two progressive waves of equal frequency and amplitude travelling in opposite directions along the same line. For example, if a rope is tied to a post at one end and made to oscillate at a particular frequency at the other end, the wave reflected by the post will overlap with the outward wave to produce a stationary wave.

Imagine two waves of equal frequency and amplitude, travelling towards each other at the same speed (see Figure 15.2):

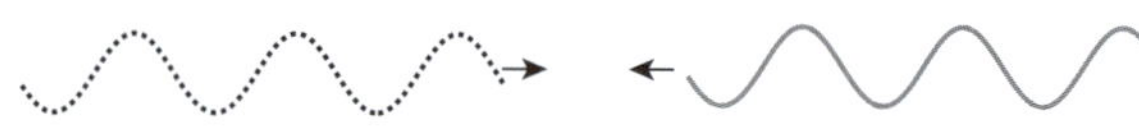

▲ **Figure 15.2** Producing a stationary wave

- At some point the two waves will be in phase and interfere constructively to form a wave with an amplitude which is twice the amplitude of one of the waves, as shown in Figure 15.3a.

- A quarter of a time period later, one wave will have moved a quarter of a wavelength to the right; the other wave will have moved the same distance to the left. The waves are now exactly half a wavelength out of step (in **antiphase**) and will cancel out completely, as shown in Figure 15.3b.

- A further time $T/4$ later the two waves are again in phase and will superpose constructively (Figure 15.3c).

Examples of stationary waves

Stretched strings

When waves are produced on a stretched string, (for example, by plucking a guitar string) several different stationary waves can be formed.

- **Fundamental frequency (first harmonic)**: The simplest stationary wave on a stretched string is shown in Figure 15.4. This is the **fundamental mode of vibration** (also called the **first harmonic**). **Nodes** are points where the amplitude of vibration is zero.

  Antinodes are points where the amplitude of vibration is a maximum.

The second- and higher-order stationary waves occur at higher frequencies of vibration.

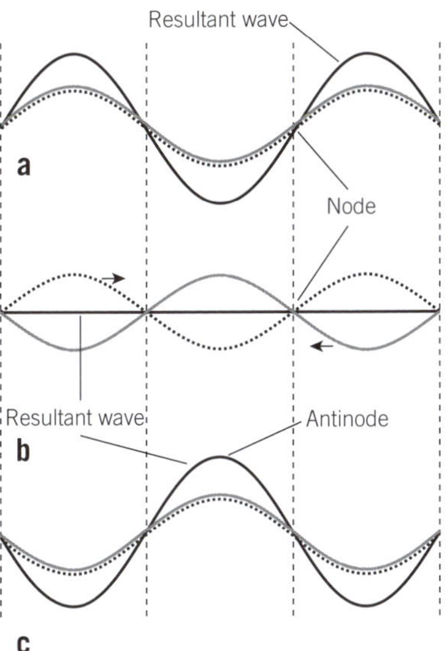

▲ **Figure 15.3** How stationary waves are formed

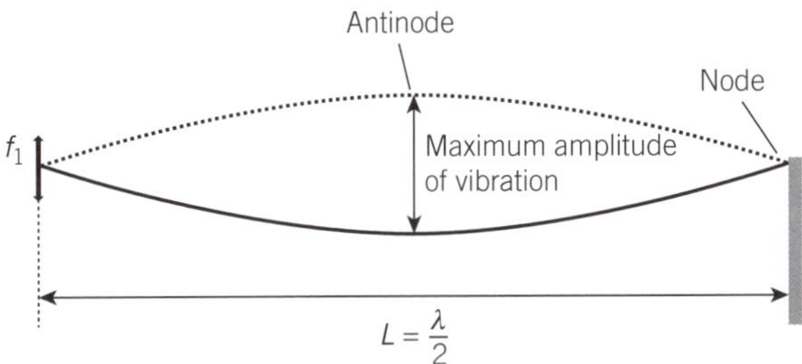

▲ **Figure 15.4** Fundamental mode of vibration

- **Second order (second harmonic)**: Points P and Q are in antiphase (π or 180° out of phase) – P is about to move up as Q is about to move down (see Figure 15.5).

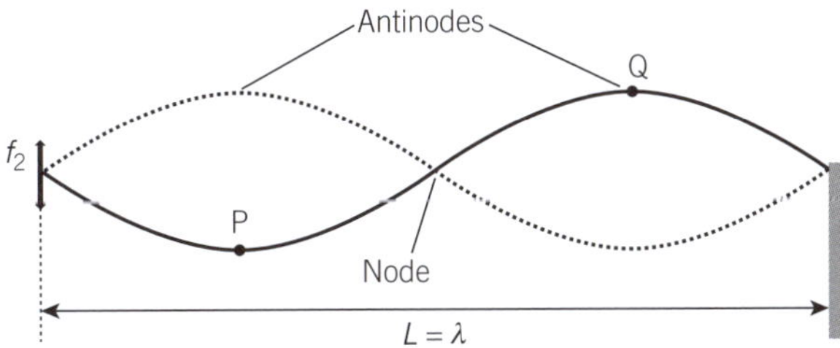

▲ **Figure 15.5** Second-order stationary wave

> **Remember**
>
> For a **one-loop** stationary wave:
>
> $$f_1 = \frac{v}{\lambda} = \frac{v}{2L}$$
>
> where L is the length of the string and v the velocity of progressive waves on the string.

> **Remember**
>
> For a **two-loop** stationary wave:
>
> $$f_2 = \frac{v}{\lambda} = \frac{v}{L}$$
>
> This frequency is twice the fundamental frequency.

Air columns

Stationary sound waves can be produced in pipes and other columns of air. A sound wave travelling down a pipe can interfere with a sound wave reflected back from the end of the pipe to form a stationary longitudinal wave. Just as with stretched strings, there are several modes of vibration.

Closed pipe

The experiment in Figure 15.6 shows how stationary waves can be produced in an air column. As the frequency of the signal generator is gradually increased a number of louder sounds are heard at specific frequencies. Figure 15.6 shows how the second-order stationary sound wave is formed.

The air molecules cannot vibrate freely at the closed end making this a **displacement node**. The air molecules have no restrictions on their movement at the open end – this is a **displacement antinode**. The displacement nodes are pressure antinodes – the air is being constantly compressed and expanded at these points. The first three stationary waves in a pipe closed at one end are shown in Figure 15.7.

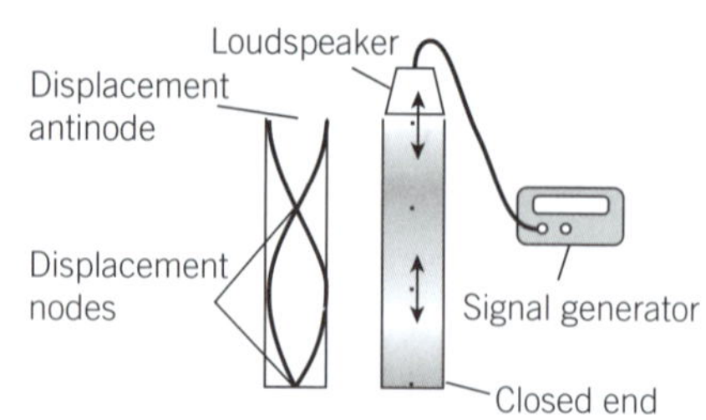

▲ Figure 15.6 Stationary longitudinal wave

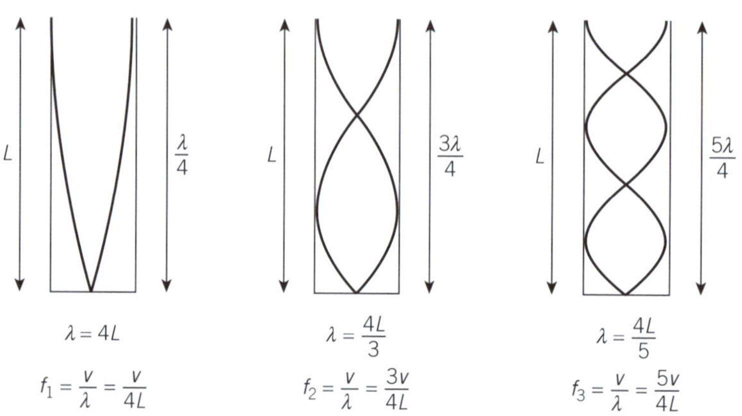

▲ **Figure 15.7** Stationary waves in a pipe closed at one end

> **Remember**
>
> The stationary wave patterns in an air column are longitudinal stationary waves. They are drawn as transverse waves to show how the amplitude of vibration of the air molecules varies along the length of the air column.

Open pipe

Stationary waves can also be produced in pipes open at both ends. In this case, both ends are displacement antinodes as the air molecules are free to vibrate with the maximum amplitude (see Figure 15.8).

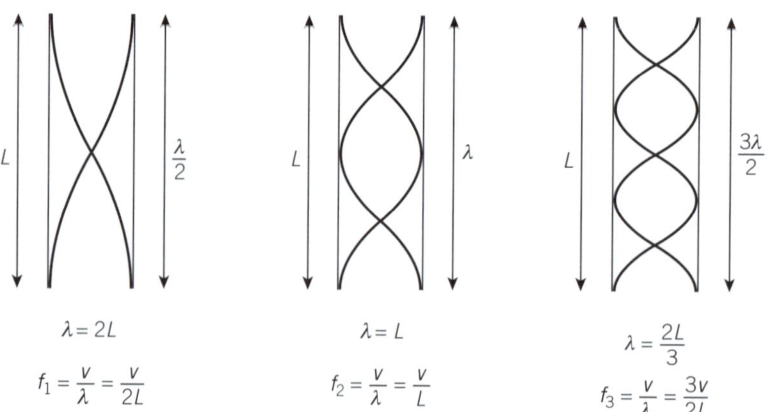

▲ **Figure 15.8** Stationary waves in open pipes

Worked example

A tuning fork is sounded above a cylinder of water, as shown in Figure 15.9. A tap is opened and the level of the water gradually falls. The sound becomes louder when the level of the water falls to certain levels.

Two successive loud sounds occur when the water level is 69.0 cm and 35.8 cm above the bench.

Determine the frequency of the tuning fork. [Speed of sound in air = $340\,\mathrm{m\,s^{-1}}$.]

Answer

The difference in the two levels of water must be the distance between two successive nodes $\left(\dfrac{\lambda}{2}\right)$, as shown in Figure 15.10:

$$\frac{\lambda}{2} = 69.0 - 35.8 = 33.2\,\mathrm{cm} \qquad \text{so} \qquad \lambda = 66.4 \times 10^{-2}\,\mathrm{m}$$

$$f = \frac{v}{\lambda} = \frac{340}{66.4 \times 10^{-2}} = 512\,\mathrm{Hz}$$

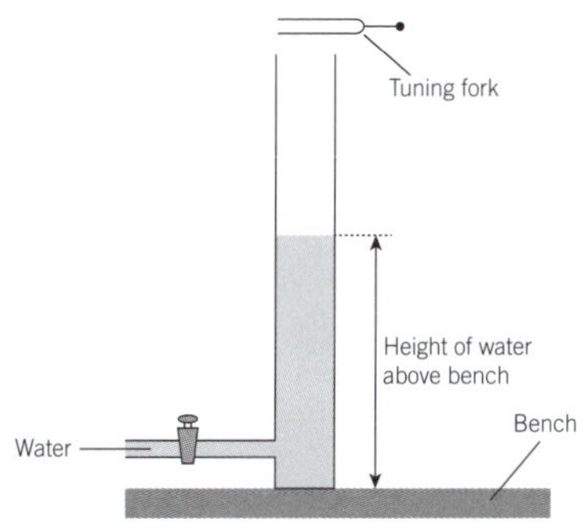

▲ **Figure 15.9** Investigating resonant frequencies

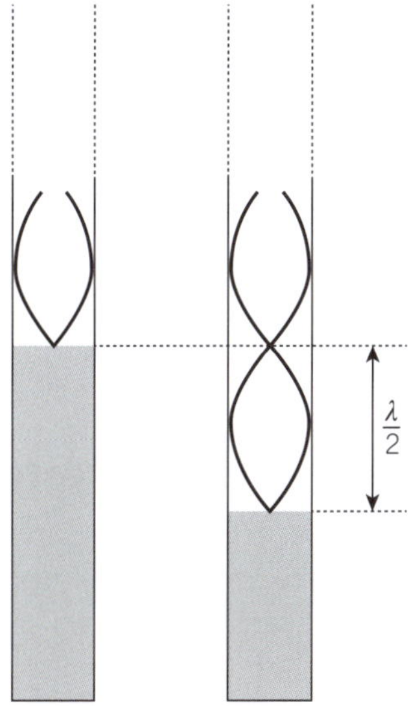

▲ **Figure 15.10**

Microwaves

Microwaves can also be used to demonstrate the properties of stationary waves. A microwave transmitter emits microwaves towards a metal plate which reflects the microwaves back towards the transmitter, as shown in Figure 15.11.

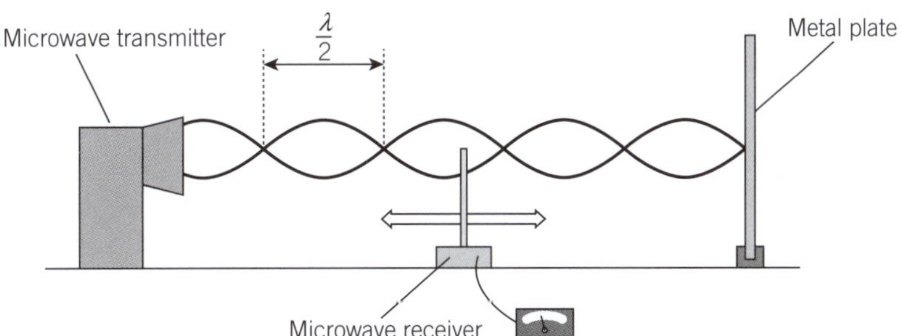

▲ **Figure 15.11** Microwave stationary waves

The transmitted and reflected waves superpose to form a stationary wave. As the microwave receiver is moved along a line directly between the microwave transmitter and the metal plate, it detects successive, equally spaced, strong, and weak signals. The distance between successive minima is the distance between two adjacent nodes – half a wavelength.

The nodes on a stationary wave have zero energy; the antinodes have maximum energy. As the positions of the nodes and antinodes do not change, no energy is transferred by a stationary wave.

Diffraction

When waves pass through a gap, or past a partial obstruction, they bend and spread out beyond the geometric 'shadow' region of the gap or obstruction, as shown in Figure 15.12. This effect is called **diffraction** and the waves are said to be 'diffracted'.

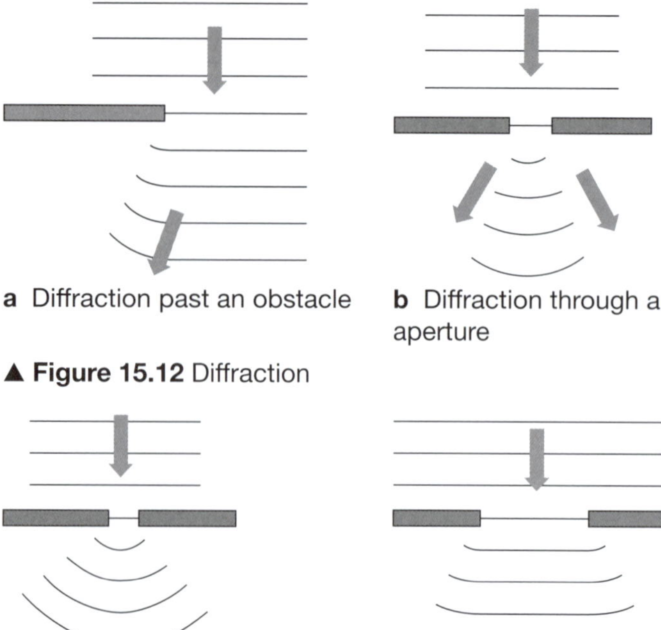

▲ **Figure 15.12** Diffraction

The amount of spreading of a wave as it passes through an aperture depends on the size of the aperture. Most diffraction occurs when the aperture is a similar size to the wavelength of the waves, as shown in Figure 15.13a.

▲ **Figure 15.13** Diffraction and aperture size

Two-source interference

If two sources of waves are in close proximity to each other (for example, two wave 'dippers' in a ripple tank, or two loudspeakers) the two sets of waves overlap. If the two sets of waves have the same frequency and similar amplitudes, and the two sources are **coherent** (have a fixed phase difference between them), they can produce an interference pattern, with points of constructive interference and points of destructive interference.

Figure 15.14 shows successive wavefronts produced by two vibrating dippers in a ripple tank (the lines represent the peaks, or crests, of waves). At points such as P, two wave peaks are interfering constructively and producing a larger amplitude; at points such as Q a wave peak from one source is overlapping with a wave trough from the other, cancelling out (destructive interference). At points such as R, two troughs are meeting and interfering constructively to produce a deeper trough.

If two loudspeakers are connected to the same signal generator and placed a short distance apart, a similar interference pattern to the one described in the ripple tank can be observed (see Figure 15.15).

A person walking from P to Q hears a series of loud and quiet sounds. Where the sound is loud, the **path difference** (the difference in the distance travelled by a sound wave from one loudspeaker compared to the other) must be a whole number of wavelengths, and so constructive interference occurs.

Where there is a quiet sound, the path difference must be an odd number of half-wavelengths so that the sound waves interfere destructively. The sound waves may not cancel out completely because the sound wave from one speaker will have travelled further than the other, so will have a smaller amplitude.

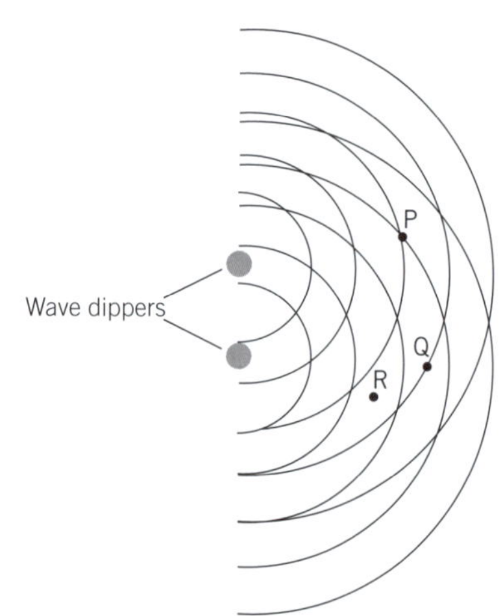

▲ **Figure 15.14** Interference patterns in a ripple tank

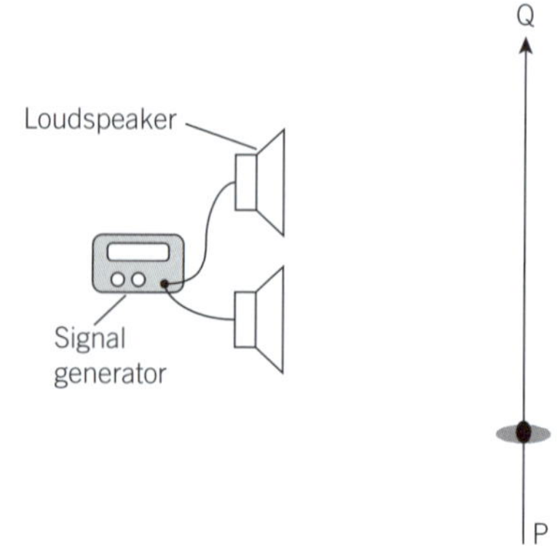

▲ **Figure 15.15** Interference patterns with sound waves

Young's double-slits experiment

Young's double-slits experiment provides evidence for the wave-like nature of light by producing interference fringes on a screen. Figure 15.16 shows how the experiment is set up.

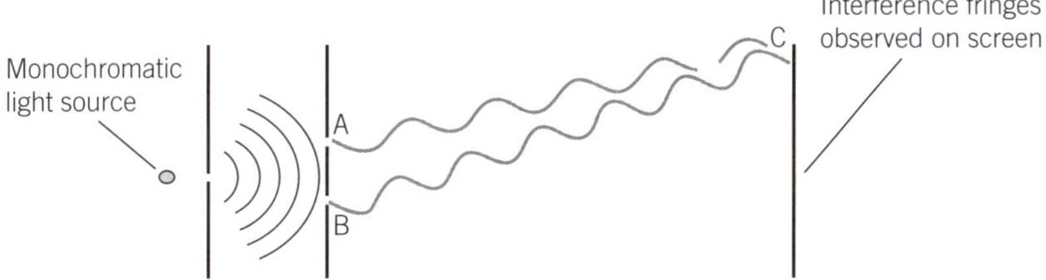

▲ **Figure 15.16** Young's double-slits experiment

Light from a monochromatic light source (e.g., a laser) passes through a single slit and diffracts (spreads out). It then passes through two narrow slits A and B. There is a fixed phase difference between the light emerging from slits A and B as the waves are part of the same wavefront; that is, the light from A and B is **coherent**.

Interference fringes (areas of light and dark) can be observed on a screen some distance away. Consider the case when the light from slits A and B emerges in phase. In Figure 15.16, the distance BC is slightly longer than the distance AC – there is a **path difference** between the light waves from slit A and slit B. If this distance is an odd number of half-wavelengths, the two waves will be out of phase (in antiphase) when they reach C and will interfere destructively – a dark fringe will be observed. If the path difference is a whole number of wavelengths the two waves will be in phase at C and interfere constructively, producing a bright fringe.

From Figure 15.17 the two waves emerging from slits A and B can be considered parallel as the distance D to the screen is much greater than the slit separation a. The path difference BP is $a\sin\theta$. The first-order bright fringe occurs when the path difference is λ, the second-order when the path difference is 2λ, and so on. The nth bright fringe occurs when the path difference is $n\lambda$. For a bright fringe:

$$n\lambda = a\sin\theta$$

> **Maths skills**
>
> For small angles:
>
> $\tan\theta \approx \sin\theta \approx \theta$ where θ is measured in radians.

Analysing Young's slits experiment

From Figure 15.17, $x_n = D\tan\theta$, but since θ is small, $\tan\theta \approx \sin\theta$, so:

$$x_n = D\sin\theta = D\frac{n\lambda}{a}$$

Similarly, the $(n+1)$th bright fringe is given by $x_{n+1} = D\frac{(n+1)\lambda}{a}$.

The distance between adjacent bright fringes is x, where:

$$x = (x_{n+1} - x_n) = \frac{D\lambda}{a} \quad \text{so} \quad \lambda = \frac{ax}{D}$$

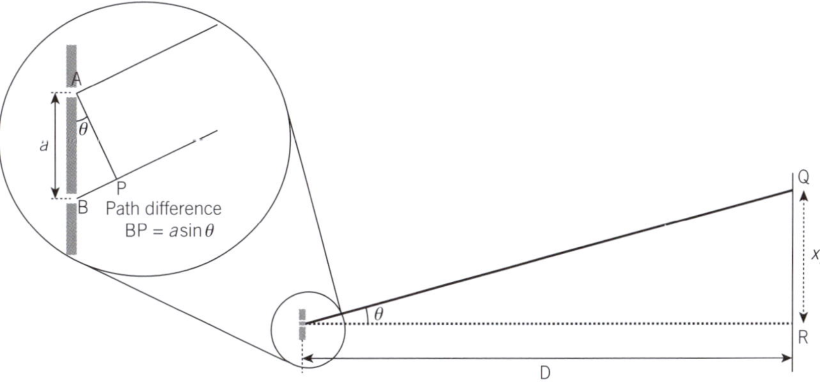

▲ **Figure 15.17** Analysing the double-slits experiment

The wavelengths of visible light are small ($\approx 5 \times 10^{-7}$ m) so the fringes are very close together. To increase the fringe separation:

- Make the slit separation a as small as possible.

- Make the distance D between the slits and the screen as large as possible.

- Use light with as long a wavelength as possible (e.g., red light rather than blue light).

Carrying out the experiment in a darkened room also makes the fringes easier to detect.

Worked example

Light of wavelength 589 nm is incident on a pair of slits, forming an interference pattern on a screen 1.40 m away. The bright fringes on the screen are 0.20 cm apart. Determine the separation of the two slits.

Answer

$$a = \frac{\lambda D}{x} = \frac{589 \times 10^{-9} \times 1.40}{0.20 \times 10^{-2}} = 4.12 \times 10^{-4}\,\text{m}\,(0.412\,\text{mm})$$

Diffraction gratings

The double-slit in Young's experiment can be replaced by multiple slits. Increasing the number of slits has the effect of making much sharper and clearer **maxima** (bright lines or 'fringes'). A diffraction grating consists of many parallel slits extremely close together, ruled on a transparent plate. The light passing through each slit is diffracted, and constructive interference occurs only at very specific angles, the light waves cancelling each other out in all other directions.

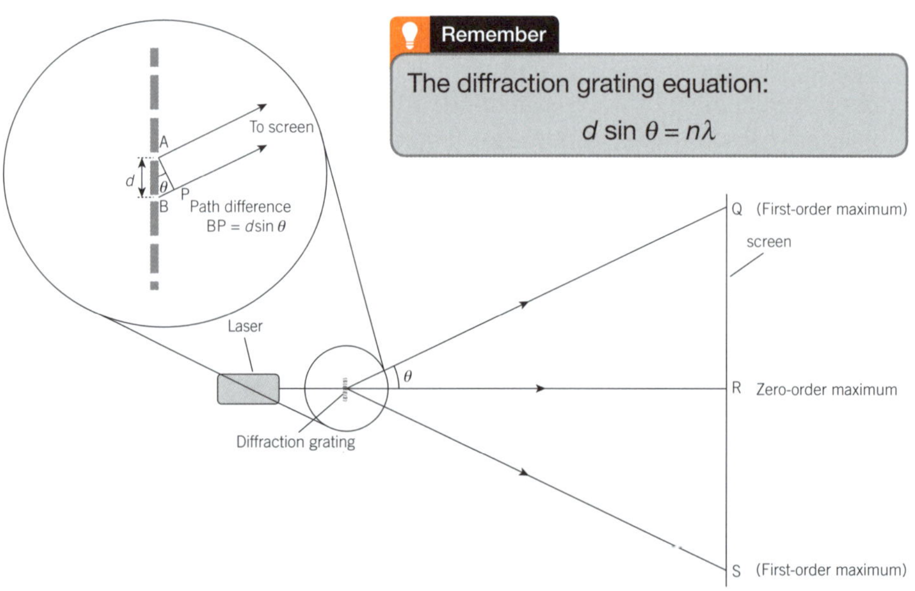

▲ **Figure 15.18** Analysing diffraction grating experiment

For the light waves from two adjacent slits to be in phase and add up constructively, the path difference BP must be a whole number of wavelengths. From Figure 15.18, if the distance between adjacent slits is d:

$$d\sin\theta = n\lambda$$

where n is an integer. There is also a **zero-order** maximum in the same direction as the incident beam (the waves coming from every slit of the grating have all travelled the same distance, so they are all in phase).

⬆ Raise your grade

(a) In relation to light waves, explain what is meant by the terms:

(i) monochromatic

A single colour ✗

A more precise, scientific definition is needed. Monochromatic light waves are waves with a single wavelength (e.g., the light from a laser).

(ii) constructive interference.

When two waves meet and make a bigger light wave ✔✗

The right idea, but a better answer would be 'When two waves overlap, and are **in phase**, so that their **amplitudes** add up to make a light wave with a larger amplitude'. [3]

(b) A student attempts to measure the wavelength of light using Young's slits, as shown below.

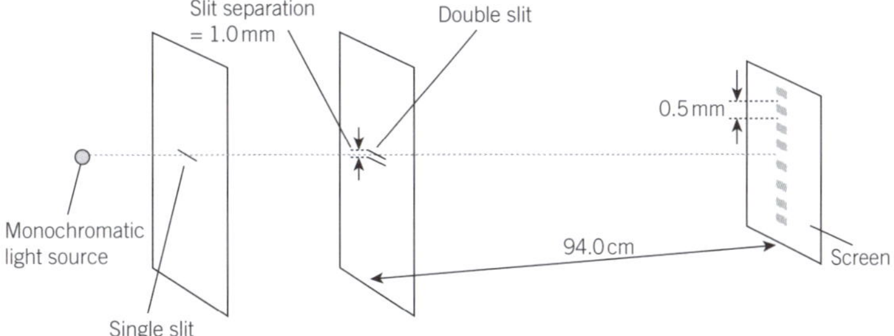

(i) The light is diffracted by the single slit. Explain what is meant by *diffraction*.

The light bends as it passes through the slit ✔✗

correct, but the candidate should have added '… and **spreads out**, beyond the geometric 'shadow' of the slit' for the second mark

(ii) Explain why the light from the double slit is *coherent*.

The light waves from the two slits are in phase. ✗

The light entering each of the two slits is from the same wavefront so there must be a fixed phase difference between the light waves emerging from them (the light waves don't need to be in phase to be coherent).

(iii) Use the student's results to calculate the wavelength of the light used.

$$\lambda = \frac{ax}{D} = \frac{1.0 \times 10^{-3} \times 0.5 \times 10^{-3}}{94.0 \times 10^{-2}} = 5.32 \times 10^{-7}\,\text{m} ✔$$

Correct calculation.

Correct equation/method.

wavelength = 5.32×10^{-7} m ✔ [5]

The student wants to make the fringes further apart so that they are easier to see and measure accurately. Suggest two changes the student could make to the experiment to achieve this.

1 Move the screen further away from the slits ✔

Correct suggestion.

2 Move the two slits further apart ✗

The opposite is true. From the two-slits equation $x = \frac{\lambda D}{a}$, so the fringe spacing x increases if a, the distance between the slits, is decreased. [2]

1 Two waves superpose as shown below.

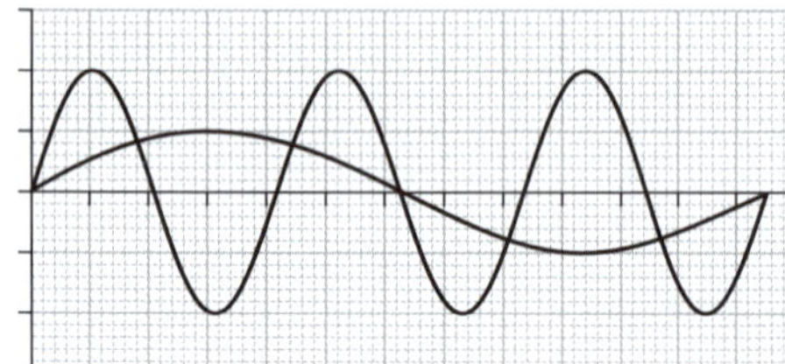

Which diagram shows the resultant of the two waves? [1]

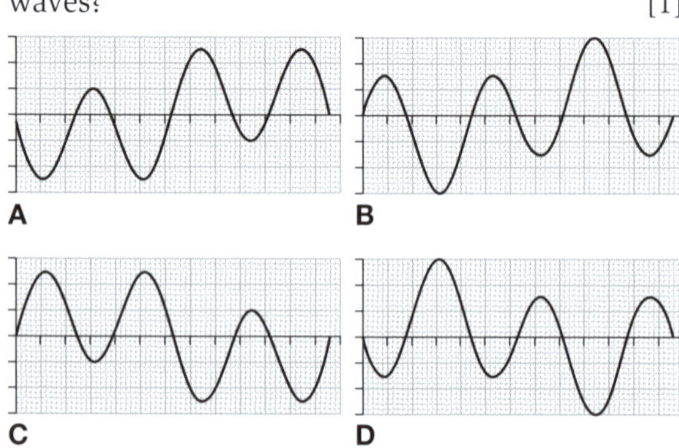

2 A string is stretched between points X and Y. One end of the string is vibrated, setting up a stationary wave, as shown below.

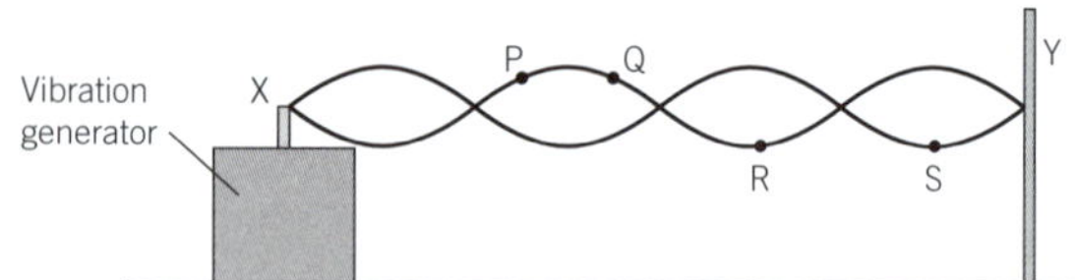

Which statement is correct?

A The string is oscillating at its fundamental frequency.

B The distance RS is one wavelength.

C Points P and Q are in phase.

D Points R is a node. [1]

3 An organ pipe of length 0.500 m is closed at one end. What are the two lowest resonant frequencies the pipe can produce?

[The speed of sound is 340 m s^{-1}.]

A 170 Hz 340 Hz

B 170 Hz 510 Hz

C 227 Hz 340 Hz

D 227 Hz 510 Hz [1]

4 Which statement correctly describes the meaning of *diffraction*?

A When waves meet in phase and their amplitudes add up.

B When waves meet out of phase and cancel out.

C When waves pass an obstacle and bend, entering the geometric shadow of the object.

D When waves from two sources have a fixed phase difference. [1]

5 A diffraction grating has 400 lines mm^{-1}. When monochromatic light passes through the grating, the third-order maxima subtend an angle of 40°, as shown below.

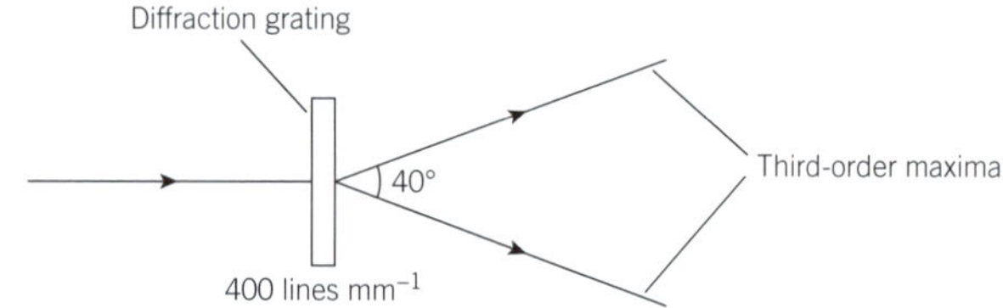

What is the wavelength of the light?

A 285 nm **B** 410 nm **C** 540 nm **D** 820 nm [1]

6 White light passes through a diffraction grating and a series of visible spectra are observed on a screen some distance away, together with a white light maximum at a point directly in line with the diffraction grating and the light source (zero order). A series of visible spectra are seen either side of the zero order.

(a) Describe what is meant by *diffraction*. [2]

(b) Explain why:

(i) the central zero-order maximum is white

(ii) the first-order maximum for green light is in a different position from the first-order maximum for red light. [2]

(c) Blue light of wavelength 460 nm produces a third-order maximum at an angle of 17.7°. A wavelength of red light produces a second-order maximum at the same angle.

Calculate:

(i) the number of lines per millimetre of the diffraction grating

(ii) the wavelength of the red light. [3]

16 Communications

Key points

- [ ] Appreciate that information may be carried by a number of different channels, including wire-pairs, coaxial cables, radio and microwave links, and optic fibres.
- [ ] Understand the term modulation, and be able to distinguish between amplitude modulation (AM) and frequency modulation (FM).
- [ ] Recall that a carrier wave, amplitude modulated by a single audio frequency, is equivalent to the carrier wave frequency together with two sideband frequencies.
- [ ] Understand the term bandwidth.
- [ ] Recall the frequencies and wavelengths used in different channels of communication.
- [ ] Demonstrate an awareness of the relative advantages of AM and FM transmissions.
- [ ] Recall the advantages of the transmission of data in digital form, compared with the transmission of data in analogue form.
- [ ] Understand that the digital transmission of speech or music involves analogue-to-digital conversion (ADC) before transmission and digital-to-analogue conversion (DAC) after reception.
- [ ] Understand the effect of the sampling rate and the number of bits in each sample on the reproduction of an input signal.
- [ ] Discuss the relative advantages and disadvantages of channels of communication in terms of available bandwidth, noise, crosslinking, security, signal attenuation, repeaters and regeneration.
- [ ] Recall the relative merits of both geostationary and polar orbiting satellites for communicating information.
- [ ] Understand and use signal attenuation expressed in dB and dB per unit length.
- [ ] Recall and use the expression 'number of dB' = $10 \lg\left(\dfrac{P_1}{P_2}\right)$ for the ratio of two powers.

Analogue signals

Music and speech are examples of **analogue signals** as they can vary continuously in both amplitude and frequency. They can be transmitted ('carried') by radio waves using either amplitude modulation (AM) or frequency modulation (FM).

Amplitude modulation (AM)

Amplitude modulation (AM) is the simplest form of modulation, and uses a carrier wave of fixed frequency. The information to be transmitted (the audio frequency) is combined with the carrier wave so that the amplitude of the radio wave varies to match the information signal, as shown in Figure 16.1.

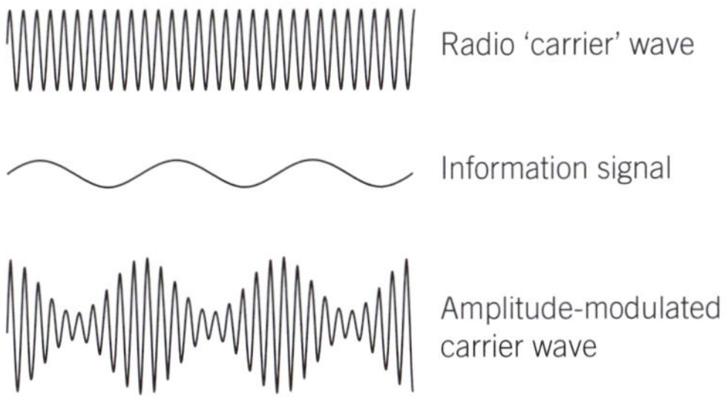

Audible signals cannot be transmitted by electromagnetic waves with audible frequencies because:

- ▶ long aerials would be needed
- ▶ radio stations would overlap (it would be impossible to 'tune in' to one particular station)
- ▶ the range would be very short.

▲ **Figure 16.1** Amplitude modulation (AM)

If the carrier-wave frequency is f_c and the audio frequency is f_a it can be shown that the amplitude-modulated carrier wave consists of:

- a wave of frequency f_c and constant amplitude
- waves of frequency $f_c - f_a$ and $f_c + f_a$ with constant amplitude

as shown in Figure 16.2.

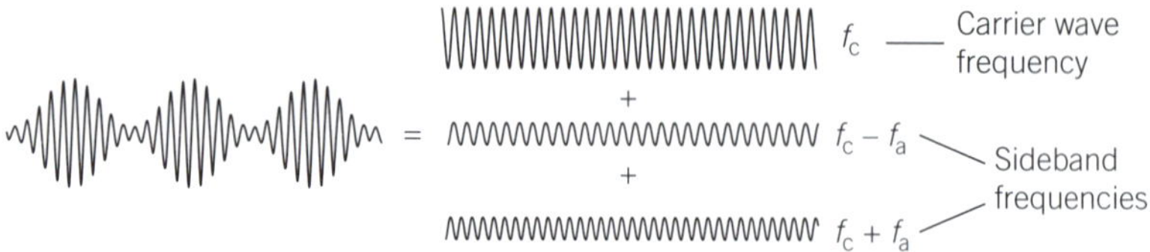

▲ **Figure 16.2** Sideband frequencies frequencies

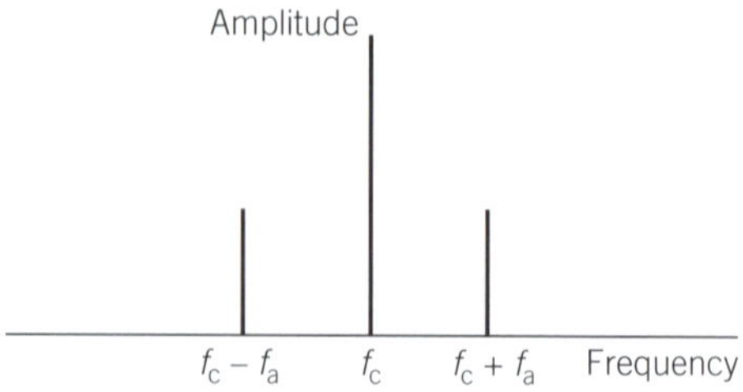

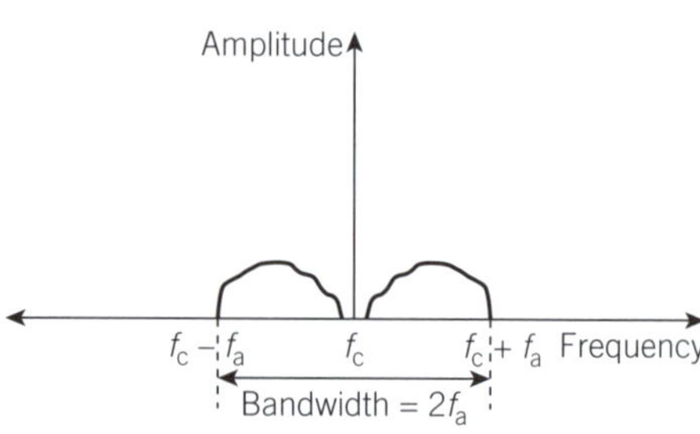

▲ **Figure 16.3** Frequency spectrum

Each audio frequency has its own pair of sideband frequencies (see Figure 16.3). When speech or music is transmitted it contains a range of frequencies up to a maximum frequency f_a and there will be a corresponding range, or band, of sideband frequencies, from $(f_c - f_a)$ to $(f_c + f_a)$. The range of sideband frequencies is called the **bandwidth** and is equal to $2f_a$, as shown in Figure 16.4.

Each radio or TV station is allocated a band of frequencies – the width of this band is the channel bandwidth. The frequencies transmitted by the station are restricted to lie within the channel bandwidth. AM carrier frequencies are in the range 535–1605 kHz, with frequencies of 540 kHz to 1600 kHz assigned at 10 kHz intervals.

▲ **Figure 16.4** Bandwidth

Frequency modulation (FM)

In **frequency modulation** (FM), the frequency of the carrier wave is modulated by the amplitude of the information signal. Where the amplitude of the information signal is high (and positive) the frequency of the carrier wave is increased; where the amplitude is negative the frequency of the carrier wave is decreased, as illustrated in Figure 16.5.

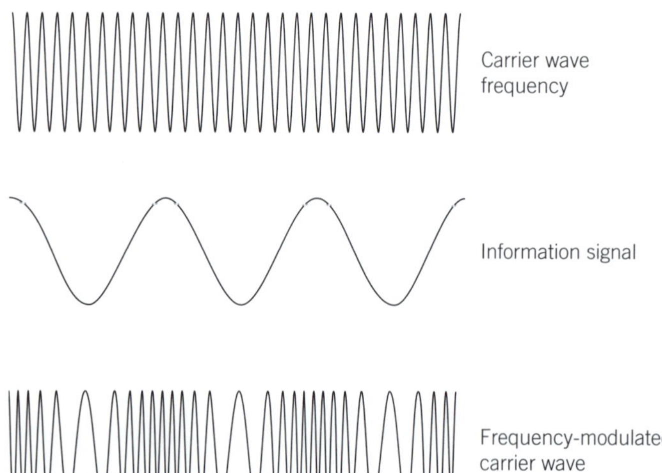

▲ **Figure 16.5** Frequency modulation

FM transmission is more expensive than AM transmission, but is less affected by **noise** – unwanted electrical interference from such things as electric motors, mobile phones, and the random thermal motion of electrons. The noise signal adds on to, and distorts, the signal being transmitted. If the signal is amplified, the noise is amplified too.

Remember

AM and FM are examples of analogue signals. They do not apply to digital signals.

Worked examples

A person is listening to music on a radio station transmitting on a frequency of 909 kHz with a bandwidth of 9 kHz.

1 **a)** Calculate:

      i) the wavelength of the radio station

      ii) the maximum audible frequency f_a.

  b) Comment on the quality of the sound heard.

Answer

a) **i)** $\lambda = \dfrac{c}{f} = \dfrac{3 \times 10^8}{909 \times 10^3} = 330\,\text{m}$

    ii) $f_a = \dfrac{9 \times 10^3}{2} = 4.5 \times 10^3\,\text{Hz}$ (4.5 kHz)

b) The human ear can detect frequencies up to 20 kHz, so the higher frequencies would not be transmitted; the quality of the sound reproduction would be poor.

2 A sinusoidal carrier wave has a frequency of 800 kHz and an amplitude of 5.0 V. The carrier wave is frequency modulated by a sinusoidal signal of frequency 10 kHz and amplitude 3.0 V.

The frequency deviation of the carrier wave is 30 kHz V^{-1}. (This means that the frequency of the carrier wave increases or decreases by 30 kHz for a change of 1.0 V in the modulating signal.)

Calculate:

  a) the amplitude of the carrier wave

  b) the maximum and minimum values of the frequency of the carrier wave.

Answer

a) amplitude = 5.0 V

b) maximum frequency = $800 + (3 \times 30) = 890\,\text{kHz}$

    minimum frequency = $800 - (3 \times 30) = 710\,\text{kHz}$

> In frequency modulation, the amplitude of the carrier wave is constant.

Comparing FM and AM transmission

▼ **Table 16.1** AM and FM transmission

AM	FM
AM transmitters and receivers are relatively cheap and simple to make.	Frequency modulation of a signal is more complex, making the cost greater than AM.
Poorer sound quality compared to FM because of narrower bandwidth – can only carry audio frequencies up to 5 kHz.	Better sound quality compared to FM due to larger bandwidth – can carry audio frequencies up to 20 kHz.
AM signals can be transmitted over long distances.	The range (~ 30 km) is smaller than AM (as higher frequency radio waves are more attenuated than low-frequency waves).
Affected by noise (e.g., from electric motors) as noise changes the amplitude of the signal and AM is amplitude modulated.	FM signals are less affected by noise – a change in the amplitude of the signal due to noise does not affect the frequency modulation of the signal.
Long wavelengths, so AM signals can diffract around obstacles such as hills and tall buildings.	Short wavelengths, so little diffraction around obstacles. Line-of-sight needed to receive an FM broadcast.

Digital signals

A digital signal is one that has been encoded as a series of '0's and '1's.
An analogue signal, such as the electrical voltage from a microphone, is
converted into a digital signal by an **analogue-to-digital converter** (ADC) in
a process called **pulse code modulation** (PCM).

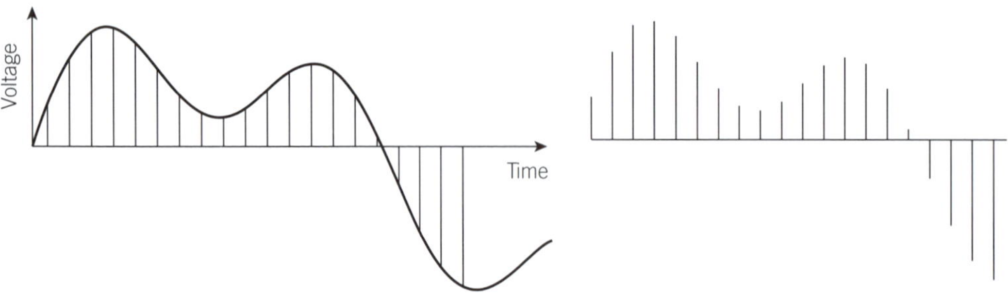

▲ **Figure 16.6** Pulse amplitude modulation (PAM)

The analogue signal is first sampled, by being measured at regular time
intervals, as shown in Figure 16.6 – a process called **pulse amplitude
modulation** (PAM). The sampling rate must be at least double the
highest frequency to be transmitted if the transmitted signal is to
be accurately reconstructed by the receiver. For speech and music
the highest frequency is 22 kHz, and so a sampling frequency of at
least 44 kHz is needed.

Each sampled voltage is assigned a quantum level corresponding
to a unique binary number. For example, in Figure 16.7 a voltage
of 3.1 V would be assigned level 8 and the binary code 1000. This
example uses 4 **bits** (binary digits) so has 2^4 (16) possible quantum
levels. A greater number of bits means the quantisation process is
more accurate, but the information takes longer to transmit and
more space to store.

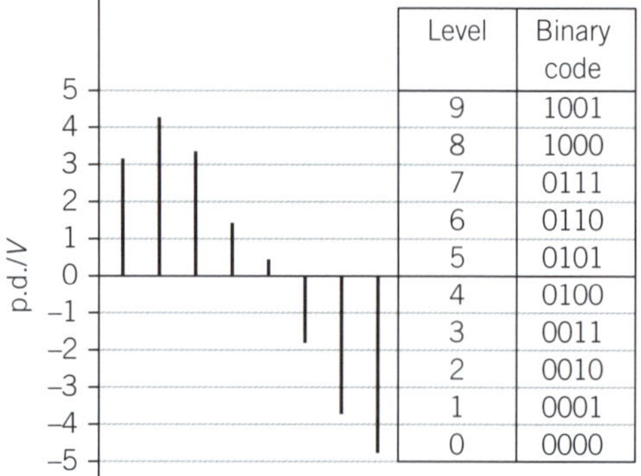

Level	Binary code
9	1001
8	1000
7	0111
6	0110
5	0101
4	0100
3	0011
2	0010
1	0001
0	0000

▲ **Figure 16.7** Quantisation

The digital signal can then be transmitted as a series of electromagnetic
wave pulses (e.g., as pulses of light sent down an optic fibre). When the
information reaches the receiver, the analogue signal is reconstructed from
the digital signal using a **digital-to-analogue converter** (DAC).

Advantages of digital transmission

Electrical noise can affect the amplitude of a digital signal in the same way
as analogue signals, but even if the transmitted signal becomes distorted by
noise, the receiving system only has to be able to recognise the separate '1's
and '0's of the signal to reconstruct the original signal accurately. In addition,
noisy signals can be 'cleaned' by regenerator amplifiers placed at regular
intervals along a transmission link, as illustrated in Figure 16.8.

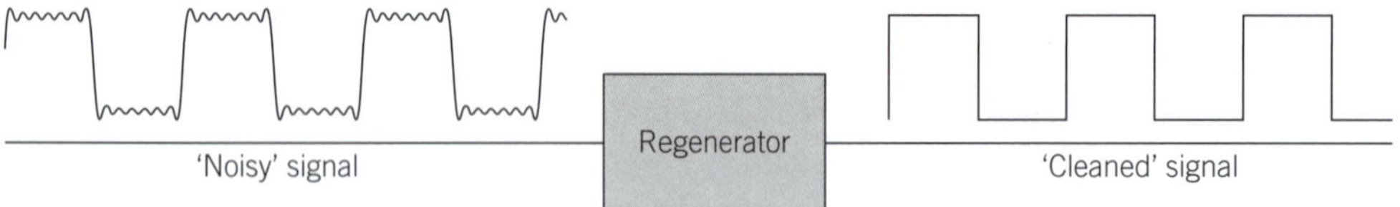

▲ **Figure 16.8** Regeneration 'cleans' a noisy signal

Digital signals can be compressed into pulses of much shorter duration than
the signal itself, and so many separate signals can be transmitted at the same
time (multiplexing).

If a large number of bits are used, some of the bits can be used to check for any transmission errors. Digital systems are also more secure (difficult to intercept) and can be encrypted for extra protection.

Transmitting information

Information can be transmitted along a range of different communication channels.

Wire pairs

Twisted wire pairs are a low-cost way of transmitting low-frequency signals over relatively short distances (up to 100 m). The heating effect of the current in the wires causes rapid attenuation of the signal over longer distances, and the wires act as aerials, radiating energy as electromagnetic waves. Twisting the wires together reduces the amount of energy radiated, but does not eliminate it completely. Wire pairs are also susceptible to cross-linking, where the signal carried by one wire pair is picked up by another, and noise from things such as electric motors and fluorescent lights.

Coaxial cables

A coaxial cable consists of a copper wire separated from an outer copper braid conductor by polythene insulation, as shown in Figure 16.9. The cable is further protected by an outer layer of plastic insulation.

Coaxial cables have much less attenuation than wire pairs, radiating less energy. There is also less noise and cross-linking, and coaxial cables are more difficult to 'tap'. They have a greater bandwidth (~ 60 MHz) than wire pairs, so can carry much more information.

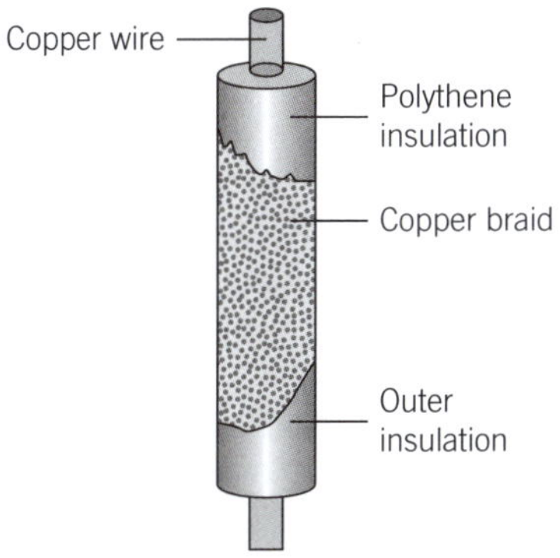

▲ **Figure 16.9** Coaxial cable

Radio waves

Radio waves are electromagnetic waves with wavelengths ranging from 0.1 m to 1 km or greater (see Table 16.2). They are used by TV and radio stations and for two-way communications, including mobile phones.

▼ **Table 16.2** Radio wavebands

	Wavelengths	Uses
Long-wave	>1 km	international long-wave AM radio
Medium-wave (MF)	100 m–1 km	medium-wave radio stations
Short-wave (HF)	10 m–100 m	AM short-wave radio
Very high frequency (VHF)	1 m–10 m	local FM radio
Ultra-high frequency (UHF)	0.1 m–1 m	TV broadcasts, mobile phones

Radio waves are affected by the atmosphere in different ways, depending on the frequency of the waves (see Figure 16.10, overleaf).

- **Surface waves:** Low-frequency (long wavelength) radio waves spread out from a transmitter because of diffraction and stay near the surface of the Earth. They also diffract around hills and tall buildings so there are few places where reception is poor.

- **Sky waves:** Medium- and high-frequency (MF and HF) waves (medium and short wavelengths) are reflected by the ionosphere and can travel around the Earth by being repeatedly reflected by the ionosphere and the ground.

- **Space waves:** Radio waves with frequencies greater than 30 MHz are used for line-of-sight communications and satellite links.

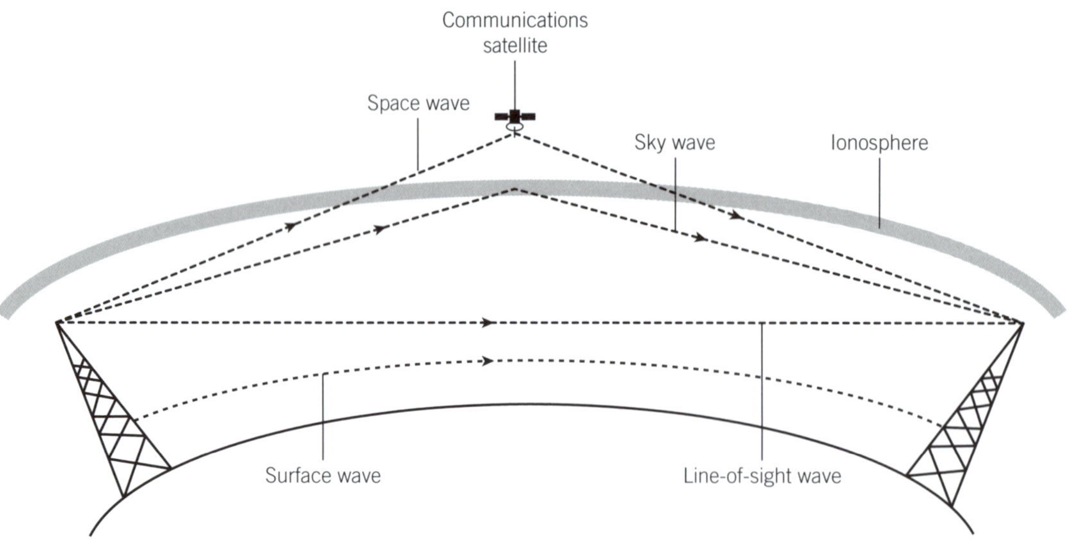

▲ **Figure 16.10** Surface waves, sky waves and space waves

Microwaves

Microwaves are electromagnetic waves with wavelengths between 1 mm and 10 cm. They have a much greater bandwidth so can carry more information. They are used for satellite communications and line-of-sight communications. For long-distance microwave links, several repeater and regenerator stations are needed, which can both amplify and 'clean' the microwave signals.

Optic fibres

Optic fibres are used to transmit digital signals as pulses of infrared light with wavelengths in the region of 1200 nm. They have several advantages over copper-wire-based communication systems, including:

- a much higher bandwidth so they can transmit much more information, at a faster rate

- less signal attenuation and so fewer regenerator/amplifier stations are needed

- greater security – they are virtually impossible to 'tap'.

They are also much less affected by noise, can be 'cleaned' by regenerator stations, and are cheaper and lighter than copper-wire systems, making them ideal for communications over long distances.

Satellite communications

Satellites are used to transmit information, such as international telephone links and satellite TV broadcasts, around the Earth, though there is a time delay between the sending and receiving of information. Two types of satellites are used:

- **Geostationary satellites**: These orbit the Earth at a height of 36 000 km above the equator once every 24 hours so that, viewed from the Earth, they appear stationary (see Figure 16.11). Microwave beams are sent from ground-based transmitters to a satellite using one carrier frequency (the **uplink**); the satellite then transmits the microwave beam back down to satellite dish receivers on the ground using a different frequency (the **downlink**).

  Satellite links for long-distance communications are better than using radio waves as they are not obstructed by hills and do not rely on reflection by the ionosphere (which varies in reflectivity during the day).

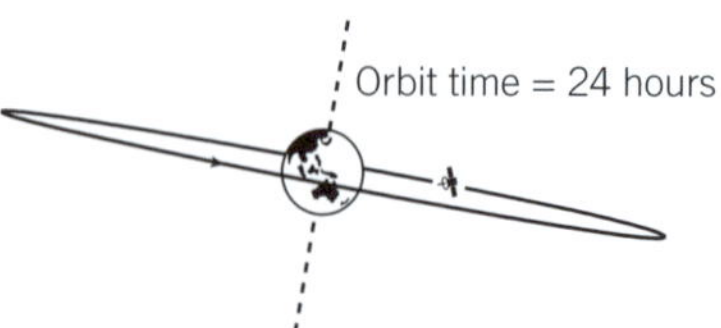

▲ **Figure 16.11** Geostationary satellite

Link

For more on geostationary orbits see Unit 8 *Gravitational fields*.

- **Polar-orbiting satellites**: Some satellites orbit the Earth above the poles at a height of 700–800 km above the Earth's surface, taking 90 minutes to complete an orbit, as shown in Figure 16.12. As polar-orbiting satellites are relatively close to the Earth's surface, they can be used for weather forecasting and surveillance.

  When used for communications, polar-orbiting satellites have much shorter delay times (because they are much nearer the Earth's surface), but the 'footprint' (the area of ground the radiation emitted by the satellite covers on the Earth's surface) is much smaller. The satellite has to be continually tracked by a ground station.

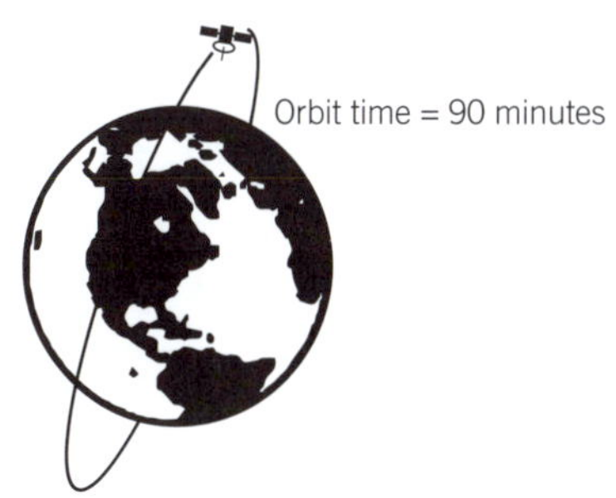

▲ **Figure 16.12** Polar orbiting satellite

Attenuation

Attenuation is the gradual decrease in the power of a signal the further it travels. Infrared waves are absorbed or scattered as they pass through an optic fibre, radio waves and microwaves are partly absorbed by the atmosphere, and electrical signals in wires lose energy through heating the wires and radiating energy.

The power may decrease by a very large factor over long distances (several powers-of-ten), and so a logarithmic scale is used to measure the attenuation. If the power falls from P_1 to P_2, the attenuation is:

$$\text{attenuation} = 10 \log_{10} \left(\frac{P_1}{P_2} \right)$$

Attenuation is measured in **decibels** (dB) where 1 dB is one-tenth of a bel (B).

The attenuation in an electrical cable or an optic fibre is often expressed in dB km^{-1}, the total attenuation divided by the length of the cable.

> **Worked example**
>
> The input power to an optic fibre of length 40 km is 14 mW. The power at the receiver is 620 μW. Calculate the attenuation per kilometre.
>
> **Answer**
>
> $$\text{attenuation} = 10 \log_{10} \left(\frac{P_1}{P_2} \right) = 10 \log_{10} \left(\frac{14 \times 10^{-3}}{620 \times 10^{-6}} \right) = 13.5 \text{ dB}$$
>
> $$\text{attenuation per km} = \frac{13.5}{40} = 0.34 \text{ dB km}^{-1}$$

The effects of attenuation can be reduced by using repeaters and regenerators:

- **Repeaters** are amplifiers used at intervals along a communication link to amplify the signal (though they amplify the noise as well as the signal).

- **Regenerators** are used in digital communications systems to 'clean' the digital signal by removing noise and restoring the digital signal to its original amplitude.

The gain of a repeater is also usually expressed in dB where:

$$\text{gain} = 10 \log_{10} \left(\frac{P_{\text{out}}}{P_{\text{in}}} \right).$$

As a signal travels it becomes weaker, and the noise becomes a greater fraction of the total signal – the signal-to-noise ratio decreases. The signal-to-noise ratio is also expressed in dB, with many communication systems setting a minimum value for this ratio.

The equation for comparing two powers P_1 and P_2 is:

$$\text{No. of dB} = 10 \log_{10} \left(\frac{P_2}{P_1} \right)$$

You must be able to recall and use this equation.

If $P_2 > P_1$ the value is positive, indicating a power **gain**.

If $P_2 < P_1$ the value is negative, indicating the power has been reduced (**attenuated**).

The negative sign can usually be ignored, but when a system has both power gains and attenuation (e.g., an FM transmission with repeaters), the overall gain can be found by adding up all the power gains and losses expressed in dB, but treating losses as negative.

Maths skills

$\log_{10}$ stands for logarithms to base 10 (written as **lg**).

$\log_e$ stands for logarithms to base e (written as **ln**).

(a) A digital communication system uses pulses of infrared radiation which pass through an optic fibre. The radiation is attenuated as it passes through the fibre.

(i) State what is meant by *attenuation*.

The infrared wave gradually gets less and less. ✗

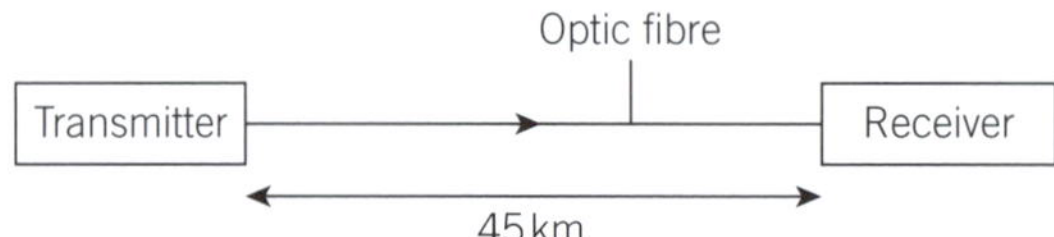

(ii) Explain why infrared radiation is used rather than visible light.

The infrared light is less scattered than visible

light because it has a shorter wavelength ✔ ✗

Infrared waves are scattered less than visible light, but because they have a **longer** wavelength than visible light. [3]

(b) The pulses are transmitted along a fibre of length 45 km. The power input from the transmitter is 11.8 mW.

Optic fibre

Transmitter ———————→ | Receiver

45 km

The signal power output at the receiver is 470 µW. The noise level at the receiver is 0.29 µW. The signal-to-noise ratio at the output must be at least 35 dB.

(i) Determine whether the signal-to-noise ratio at the receiver is greater than 35 dB.

$$\text{Signal-to-noise ratio} = 10 \log_{10}\left(\frac{P_{signal}}{P_{noise}}\right) = 10 \log_{10}\left(\frac{470}{0.29}\right) = 32\,\text{dB} \checkmark$$

The signal-to-noise ratio is less than 35 dB

The correct method and values substituted correctly.

(ii) Calculate the attenuation per kilometre of the optic fibre. ✔

$$\text{Attenuation} = 10 \log_{10}\left(\frac{P_{transmitter}}{P_{receiver}}\right) = 10 \log_{10}\left(\frac{118 \times 10^{-3}}{470 \times 10^{-6}}\right) = 14.0\,\text{dB}$$

Correct calculation.

✔ The attenuation along the whole length of the optic fibre has been calculated correctly ✗ but the question asks for the attenuation **per km**. The correct answer is $\frac{14.0}{45} = 0.31\,\text{dB km}^{-1}$

attenuation per km = ___14.0___ dB km⁻¹ [4]

(c) To improve the system, a regenerator is connected halfway along the cable.

Describe the purpose of a regenerator.

The signal is amplified by the regenerator ✗ ✗

The regenerator does not amplify the signal – it reads the '0's and '1's of the distorted signal and recreates them as sharp '0's and '1's – i.e, it removes or 'cleans' the noise from the signal. [2]

? Exam-style questions

1 (a) Describe what is meant by *amplitude modulation* (AM). [2]

(b) The graph shows how the amplitude of the signal from a radio station varies with frequency.

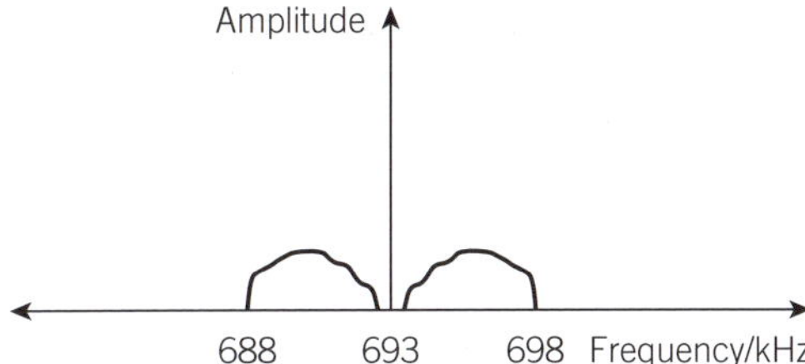

State, for this signal:

(i) the bandwidth

(ii) the carrier frequency

(iii) the maximum audio frequency that can be transmitted. [3]

(c) Use your answer to (b)(iii) to comment on the quality of the sound that would be heard listening to this radio station. [2]

2 (a) Suggest two reasons why audio signals are not transmitted using electromagnetic waves with audio frequencies. [2]

(b) Describe what is meant by *frequency modulation*. [2]

(c) State two advantages and two disadvantages of FM radio transmission compared to AM radio transmission. [4]

3 (a) State two advantages of digital transmission of information compared to analogue transmission. [2]

(b) In the recording of music using digital technology, describe the effect of increasing:

(i) the sampling rate

(ii) number of bits. [2]

(c) A 4 gigabit music CD records music using a sampling rate of 44 kHz and 16 bits.

(i) Explain why 44 kHz is a sufficiently high sampling rate.

(ii) Determine how many quantisation levels are possible using 16 bits. [3]

(d) Two (stereo) channels are recorded on the CD. Estimate the playing time available on the CD. [3]

4 An electrical signal is transmitted along a cable of length 200 km. The input power to the cable is 500 mW and the attenuation of the cable is 0.3 dB km^{-1}.

(a) Calculate:

(i) the attenuation of the cable

(ii) the output power of the receiver. [3]

(b) A repeater with a gain of 20 dB is placed halfway along the cable.

(i) Explain the purpose of a repeater.

(ii) Calculate the new output power at the receiver. [3]

5 An Earth station sends a microwave signal to a satellite in geostationary orbit. The attenuation of the signal is 5.4×20^{-3} dB km^{-1}. The power transmitted from the Earth station 16 kW.

(a) State one reason why microwaves are used for satellite communications. [1]

(b) Calculate:

(i) the attenuation of the signal by the time it reaches the satellite

(ii) the power received by the satellite. [3]

(c) The signal-to-noise ratio at the satellite must be greater than 40 dB. Calculate the maximum value of the noise power at the satellite. [2]

6 The minimum acceptable signal-to-noise ratio in an optic fibre cable is 25 dB. The noise power is 3.5×10^{-10} W.

(a) Calculate the minimum power of the signal. [2]

(b) The input power to the cable is 400 μW and the attenuation of the cable is 0.65 dB km^{-1}.

Calculate the maximum permissible length of cable. [2]

17 Electric fields

Key points

- ☐ Understand that an electric field is an example of a field of force, and define electric field strength as force per unit positive charge acting on a stationary point charge.
- ☐ Represent an electric field using field lines.
- ☐ Recall, and use, $E = \dfrac{\Delta V}{\Delta d}$ to calculate the field strength of the uniform field between charged parallel plates in terms of potential difference and separation.
- ☐ Calculate the forces on charges in uniform electric fields.
- ☐ Describe the effect of a uniform electric field on the motion of charged particles.
- ☐ Understand that, for any point outside a metal sphere, the charge on the sphere acts as a point charge at its centre.
- ☐ Recall, and use, Coulomb's law, $F = \dfrac{Q_1 Q_2}{4\pi\varepsilon_0 r^2}$, for the force between two point charges in free space or air.
- ☐ Recall, and use, $E = \dfrac{Q}{4\pi\varepsilon_0 r^2}$ for the field strength of a point charge in free space or air.
- ☐ Define potential at a point as the work done to bring a unit positive charge from infinity to the point.
- ☐ State that the field strength at a point is equal to the negative of the potential gradient at that point.
- ☐ Use the equation $V = \dfrac{Q}{4\pi\varepsilon_0 r}$ for the potential in the field of a point charge.
- ☐ Recognise the similarities between aspects of electric fields and gravitational fields.

What is an electric field?

An **electric field** is a region where charged particles experience forces. If a small positive charge Q is placed at a point in the field and it experiences a force F, then the **electric field strength** E at that point is defined as:

$$E = \frac{F}{Q}$$

Electric field strength is a vector quantity – it has direction as well as magnitude. The direction of the electric field is the direction of the force on a unit **positive** charge, and can be shown by electric field lines (see Figure 17.1).

> **Remember**
>
> $$E = \frac{F}{Q}$$
>
> SI units are N C^{-1}

> **Remember**
>
> ▶ Field lines point away from positive charges towards negative charges.
>
> ▶ The closer together the field lines, the greater the electric field strength.
>
> ▶ Electric field lines do not cross or touch each other.

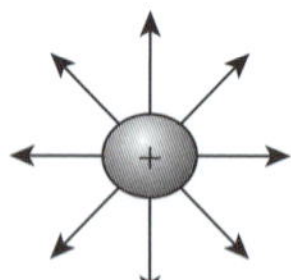

a Near an isolated positive charge

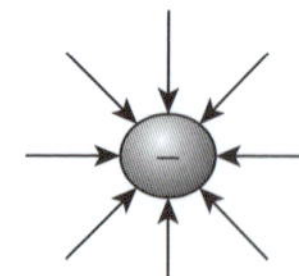

b Near an isolated negative charge

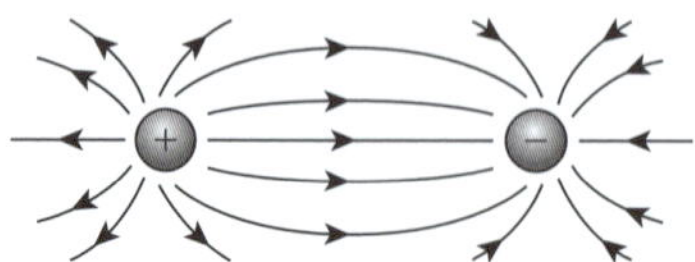

c Between two opposite charges

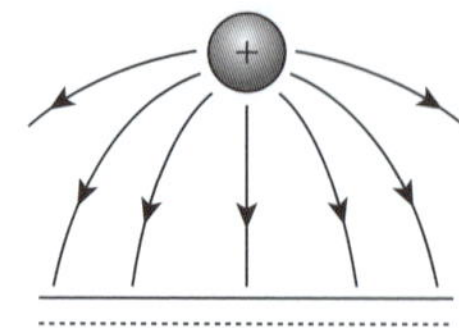

d Between a point positive charge, and an oppositely-charged plate

▲ **Figure 17.1** Examples of electric field lines

Uniform electric fields

Electric field between two parallel plates

If there is a potential difference across two large parallel plates, the electric field between the plates is **uniform** – it is constant both in magnitude and direction (see Figure 17.2a).

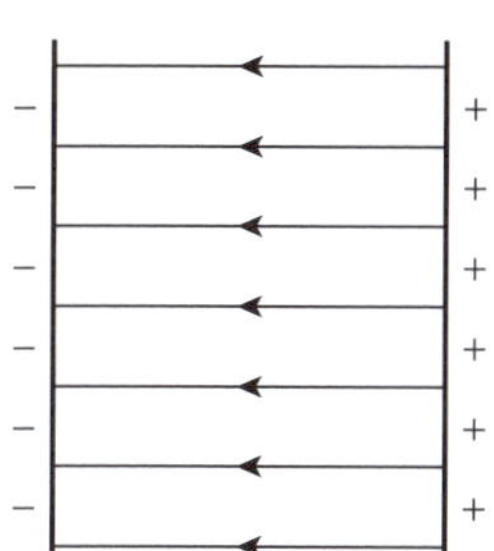

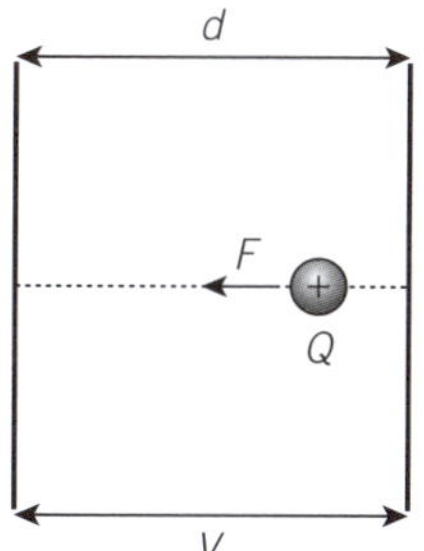

a Field between two parallel plates   **b** Force between two parallel plates

▲ **Figure 17.2** Electric field and force between two parallel plates

For two metal plates that are a distance d apart with a p.d. V between them, when a charge $+Q$ moves from the positively charged plate to the negatively charged plate (see Figure 17.2b), the work done on the charge is:

$$Fd = QV$$

$$E = \frac{F}{Q} = \frac{V}{d}$$

> **Remember**
>
> $E = \dfrac{V}{d}$ for parallel plates
>
> SI units are $V\,m^{-1}$
>
> (equivalent to $N\,C^{-1}$)
>
> This equation can also be written:
>
> $$E = \frac{\Delta V}{\Delta d}$$

Worked examples

1 An electron is between two parallel plates that are 2.0 cm apart. The p.d. between the plates is 5.0 kV. Calculate:

   a) the electric field strength between the plates   **b)** the force on the electron.

Answer

   a) $E = \dfrac{V}{d} = \dfrac{5 \times 10^{3}}{2.0 \times 10^{-2}} = 2.5 \times 10^{5}\,V\,m^{-1}$ (or $N\,C^{-1}$)

   b) $F = EQ$
   $= Ee = 2.5 \times 10^{5} \times 1.6 \times 10^{-19} = 4.0 \times 10^{-14}\,N$

2 A small charged ball, of mass 5.0 g, is suspended by a nylon thread between two parallel plates, and hangs at an angle of 40° to the vertical (see Figure 17.3). The p.d. between the plates is 2.0 kV, and the distance between the plates is 10.0 cm.

   How much charge is on the ball?

Answer

Resolving vertically: $T\cos 40° = mg = 5.0 \times 10^{-3} \times 9.81 = 4.91 \times 10^{-2}$

Resolving horizontally: $T\sin 40° = F_e = \dfrac{V}{d} \times Q = \dfrac{2.0 \times 10^{3}}{10.0 \times 10^{-2}} \times Q = 2 \times 10^{4} Q$

Dividing the second equation by the first:

$$\frac{2 \times 10^{4} Q}{4.91 \times 10^{-2}} = \frac{\sin 40}{\cos 40} = \tan 40$$

$$Q = \frac{4.91 \times 10^{-2} \times 0.839}{2 \times 10^{4}} = 2.1 \times 10^{-6}\,C \ (2.1\,\mu C)$$

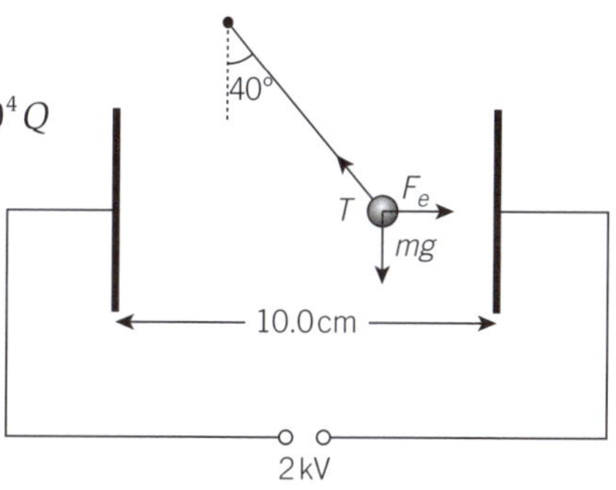

▲ **Figure 17.3**

Motion of charged particles in a uniform electric field

Uniform electric fields can change the speed and direction of charged particles.

> **Worked example**
>
> The p.d. between two metal plates, S and T, is 3.0 kV. The two plates are 6.0 cm apart. An electron is at rest next to the negatively charged plate S. It accelerates towards the positively charged plate T, a metal disc with a hole in the middle (see Figure 17.4).
>
> Calculate:
>
> **a)** the electric field strength between the two plates
>
> **b)** the force on the electron
>
> **c)** the acceleration of the electron
>
> **d)** the speed of the electron when it reaches T.
>
> **Answer**
>
> **a)** $E = \dfrac{V}{d} = \dfrac{3.0 \times 10^3}{6.0 \times 10^{-2}} = 5.0 \times 10^4 \, \text{N C}^{-1}$
>
> **b)** $F = Ee = 5.0 \times 10^4 \times 1.6 \times 10^{-19} = 8.0 \times 10^{-15} \, \text{N}$
>
> **c)** Using $F = ma$:
>
> $a = \dfrac{F}{m} = \dfrac{8.0 \times 10^{-15}}{9.1 \times 10^{-31}} = 8.8 \times 10^{15} \, \text{m s}^{-2}$
>
> **d)** Using $v^2 = u^2 + 2as$:
>
> $v^2 = 0^2 + 2 \times (8.8 \times 10^{15}) \times (6.0 \times 10^{-2}) = 1.1 \times 10^{15}$
>
> $v = 3.2 \times 10^7 \, \text{m s}^{-1}$

This example shows how electrons are accelerated in, for example, cathode-ray oscilloscopes.

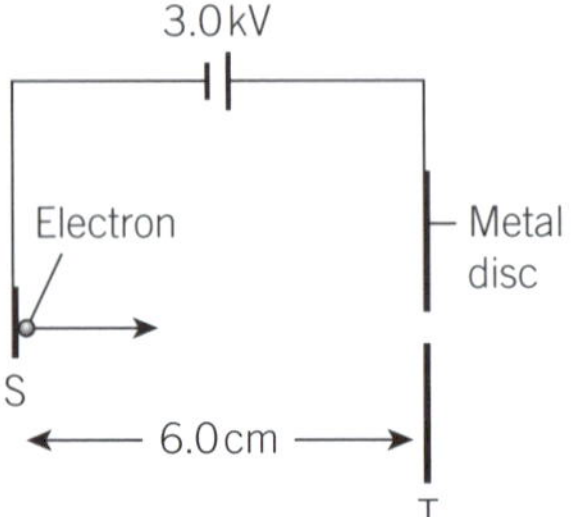

▲ **Figure 17.4**

> ★ **Exam tip**
>
> The mass of an electron m_e and the electronic charge e are provided in Exam Papers 1, 2, and 4.

Electric field strength between point charges

Coulomb's law

Coulomb's law says that if two **point charges**, Q_1 and Q_2, are a distance r apart (see Figure 17.5), the force F between them is:

$$F = \frac{Q_1 Q_2}{4\pi \varepsilon_0 r^2}$$

ε_0 is a constant, called the permittivity of free space. Its value is $8.85 \times 10^{-12} \, \text{F m}^{-1}$.

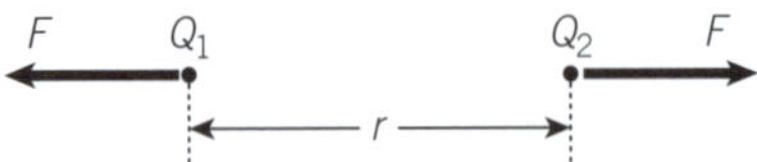

▲ **Figure 17.5** The force between two point charges

If the two charges are both positive, or both negative, the force is repulsive (by convention, a positive force). If one charge is positive, and the other negative, the force is attractive (by convention, a negative force).

Electric field strength of a point charge

The **electric field strength** at a point is the force on a unit positive charge at that point, and so for a point charge Q, by replacing Q_2 with '1' in Coulomb's law:

$$E = \frac{Q}{4\pi \varepsilon_0 r^2}$$

> ★ **Exam tip**
>
> If you're asked to define Coulomb's law, make sure that you refer to the force between two **point charges**.

> ★ **Exam tip**
>
> $\dfrac{1}{4\pi \varepsilon_0} = 8.99 \times 10^9 \, \text{N m}^2 \text{C}^{-2}$
>
> This value, and the value of ε_0, are provided in Exam Papers 1, 2, and 4.

> 💡 **Remember**
>
> Coulomb's law is defined for **point** charges, but a sphere of charge, with the charges equally spread over the sphere, acts as if all the charge were concentrated at the centre of the sphere.

Worked examples

1   Calculate the force between two protons $2.0 \times 10^{-12}\,$m apart.
    (The charge on a proton $= +e = 1.6 \times 10^{-19}\,$C)

Answer

$$F = \frac{Q_1 Q_2}{4\pi\varepsilon_0 r^2} = 8.99 \times 10^9 \times \frac{(1.6 \times 10^{-19})^2}{(2 \times 10^{-12})^2} = 5.8 \times 10^{-5}\,\text{N}$$

2   A metal sphere of radius 50 mm carries a charge of $+200\,\mu$C.

   a)   What is the strength of the electric field 20 mm from the surface of
            the sphere?

   b)   In which direction is the electric field?

Answer

   a) $E = \dfrac{Q}{4\pi\varepsilon_0 r^2} = 8.99 \times 10^9 \times \dfrac{200 \times 10^{-6}}{(70 \times 10^{-3})^2} = 3.7 \times 10^8\,\text{NC}^{-1}$

   b)   The direction is radially, away from the sphere (see Figure 17.6).

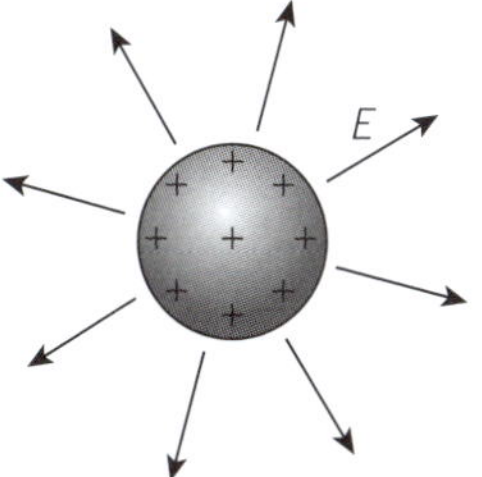

▲ **Figure 17.6**

Electric potential

The **electric potential** at a point is defined as the work done in bringing
unit positive charge ($+1\,$C) from infinity to that point. The work done is
independent of the path taken (see Figure 17.7).

For a point charge $+Q$, the energy needed to bring $+1\,$C from infinity to a
point P that is a distance r from Q (i.e., the potential at point P) is inversely
proportional to r (see Figure 17.8). The potential V is calculated using the
equation:

$$V = \frac{Q}{4\pi\varepsilon_0 r}$$

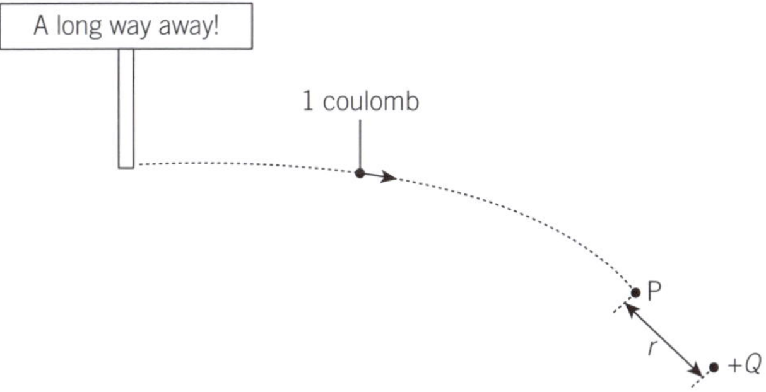

▲ **Figure 17.7** The meaning of electric potential

Moving a unit positive charge towards $+Q$ requires work to be done as the
positive charges are trying to push each other apart. The smaller the value
of r (the closer the test charge gets) the greater the amount of work required.
The unit positive charge gains electrical potential energy as it approaches Q,
and so V is positive.

No work needs to be done to move the positive test charge towards a charge $-Q$.
Instead, the unit positive charge is pulled towards $-Q$ and gains kinetic energy.
The test charge loses potential energy as it approaches $-Q$, and so V is negative.

Electric potential is useful for calculating the energy needed or gained when
charged particles move from one place to another. If the potential is known
at two different points, then the energy needed for a unit positive charge to
move from one point to the other is the potential difference.

> **Remember**
>
> Electric force and electric
> field strength are both
> vectors because they
> have direction as well as
> magnitude.
>
> Electric potential is a
> scalar quantity – it only
> has magnitude.

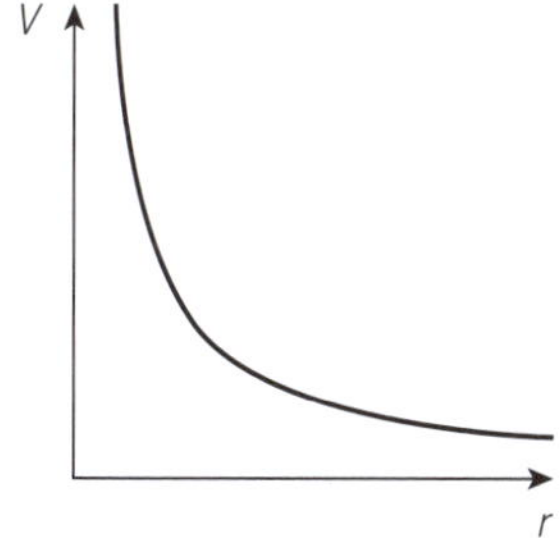

▲ **Figure 17.8** Electric
potential V at a distance r
from a point charge Q

> ★ **Exam tip**
>
> The formula
> $$V = \frac{Q}{4\pi\varepsilon_0 r}$$
> is provided in Exam
> Papers 1, 2, and 4, but
> remember it has units of
> J C^{-1}.

Electric field and potential gradient

For two parallel plates, with a potential difference V between them, the potential V increases linearly from 0 V to + V (Figure 17.9). As shown earlier, the field strength is equal to $-\frac{\Delta V}{\Delta x}$ (the negative of the potential gradient), and is constant.

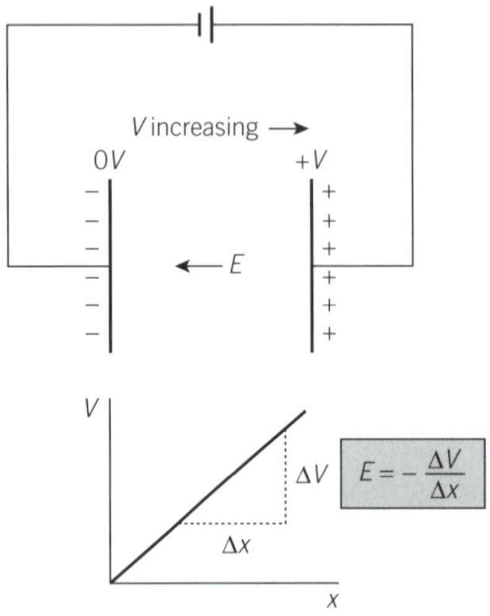

▲ **Figure 17.9** Electric field E between two parallel plates

It can be shown that the equation:

electric field strength = − potential gradient

applies more generally. If a graph of electric potential against distance is plotted, the field strength at any point is the negative of the gradient of the graph at that point (Figure 17.10).

Remember

The negative sign is needed because the field strength E is in the opposite direction to the displacement Δx.

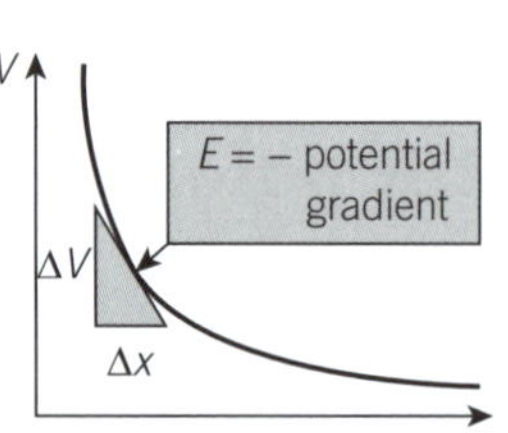

▲ **Figure 17.10** Field strength is the negative of the potential gradient

Comparing electrical and gravitational fields

The gravitational force between two masses and the electrical force between two charges both obey inverse-square laws (see Table 17.1).

▼ **Table 17.1** Comparison of electrical and gravitational fields

	Force	Field	Potential	Potential energy
Gravity	$F = \frac{Gm_1m_2}{r^2}$	$g = \frac{GM}{r^2}$	$\phi = -\frac{GM}{r}$	P.E. $= \frac{Gm_1m_2}{r}$
Electricity	$F = \frac{Q_1Q_2}{4\pi\varepsilon_0 r^2}$	$E = \frac{Q}{4\pi\varepsilon_0 r^2}$	$V = \frac{Q}{4\pi\varepsilon_0 r}$	P.E. $= \frac{Q_1Q_2}{4\pi\varepsilon_0 r}$

- The electrical force can be attractive or repulsive, depending on the sign of the charges, but the gravitational force is always attractive.
- Electrical potential can be positive or negative, depending on the sign of the charge. Gravitational potential must always be negative as work needs to be done to move a test mass to infinity where, by convention, it has zero potential energy.
- As two masses are moved apart, the potential energy of the two masses increases (work has to be done to separate the two masses).
- As two positive charges are moved apart, the potential energy of the two charges decreases (work has to be done to push the two positive charges together).

⬆ Raise your grade

A beam of electrons is fired from an electron 'gun' inside a vacuum tube. The electrons pass between two metal plates, a distance d apart. A potential difference V is applied between the two plates, and the electron beam is deflected upwards, as shown.

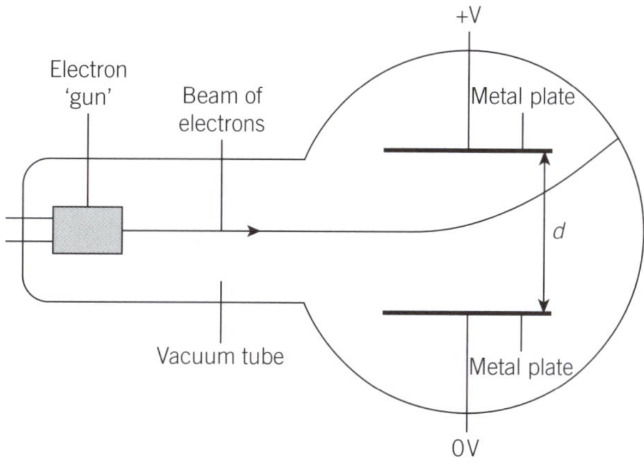

(a) Explain why the electrons are deflected upwards.

Electrons are negatively charged. ✗

> Not sufficient explanation – should add 'and are attracted to the positive plate because opposite charges attract'.

[1]

(b) State, and explain, the effect on the deflection of the electrons for each of the following changes: **(i)** the distance d between the plates is decreased

The electric field strength E (= V/d) will increase, increasing the upwards force on the electrons, so the electrons will be deflected more. ✔

> A good answer.

[2]

(ii) the potential difference V between the plates is decreased

The electric field strength E will decrease, decreasing the upwards force on the electrons, so the electrons will be deflected less. ✔

> A good answer.

[1]

(iii) the speed of the electrons is decreased.

The force on the electrons has not changed so the deflection will be the same. ✗

[1]

> An incorrect answer. The electrons spend more time between the two plates (because they are moving more slowly) so the upwards force acts on them for longer – the deflection would be larger.

(c) Describe, and explain, what would happen if the beam of electrons was replaced by a beam of α-particles travelling at the same speed as the electrons.

Alpha particles are positively charged so they would be attracted to the negative plate. They would deflect downwards. ✔ ✗

[2]

> For the second mark the candidate should add that the deflection will be smaller because the mass of an alpha particle is much greater than the mass of an electron. The force acting on an alpha particle will be double that on an electron, but the mass of an alpha particle is several thousand times greater than the mass of an electron.

1 A charged oil drop, of mass 5.0×10^{-15} kg, is held stationary between two parallel, charged plates, 5.0 mm apart.

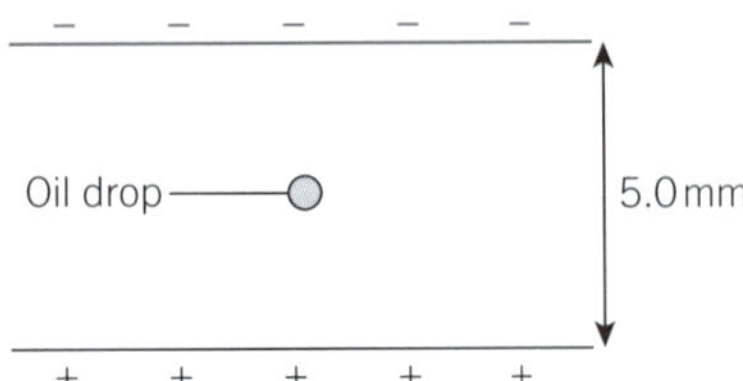

The p.d. between the two plates is 516 V.

What is the charge on the oil drop?

A 4.8×10^{-20} C   **C** 4.8×10^{-17} C

B 4.8×10^{-19} C   **D** 4.8×10^{-16} C   [1]

2 Two parallel metal plates, 20.0 cm apart, are connected to the terminals of a 5.0 kV supply.

What is the field strength between the plates?

A $0.25 \, \text{N C}^{-1}$

B $25 \, \text{N C}^{-1}$

C $2.5 \times 10^4 \, \text{N C}^{-1}$

D $2.5 \times 10^6 \, \text{N C}^{-1}$ [1]

3 An electron is positioned midway between two oppositely charged, parallel plates connected to a d.c. supply. Which one of the graphs shows how the force F on the electron varies with the separation d of the plates?   [1]

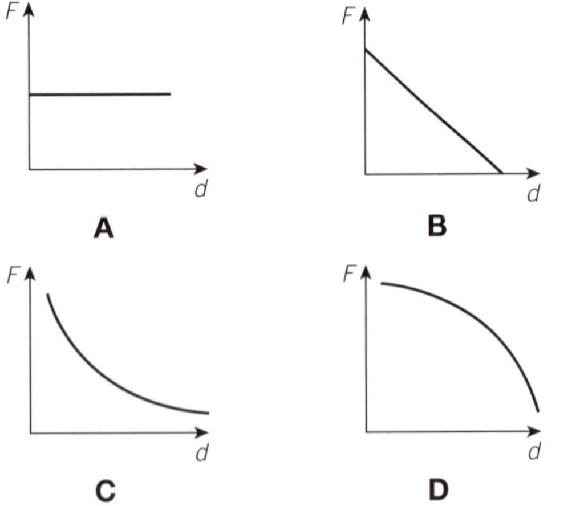

4 (a) Sketch the electric field pattern between two negatively charged particles. Each particle carries the same amount of charge.

    [1]

(b) How would the diagram change if one of the particles had double the charge of the other?   [2]

5 A sphere of diameter 4.0 cm carries a negative charge of 6.0 nC.

(a) Calculate the electric field strength:
    (i) on the surface of the sphere
    (ii) at a point P, 10.0 cm from the centre of the sphere.   [3]

(b) Calculate the electric potential:
    (i) on the surface of the sphere
    (ii) at a point P, 10.0 cm from the centre of the sphere.   [3]

(c) Calculate the work done in moving a charge of + 4 pC from the surface of the sphere to point P.   [2]

6 A metal sphere is suspended from an insulating wire. The sphere is connected to the positive output of a 5 kV d.c. supply.

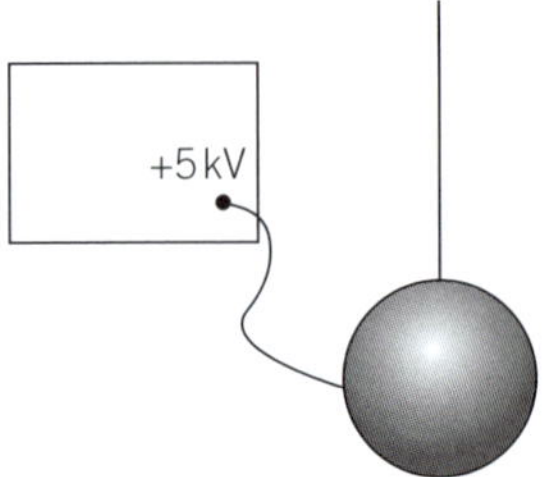

The potential 500 mm from the centre of the sphere is 3.0 kV.

Calculate:

(a) the charge on the sphere   [2]

(b) the radius of the sphere   [2]

(c) the potential 800 mm from the centre of the sphere.   [2]

7 The force between two charged particles is inversely proportional to the square of the distance between them – an example of an *inverse square law*. The variation of the gravitational force between two masses with their distance apart is also an inverse square law.

(a) Describe two ways in which gravitational and electrostatic forces are similar.   [2]

(b) Describe one way in which the two forces are not similar.   [1]

18 Capacitance

Key points

- ☐ Define capacitance and the farad for isolated conductors and parallel-plate capacitors.
- ☐ Recall and use $C = \dfrac{Q}{V}$.
- ☐ Derive and use the formula for the capacitance of capacitors connected in series.
- ☐ Derive and use the formula for the capacitance of capacitors connected in parallel.
- ☐ Deduce, from the area under a potential–charge graph, the equation $W = \dfrac{1}{2}QV = \dfrac{1}{2}CV^2$.
- ☐ Understand some of the functions of capacitors in simple circuits.

Parallel-plate capacitors

Capacitors are devices for storing charge and energy in electric circuits. They usually consist of two metal plates a short distance apart, separated by an insulating material. Figure 18.1 shows the circuit symbol for a capacitor.

▲ **Figure 18.1** Symbol for a capacitor

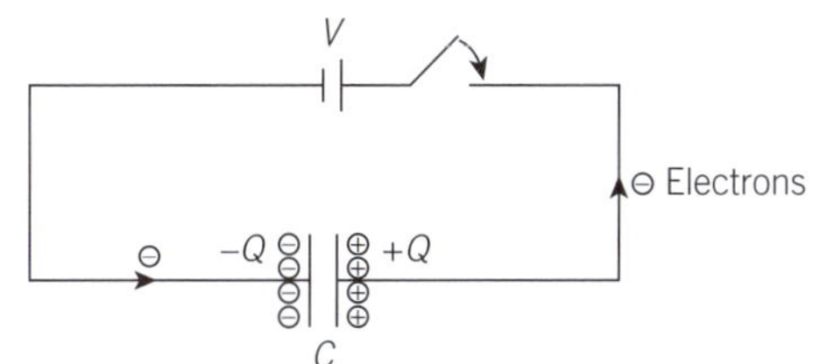

▲ **Figure 18.2** Charging a capacitor

In the circuit shown in Figure 18.2, when the switch is closed, electrons flow from the negative terminal of the cell onto the left-hand plate of the capacitor, and an equal number of electrons leave the right-hand plate and move to the positive terminal of the cell. The left-hand plate then has a negative charge $-Q$ and the right-hand plate has an equal positive charge $+Q$ (the **net** charge stored in a capacitor is zero).

The greater the value of V, the potential difference (p.d.) across the capacitor, the greater the charge Q. Doubling V doubles the charge Q (see Figure 18.3):

▲ **Figure 18.3** $Q \propto V$

$$Q \propto V$$

or

$$Q = CV$$

where C is the capacitance of the capacitor. The units of capacitance are **farads** (F). **Capacitance** is the charge stored per unit potential difference.

If a p.d. of 1V produces a charge of 1C (coulomb) on the plates of a capacitor (positive charge on one plate, negative on the other) the capacitor has a capacitance of 1F.

> **Remember**
>
> The capacitance C of a capacitor is the charge Q stored per unit potential difference:
> $$C = \frac{Q}{V}$$

Worked example

What are the SI base units of capacitance?

Answer

Units are: $\dfrac{C}{V} = \dfrac{\text{As}}{\text{JC}^{-1}} = \dfrac{\text{As}}{\text{kg m}^2\,\text{s}^{-2}\,\text{A}^{-1}\,\text{s}^{-1}} = \text{kg}^{-1}\,\text{m}^{-2}\,\text{s}^4\,\text{A}^2$

Most capacitors have capacitances very much smaller than 1F.

Table 18.1 gives the prefixes often used for capacitances.

▼ **Table 18.1** Prefixes used with capacitance

Microfarads	μF	10^{-6} F
Nanofarads	nF	10^{-9} F
Picofarads	pF	10^{-12} F

Worked examples

1   A 500 nF capacitor is connected to a 12 V d.c. supply.
    What is the charge on either plate of the capacitor?

Answer

$$Q = CV = 500 \times 10^{-9} \times 12 = 6\,\mu C$$

2   A potential difference of 5.0 kV is connected across a 470 μF capacitor.

   a) What is the charge stored?

   b) How many extra electrons are there on the negative plate of the
           capacitor? [$e = 1.6 \times 10^{-19}$ C]

Answer

   a) $Q = CV = 470 \times 10^{-6} \times 5 \times 10^{3} = 2.35$ C

   b) number of 'extra' electrons on negative plate $= \dfrac{2.35}{1.6 \times 10^{-19}}$

   $$= 1.47 \times 10^{19}$$

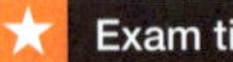

Take care not to confuse
C (the symbol for the unit
of charge, the coulomb)
and C (the symbol for
the capacitance of a
capacitor).

This means there will be
+2.35 C of charge on
one plate, and −2.35 C of
charge on the other plate.

Capacitors in series

Figure 18.4 shows three capacitors, with capacitances C_1, C_2, and C_3, connected
in series to a d.c. supply with a terminal p.d. V. As the capacitors are in
series, the charges on each plate of all the capacitors must be the same
(charge is conserved).

From Kirchhoff's second law:

$$V = V_1 + V_2 + V_3$$

$$= \frac{Q}{C_1} + \frac{Q}{C_2} + \frac{Q}{C_3}$$

$$= Q\left(\frac{1}{C_1} + \frac{1}{C_2} + \frac{1}{C_3}\right)$$

The three capacitors are equal to one capacitor of capacitance C, where:

$$\frac{1}{C} = \left(\frac{1}{C_1} + \frac{1}{C_2} + \frac{1}{C_3}\right)$$

Capacitors add up in series in the same way as resistors add up in parallel.
Adding capacitors up in series makes a capacitor with **smaller capacitance**.

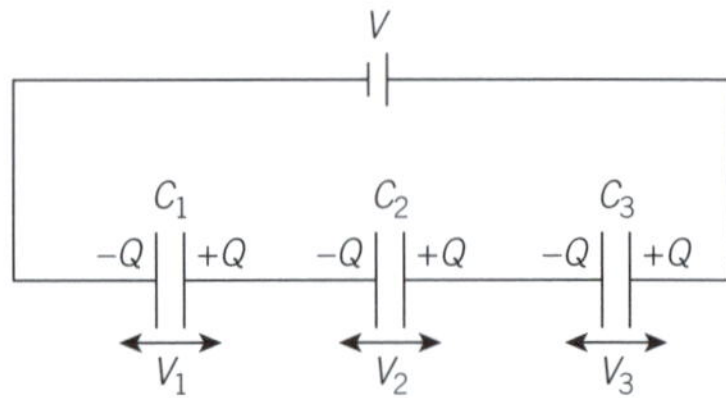

▲ **Figure 18.4** Capacitors in series

🔗 **Link**

See Unit 20 *Direct current
circuits (d.c.)* for more
about Kirchhoff's laws.

★ **Exam tip**

$$\frac{1}{C} = \left(\frac{1}{C_1} + \frac{1}{C_2} + \frac{1}{C_3}\right)$$

Capacitors in series add
up like resistors in parallel.

Worked examples

1 Three capacitors with capacitance $3\,\mu\text{F}$, $4\,\mu\text{F}$, and $5\,\mu\text{F}$ are connected in series. What is their combined capacitance?

Answer

$$\frac{1}{C} = \left(\frac{1}{3} + \frac{1}{4} + \frac{1}{5}\right)$$

$$= \frac{20 + 15 + 12}{60}$$

$$\frac{1}{C} = \frac{47}{60}$$

$$C = 1.3 \ \text{F}$$

2 A $5\,\mu\text{F}$ capacitor is connected in series to a $10\,\mu\text{F}$ capacitor. The combination of capacitors is connected to a $12\,\text{V}$ d.c. supply (Figure 18.5).

Calculate:

a) the p.d. across the $10\,\mu\text{F}$ capacitor

b) the charge on the $5\,\mu\text{F}$ capacitor.

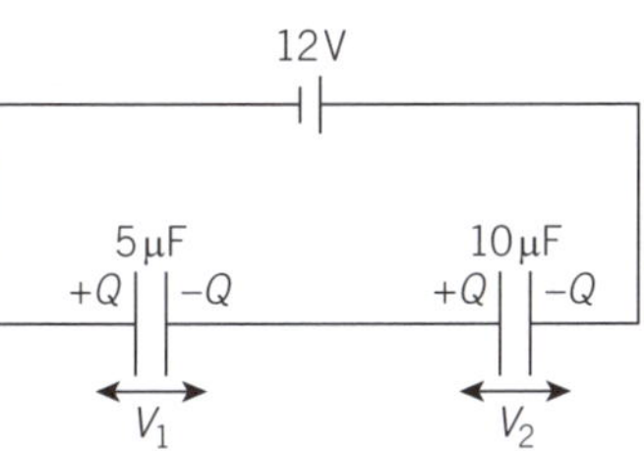

▲ **Figure 18.5**

Answer

a) From Kirchhoff's second law: $V_1 + V_2 = 12$ (eqn 1)

Charge on both capacitors is the same:

$$Q = CV = 5 \times 10^{-6}\, V_1 = 10 \times 10^{-6}\, V_2$$

$$V_1 = 2V_2 \qquad \text{(eqn 2)}$$

From eqn 1 and eqn 2: $V_1 = 8\,\text{V}$, $V_2 = 4\,\text{V}$

b) $Q = CV = 5 \times 10^{-6} \times 8 = 40\,\mu\text{C}$

Capacitors in parallel

In Figure 18.6 three capacitors, $(C_1, C_2, \text{and } C_3)$ are connected in parallel to a d.c. supply with a terminal p.d. V.

The potential difference across each capacitor is the same as the potential difference across the d.c. supply (V), as they are all in parallel.

The total charge 'stored' is Q, where:

$$Q = Q_1 + Q_2 + Q_3$$

$$= C_1 V + C_2 V + C_3 V$$

$$Q = V (C_1 + C_2 + C_3)$$

$$\frac{Q}{V} = C = C_1 + C_2 + C_3$$

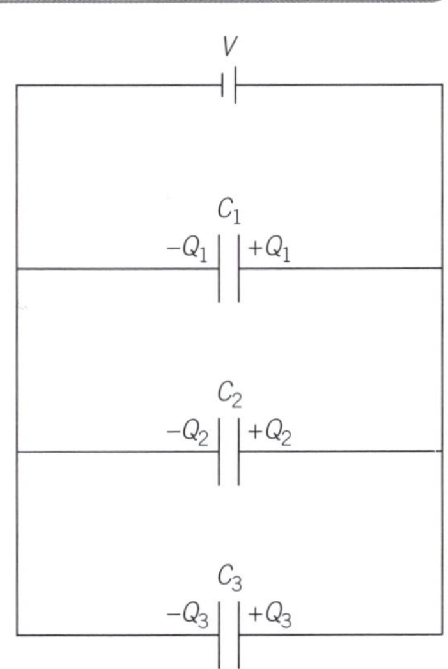

▲ **Figure 18.6** Capacitors in parallel

Capacitors add up in parallel in the same way as resistors add up in series. Adding capacitors up in parallel makes a capacitor with **larger capacitance**.

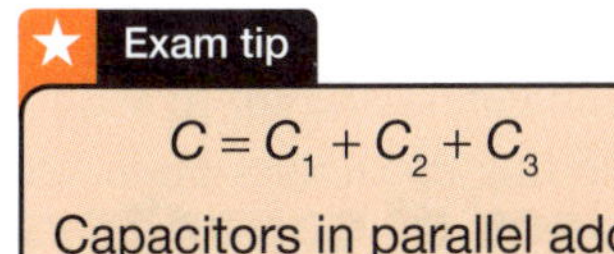

Worked examples

1 Three capacitors, with capacitance 1000 µF, 2200 µF, and 4700 µF are connected in parallel.

    a) What is the total effective capacitance?

    b) If the parallel combination of capacitors is connected to a 12 V d.c. power supply, what is the total charge stored?

Answer

    a) $C = C_1 + C_2 + C_3 = 1000 + 2200 + 4700 = 7900 \, \mu F$

    b) $Q = CV = 7900 \times 10^{-6} \times 12 = 95 \, mC$

2 Three capacitors are connected as shown in Figure 18.7. The total effective capacitance is 20 µF. What is the value of C?

Answer

$$\frac{1}{(20+40)} + \frac{1}{C} = \frac{1}{20}$$

$$\frac{1}{C} = \frac{1}{20} - \frac{1}{60} = \frac{2}{60}$$

$$C = 30 \, \mu F$$

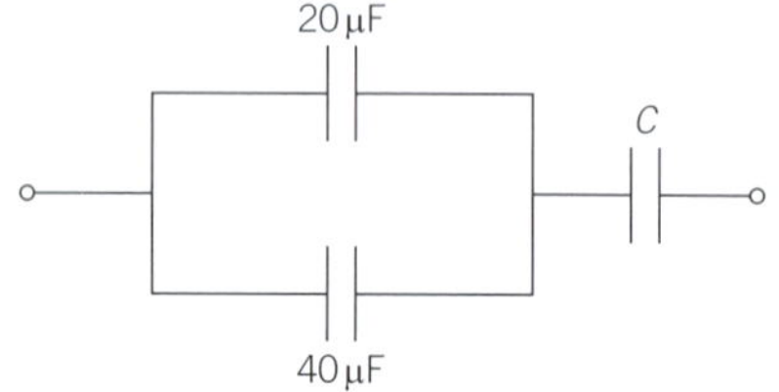

▲ **Figure 18.7**

Energy stored in capacitors

When a capacitor is being charged, work is done in pushing charge onto one plate and 'pulling' it off the other. For a capacitor that is initially uncharged, the work done in pushing the first few charges onto one of the plates is very small, but as the charge on the plates builds up it gets harder and harder to push more charge onto the plate (because the charges already there are repelling them).

The energy stored W in a capacitor charged to a potential difference V is equal to the area under the graph of p.d. against charge (see Figure 18.8).

Calculating the energy stored

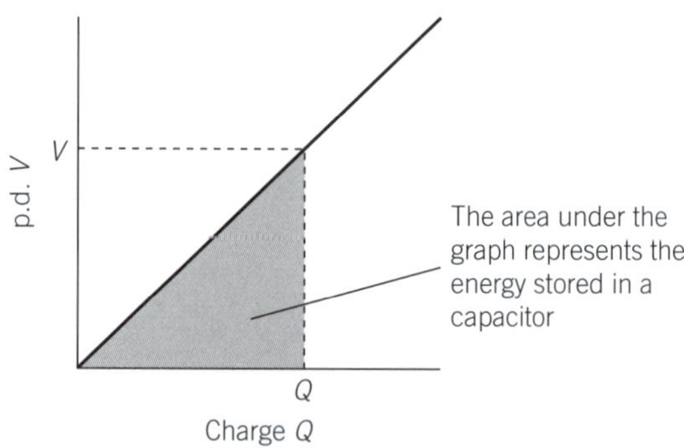

▲ **Figure 18.8** Energy stored in a capacitor

From Figure 18.8:

Energy stored: $\qquad\qquad W = \dfrac{1}{2}QV$

Using $Q = CV$: $\qquad\qquad = \dfrac{1}{2}CV^2$

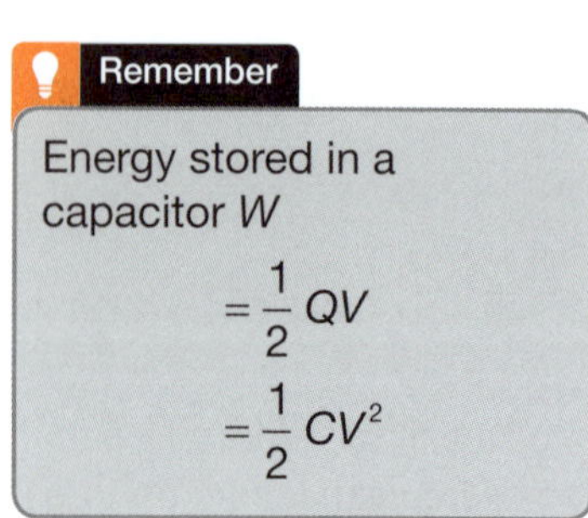

Uses of capacitors

Capacitors have a wide range of uses, including defibrillators, flash photography and back-up power supplies for devices such as computers.

Worked examples

1 The graph (Figure 18.9) shows the charge Q on a capacitor as the potential difference V across it is gradually increased.

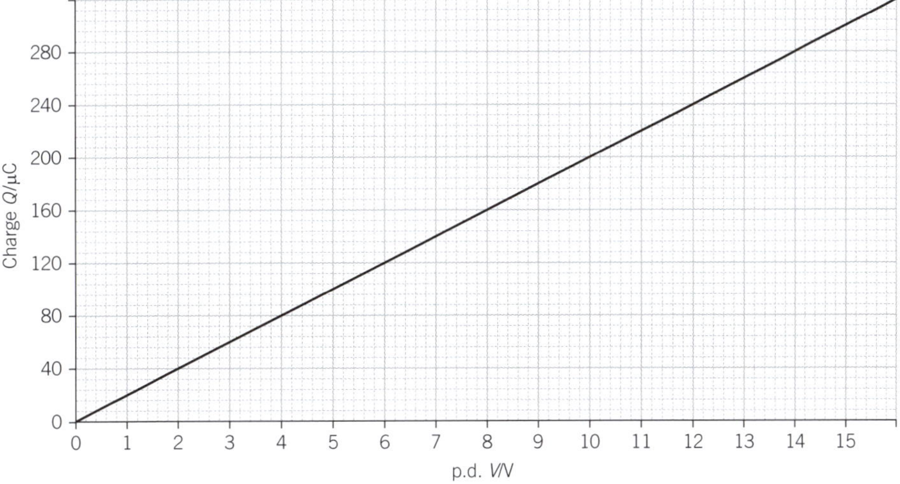

▲ **Figure 18.9**

   a) What is the capacitance of the capacitor?

   b) What is the energy stored in the capacitor when the p.d. across it is 14.0 V?

   c) The capacitor is partially discharged so that the p.d. across it falls from 14.0 V to 9.0 V. how much energy is lost by the capacitor?

Answer

   a) $C = \dfrac{Q}{V} = \dfrac{240 \times 10^{-6}}{12} = 20 \times 10^{-6}\,\text{F}\ (20\,\mu\text{F})$

   b) energy stored $= \dfrac{1}{2}CV^2 = 0.5 \times 20 \times 10^{-6} \times 14.0^2 = 1.96 \times 10^{-3}\,\text{J}$

   c) energy lost = area under graph between 14.0 V and 9.0 V
   $$= \dfrac{1}{2}(280 + 180) \times 10^{-6} \times (14.0 - 9.0) = 1.15 \times 10^{-3}\,\text{J}$$

2 A defibrillator (see Figure 18.10) is a device that provides an electric shock to the heart of someone who is in cardiac arrest. A capacitor of capacitance 15 μF is first charged to a potential difference of 8.0 kV. It is then discharged through the patient using two metal disks connected to insulated handles.

   Calculate:

   a) the charge on either plate of the capacitor

   b) the energy stored in the capacitor.

Answer

   a) $Q = CV = 15 \times 10^{-6} \times 8.0 \times 10^{3} = 0.12\,\text{C}$

   b) $W = \dfrac{1}{2}CV^2 = \dfrac{1}{2} \times 15 \times 10^{-6} \times (8.0 \times 10^{3})^2 = 480\,\text{J}$

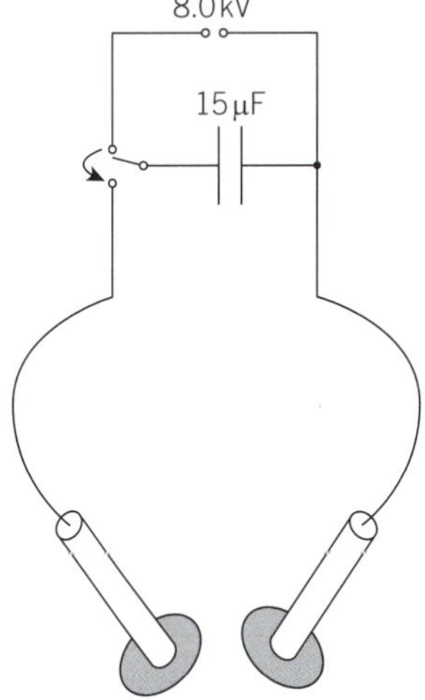

▲ **Figure 18.10** Defibrillator

3 The top and bottom layers of a thundercloud can be treated as a capacitor of approximate capacitance 10^{-8} F. Just before a lightning strike the p.d. between the top and bottom of the thundercloud is approximately 3×10^{9} V. Calculate the energy stored in a thundercloud.

Answer

$$W = \dfrac{1}{2}CV^2 = \dfrac{1}{2} \times 10^{-8} \times (3 \times 10^{9})^2 = 4.5 \times 10^{10}\,\text{J}$$

 **Raise your grade**

A capacitor of capacitance 200 μF is connected to a 12 V d.c. supply.

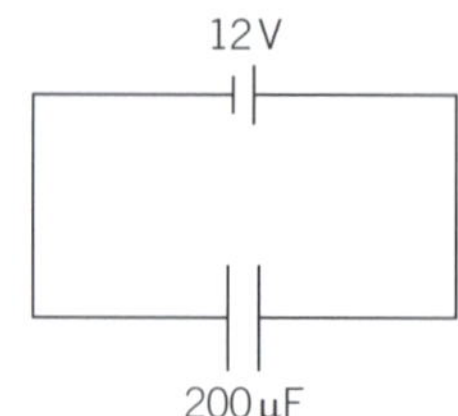

(a) Define *capacitance*.

Something has capacitance if it stores charge when a voltage is across it ...... ✗ ....

.... This is a description of what capacitors do, rather than a definition. A good .... [1]
answer would be 'The capacitance of a capacitor is defined as the charge
stored per unit potential difference'.

(b) (i) Show that the charge on the capacitor is 2.4 mC.

$$Q = CV = 12 \times 200 = 2400\ \mu C = 2.4\ mC \qquad ✔$$

A good answer. [1]

(ii) Calculate the energy stored in the capacitor.

$$\text{Energy stored} = \tfrac{1}{2}CV^2 = 0.5 \times 200 \times 12^2 = 1.44 \times 10^4\ J. \qquad ✔ ✗$$

The candidate has substituted the correct values into the correct equation
(method mark), but has not taken account of the capacitance being in μF, not F.
The answer should be $0.5 \times 200 \times 10^{-6} \times 12^2 = 1.44 \times 10^{-2}\ J$. [2]

(c) The 200 μF capacitor, still charged, is disconnected from the power supply and connected
to an uncharged 100 μF capacitor.

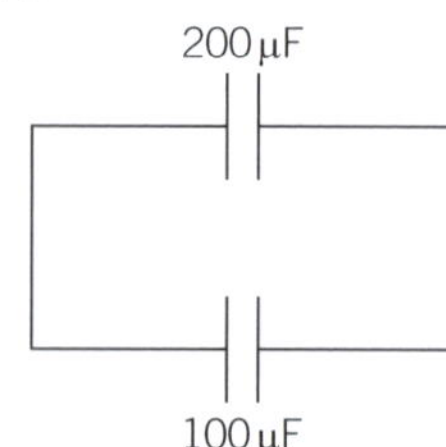

(i) Show that the potential difference across the 200 μF capacitor is now 8 V.

The 2400 μC charge is 'shared' between the 2 capacitors so that the p.d. across each
capacitor is the same (the capacitors are in parallel).

8 V across 100 μF means 800 μC; 8 V across 200 μF capacitor means 1600 μC.
Total charge = 800 + 1600 = 2400 μC (charge is conserved) ✔ ✔ ✔

A good answer. [3]

(ii) Calculate the total energy stored in the two capacitors.

$$\text{Total energy stored} = \tfrac{1}{2}C_1V^2 + \tfrac{1}{2}C_2V^2 = \tfrac{1}{2} \times 100 \times 8^2 + \tfrac{1}{2} \times 200 \times 8^2$$

$$= 0.96 \times 10^4\ J \qquad ✔ ✔$$

Correct method and substitution. Calculation is correct, allowing for error carried forward –
already penalised for using μF rather than converting to F. The answer should be $0.96 \times 10^{-2}\ J$. [2]

(d) Explain why your answers to (b)(ii) and (c)(ii) are not the same. ✗

Electrical resistance ....

Not enough for a mark. The candidate should have written 'Charge must flow from one capacitor ... [1]
to the other (an electric current) so energy is lost as heat because of electrical resistance'.

❓ Exam-style questions

1 When a capacitor is charged to 400 V the charge stored on the capacitor is 800 mC.

   (a) What is the capacitance of the capacitor? [1]

   (b) Why is it ambiguous to refer to the charge *stored* on a capacitor? [1]

2 A 400 µF and a 600 µF capacitor are connected in parallel and the combination is connected in series to a 1000 µF capacitor. The three capacitors are then connected to a 10 V d.c. power supply.

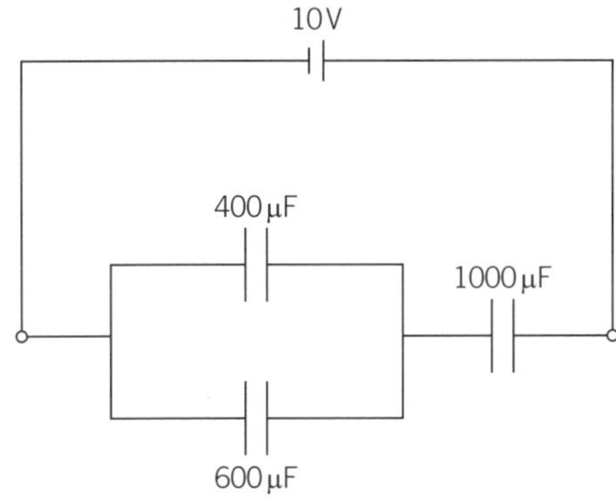

   (a) Determine the combined capacitance of:

     (i) the two capacitors in parallel

     (ii) the three capacitors in combination. [2]

   (b) Calculate the charge:

     (i) on the 400 µF capacitor

     (ii) on the 600 µF capacitor. [2]

   (c) Calculate the p.d. across the 1000 µF capacitor. [1]

3 A photoflash capacitor fitted to a camera has a capacitance of 500 µF. It is charged to a voltage of 300 V.

   (a) Determine the energy stored in the capacitor. [2]

   (b) The capacitor discharges completely in 3 ms. Calculate the average power output of the flash. [2]

4 A student has a 20 µF capacitor, a 47 µF capacitor, and an 82 µF capacitor.

   (a) (i) What is the largest capacitance that he can make from the three capacitors?

     (ii) Draw a diagram to show how the three capacitors should be connected. [2]

   (b) (i) How can the three capacitors be combined to have an overall capacitance of 50 µF?

     (ii) Draw a diagram to show how the three capacitors should be connected. [2]

5 A 1000 µF capacitor is connected to a 6 V d.c. supply, as shown.

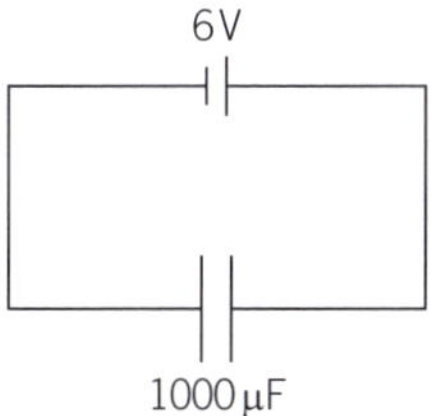

   (a) Calculate:

     (i) the charge on the capacitor

     (ii) the energy stored in the capacitor. [2]

The 1000 µF capacitor, still charged, is disconnected from the battery. A 500 µF capacitor is now charged to 6 V by the d.c. supply and then disconnected. The two capacitors are then joined together as shown below; the positively-charged plate of one capacitor being connected to the negatively-charged plate of the other.

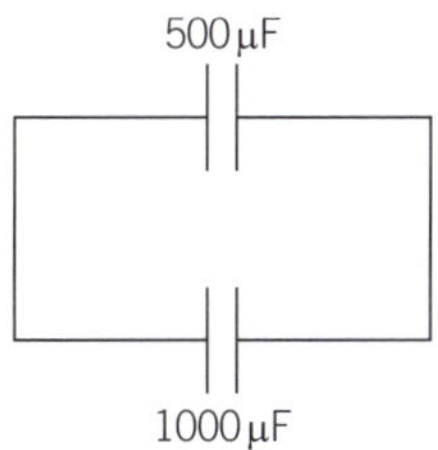

   (b) Calculate the charge on either plate of:

     (i) the 500 µF capacitor

     (ii) the 1000 µF capacitor. [2]

   (c) Determine the potential difference across:

     (i) the 500 µF capacitor

     (ii) the 1000 µF capacitor. [2]

   (d) (i) Calculate the total energy stored in the two capacitors.

     (ii) Explain why this is less than your value in (a)(ii). [2]

Key points

- ☐ Understand that electric current is a flow of charge carriers and that the charge on the carriers is quantised.
- ☐ Define the coulomb, and be able to recall and use $Q = It$.
- ☐ Derive and use the equation $I = Anvq$, where n is the number density of charge carriers.
- ☐ Define potential difference and the volt, and be able to recall and use $V = W/Q$ and $P = VI$.
- ☐ Define resistance and the ohm, and recall and use $V = IR$.
- ☐ Be able to recall and use $P = I^2R$.
- ☐ State Ohm's law.
- ☐ Sketch and explain the I-V characteristics of: a metallic conductor at constant temperature, a filament lamp, a semiconductor diode.
- ☐ Recall and use $R = \rho L/A$.
- ☐ Link the change in resistance of a light-dependent resistor with changing light intensity.
- ☐ Sketch the graph of resistance against temperature for a negative temperature coefficient thermistor.
- ☐ Understand the action of a piezo-electric transducer and its application in a simple microphone.
- ☐ Describe how a metal-wire strain gauge works, and relate the extension of a strain gauge to the change in its resistance.

Current and charge

An **electric current** is any flow of electrically-charged particles. In metal conductors such as copper wires the charge carriers are electrons; in liquids the charge carriers are ions.

The amount of **charge** on a charged particle is **quantised** – it is always a multiple of the electronic charge $e = 1.6 \times 10^{-19}\,$C, where C stands for **coulomb**, the unit of electric charge.

The **current** I is the amount of charge flowing past a point each second. If a charge Q flows in a time t, the current I is given by:

$$I = \frac{Q}{t}$$

> **Remember**
>
> The **coulomb** (C) is the SI unit of electric charge and is defined as the charge conveyed by a current of 1 ampere in 1 second.
>
> charge = current × time
>
> $$Q = It$$

Worked example

A current of 60 mA flows in a wire for 1 minute.

Calculate:

a) how much charge passes

b) how many electrons pass. [$e = 1.6 \times 10^{-19}\,$C.]

Answer

a) $Q = It = 60 \times 10^{-3} \times 60 = 3.6\,$C

b) number of electrons $= \dfrac{3.6}{1.6 \times 10^{-19}}$

$= 2.25 \times 10^{19}$ electrons

When an electric current passes through a metal wire the charge is carried by electrons which 'drift' from one end of the wire to the other. Figure 19.1 shows a wire of cross-sectional area A carrying a current I. The **number density** n of the material is the number of free charge carriers per cubic metre. It is sometimes also called the charge density.

>  **Link**
>
> See Unit 10 *Ideal gases* for more about Brownian motion.

As well as their random thermal ('Brownian') motion, the electrons have a drift velocity v, as shown in Figure 19.1.

In time t, the charge carriers travel a distance $L = vt$, and so the number of charge carriers passing any given point in time t is nAL, where n is the number density of charge carriers and A is the cross-sectional area of the wire.

If the charge on each charge carrier is q, the current I is:

$$I = \frac{nALq}{t} = nAqv \qquad \left(\text{as } v = \frac{L}{t}\right)$$

For conductors such as copper, the number density of charge carriers is typically $10^{29}\,\text{m}^{-3}$. For semiconductors it is much smaller, typically $10^{9} - 10^{13}\,\text{m}^{-3}$. **Semiconductors** are materials with resistances that lie between metals (low resistance) and insulators (high resistance).

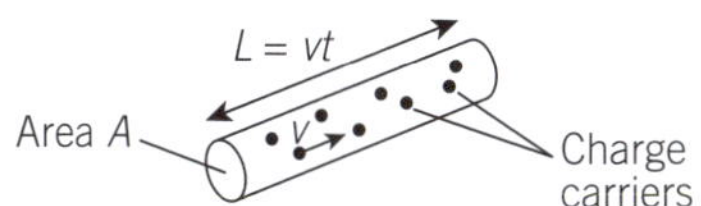

▲ **Figure 19.1** Charge carriers in a conductor

Remember

$$I = nAqv$$

You need to be able to derive this equation.

Worked example

Estimate the drift velocity of electrons in a copper wire of diameter 0.56 mm when a current of 60 mA passes through it. Assume the charge density of copper is $10^{29}\,\text{m}^{-3}$ and use $e = 1.6 \times 10^{-19}\,\text{C}$.

Answer

a) $\quad v = \dfrac{I}{nAq} = \dfrac{60 \times 10^{-3}}{10^{29} \times \pi \times \left(0.28 \times 10^{-3}\right)^{2} \times 1.6 \times 10^{-19}}$

$\qquad = 1.5 \times 10^{-5}\,\text{m s}^{-1}\ (0.015\,\text{mm s}^{-1})$

Remember

Conventional current
I is shown flowing from positive to negative; in reality it is electrons which flow the other way – from the negative terminal to the positive terminal.

Potential difference and the volt

In the circuit shown in Figure 19.2, the electrons effectively 'pick up' energy as they pass through the cell and 'deposit' it at the lamp. The amount of energy delivered to the lamp per **coulomb** of charge passing through the lamp is called the **potential difference** (p.d.), or voltage V, across the bulb:

$$V = \frac{W}{Q}$$

where W is the work done when a charge Q passes.

In Figure 19.2 a **voltmeter** connected across the lamp measures how much energy per coulomb is lost, in volts (V). 1 volt is 1 joule per coulomb. A voltmeter is connected in parallel with ('across') a component because it is comparing the energy per coulomb on either side of the component.

In Figure 19.3, if the current flowing through a component is I and the p.d. across the component is V, the **power** P dissipated in the component is:

$$P = \frac{\text{energy per}}{\text{second}} = \frac{\text{no of coulombs}}{\text{per second}} \times \frac{\text{energy lost per}}{\text{coulomb}}$$

or $\qquad\qquad\qquad P = IV$

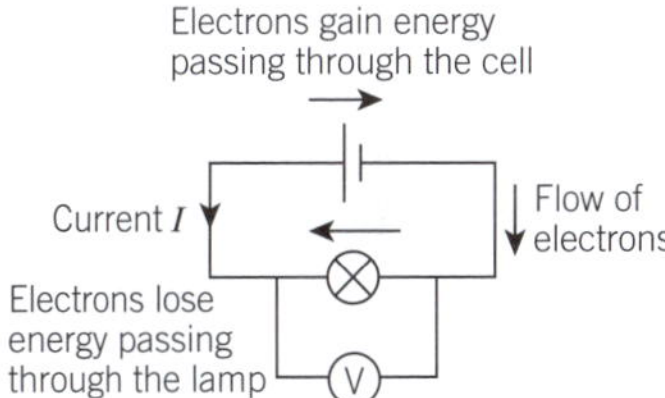

▲ **Figure 19.2** Electron flow

Remember

The **potential difference** is defined as the work done (or energy transferred) per unit charge.

The **volt** (V) is the SI unit of potential difference and is equal to 1 joule per coulomb.

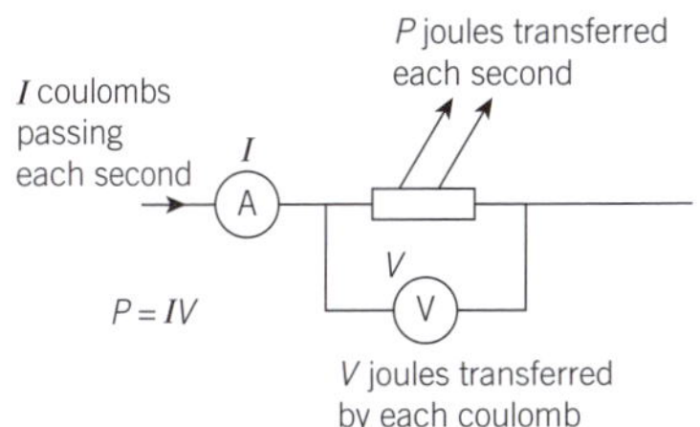

▲ **Figure 19.3** Energy transfer

> **Worked example**
>
> An electric kettle is labelled '230 V, 3 kW'.
>
> **a)** Calculate the electric current in the kettle when in use.
>
> **b)** Determine the energy transferred to the kettle if it is switched on for 5 minutes.
>
> **Answer**
>
> **a)** $I = \dfrac{P}{V} = \dfrac{3000}{230} = 13.0\,\text{A}$
>
> **b)** energy = power × time = $3000 \times (5 \times 60) = 9 \times 10^5\,\text{J}$ (900 kJ)

Resistance and the ohm

The **resistance** of an electrical component is a measure of how difficult it is to pass a current through the component (see Figure 19.4).

Resistance is defined by the equation:

$$R = \frac{V}{I}$$

If 1 A flows in a component when the potential difference across it is 1 V, then the resistance of the component is 1 ohm (Ω).

Combining the equations $P = IV$ and $V = IR$ gives:

$$P = I^2R$$

I–V characteristics

The circuit shown in Figure 19.5 can be used to investigate how the electrical resistance of a component changes. By adjusting the variable resistor, the current can be measured for different values of p.d. across the resistor.

I–V characteristic of an ohmic resistor

For some resistors, including metal wires at constant temperature, the current passing through the resistor is directly proportional to the applied voltage, as shown in Figure 19.6.

$$R = \frac{V}{I} = \text{constant}$$

This is known as **Ohm's law**. A resistor that obeys Ohm's law is an **ohmic resistor**. Ohm's law is more of a 'rule' than a law, which only some resistors obey, and then only under certain conditions; for example, constant temperature.

Other electrical components generally do not obey Ohm's law.

> ★ **Exam tip**
>
> When describing Ohm's law, it is not enough to write:
>
> $$R = \frac{V}{I}$$
>
> This is just the definition of resistance; for Ohm's law to be obeyed the resistance must be **constant**.
>
> To calculate the resistance, read off a value of V and the corresponding value of I on the graph and then calculate V/I.

> 💡 **Remember**
>
> Power (W) = current (A) × p.d.(V)
>
> $$P = IV$$
>
> Power is measured in **watts** (W).

▲ **Figure 19.4** Electrical resistance

> ★ **Exam tip**
>
> The **ohm** (Ω) is the SI unit of electrical resistance and is equal to $1\,\text{V A}^{-1}$.

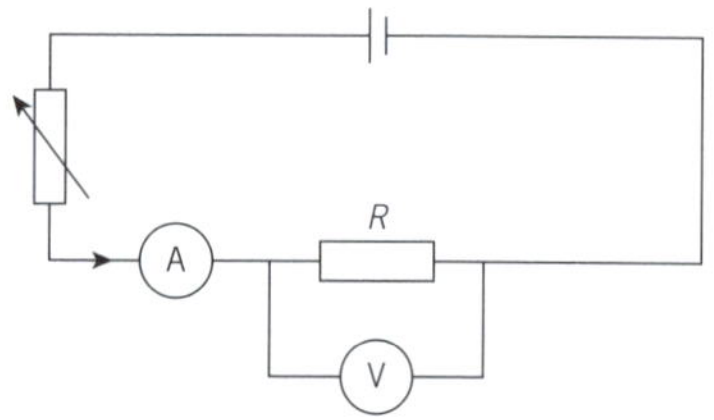

▲ **Figure 19.5** Measuring resistance

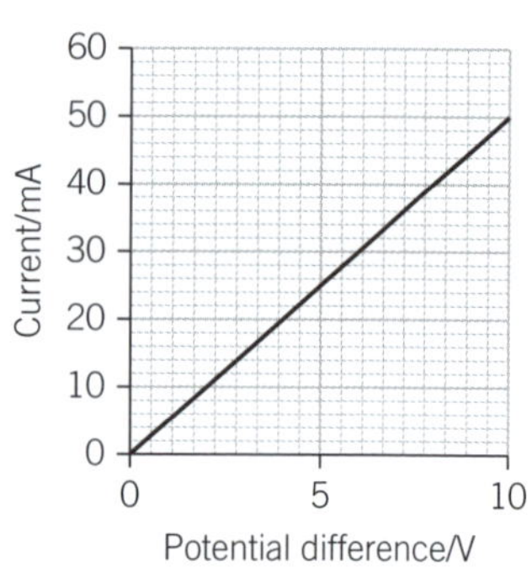

▲ **Figure 19.6** I–V characteristic of an ohmic resistor

I–V characteristic of a filament lamp

For a filament lamp doubling the voltage from 5 V to 10 V increases the current, but doesn't double it, as shown in Figure 19.7. Since $R = V/I$ the resistance of the lamp must be **increasing**. This is because as more current passes through the lamp, the filament (usually made of tungsten metal) gets hotter, and so the metal atoms vibrate much more vigorously, making it more difficult for electrons to pass through.

I–V characteristic of a semiconductor diode

When a semiconductor diode is reverse-biased (connected in its reverse direction – see Figure 19.8a) it has a very large resistance – almost no current flows as shown in Figure 19.9.

If the diode is forward-biased (connected in its forward direction – see Figure 19.8b.) a small p.d (~0.6 V) causes the diode to conduct with almost no resistance as shown in Figure 19.9.

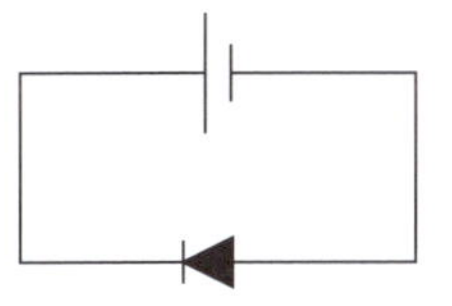

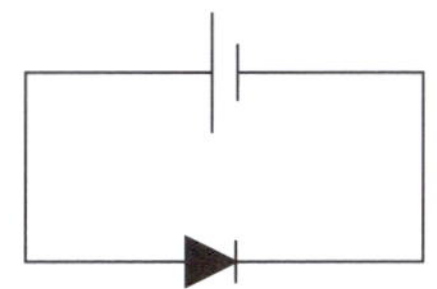

a Reverse biased   **b** Forward biased

▲ **Figure 19.8** Forward- and reverse-biased diodes

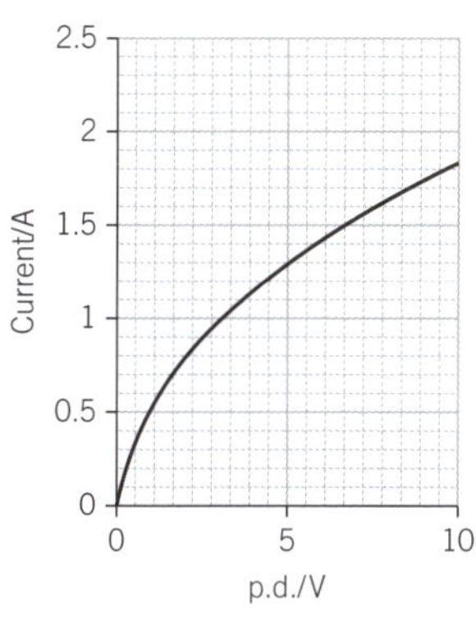

▲ **Figure 19.7** *I–V* characteristic of a filament lamp

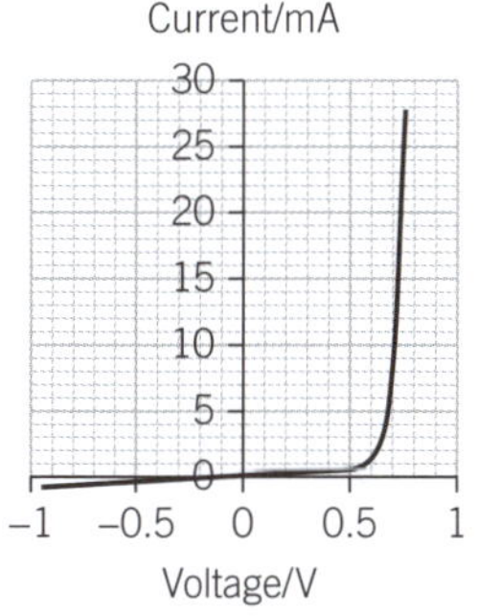

▲ **Figure 19.9** *I–V* characteristic of a semiconductor diode

Resistivity

For a conductor such as a metal wire or a carbon resistor (see Figure 19.10), the resistance of the conductor depends on its dimensions and on the material it is made from. Doubling the length of a wire doubles the resistance. Doubling the diameter of a wire increases the cross-sectional area of the wire by a factor of four which **decreases** the resistance by the same factor (it is as if four of the original wires have been connected in parallel).

For a conductor of length L and cross-sectional area A, we can write:

$$R = \frac{\rho L}{A}$$

where ρ is the resistivity of the material. ρ is measured in $\Omega\,$m. Good conductors such as copper have very low resistivities (about $10^{-8}\,\Omega\,$m). Good insulators have very high resistivities (about 10^{12}–$10^{16}\,\Omega\,$m). Semiconductors have resistivities between 10^{-3} and $10^{8}\,\Omega\,$m.

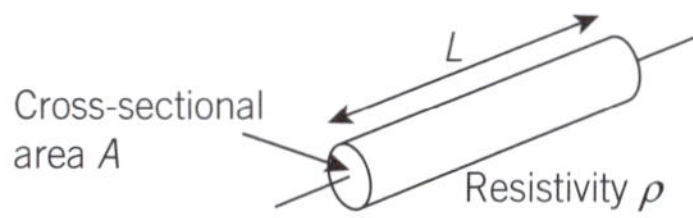

▲ **Figure 19.10** Resistivity

> **Remember**
>
> Resistance, $R = \dfrac{\rho L}{A}$
>
> The units of resistivity ρ are $\Omega\,$m.

Worked example

Constantan wire is an alloy of copper and nickel with a resistivity of $49.0 \times 10^{-5}\,\Omega\,$mm. What is the resistance of a constantan wire of length 50.0 cm and diameter 0.46 mm?

Answer

$$R = \frac{\rho L}{A} = \frac{4.90 \times 10^{-7} \times 0.500}{\pi \times \left(0.23 \times 10^{-3}\right)^{2}} = 1.47\,\Omega$$

> ★ **Exam tip**
>
> Always check that the units are **consistent**. In this example, convert the units of diameter and length to metres and the resistivity to $\Omega\,$m.
>
> Also check that you've halved the diameter to find the radius when using πr^{2} to find the cross-sectional area of the wire, and don't forget to square r!

Sensing devices

Light-dependent resistor (LDR)

The resistance of some semiconductor materials (e.g., cadmium sulphide) is altered by the amount of light falling on them (Figure 19.11). The light energy is absorbed by the material, releasing electrons in the material to become 'conduction electrons'. A typical **light-dependent resistor** (LDR) has a resistance of $10\,M\Omega$ in complete darkness but this falls to a few hundred ohms in bright light.

LDRs are used in a range of electronic devices to operate light-sensitive switches including automatic street lights and camera light-meters. Its symbol is shown in Figure 19.12.

Negative temperature coefficient thermistor

The resistance of thermistors changes with temperature. A **negative temperature coefficient** thermistor has a smaller resistance at higher temperatures, as shown in Figure 19.13 – the rise in temperature increases the number of electrons that are 'free' to conduct.

The change of resistance can be very large for a small increase or decrease in temperature. Thermistors are used in thermostats, fire alarms and digital thermometers. They monitor the oil and water temperature in cars and prevent high currents flowing through computers and electric motors when first switched on. Its symbol is shown in Figure 19.14.

Potential divider circuits

Thermistors and LDRs are often used as part of a potential divider circuit as shown in Figures 19.15 and 19.16.

Piezo-electric transducer

The piezo-electric transducer (Figure 19.17) generates a potential difference when it is compressed (i.e., subjected to mechanical stresses). The piezo-electric material is crystalline and contains positive and negative ions. When a stress is applied to the material, the centres of charge of the positive and negative ions move in opposite directions, creating a small potential difference. Sound waves directed at a piezo electric microphone cause the crystal to compress and expand, generating an alternating p.d. which can be connected to an amplifier.

The principle can also be applied in reverse, as a loudspeaker – an alternating voltage applied across a piezo-electric crystal will cause the distance across the material to increase and decrease. If the crystal is attached to a diaphragm, sound waves are produced.

Strain gauge

A strain gauge consists of a long thin wire mounted on a thin plastic strip. The gauge can be fixed to a building or other structures such as bridges – as the structure moves or bends the wire stretches, and so its resistance increases due to its length increasing and its diameter decreasing. See page 151 for more details of how strain gauges are used.

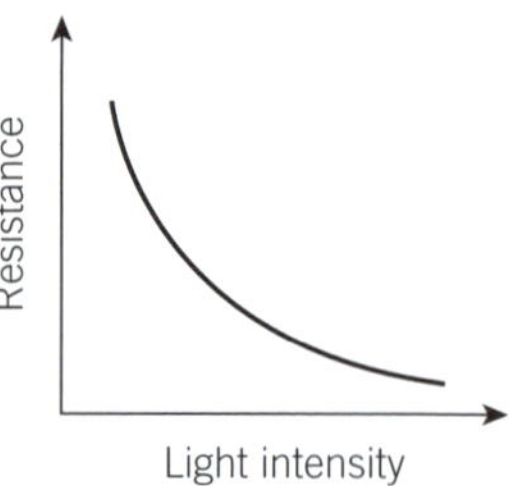

▲ **Figure 19.11** Resistance of an LDR against light intensity

▲ **Figure 19.12** Symbol for an LDR

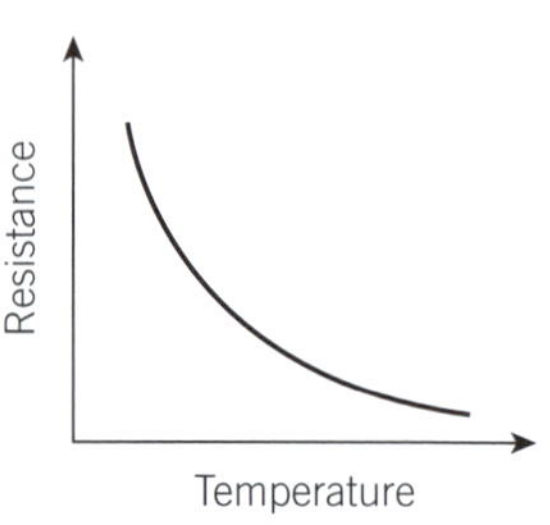

▲ **Figure 19.13** Resistance of a thermistor against temperature

▲ **Figure 19.14** Symbol for a thermistor

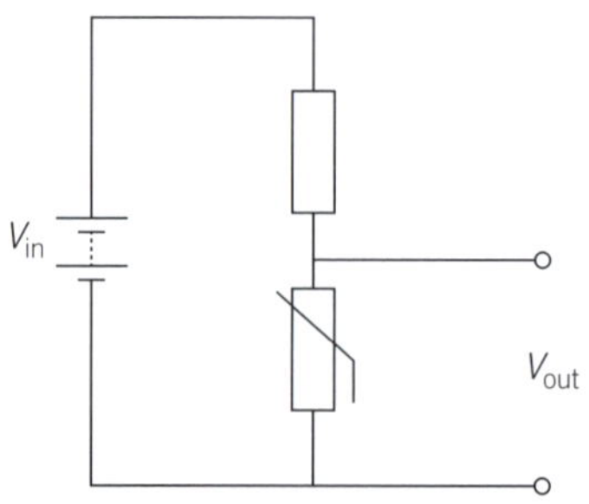

▲ **Figure 19.15** Temperature control: As the temperature decreases, V_{out} increases

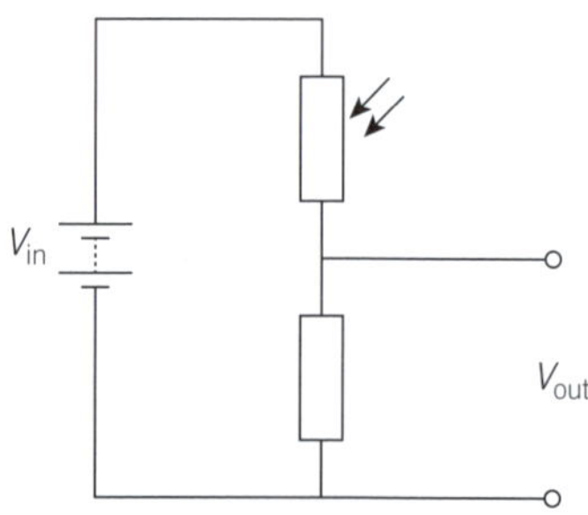

▲ **Figure 19.16** Monitoring light level: As the light level increases, V_{out} increases

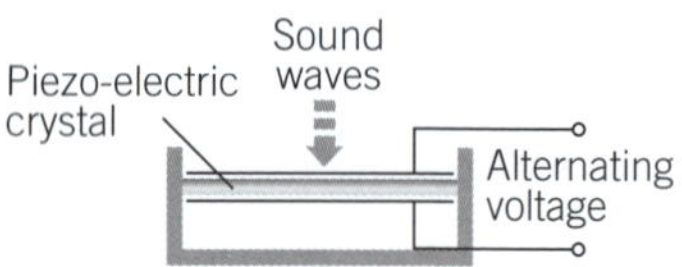

▲ **Figure 19.17** Piezo-electric microphone

A **transducer** is a device which converts one form of energy into another.

Raise your grade

(a) Define the *ohm*.

The ohm is the S.I. unit of resistance ✗ .... [1]

> The statement is true but is not a definition of the ohm.
>
> A better answer would be '1 ohm is 1 volt per ampere'.

(b) A three-way light bulb has two filaments and power outputs of 60 W, 120 W, and 180 W. It is connected to a 120 V mains supply as shown.

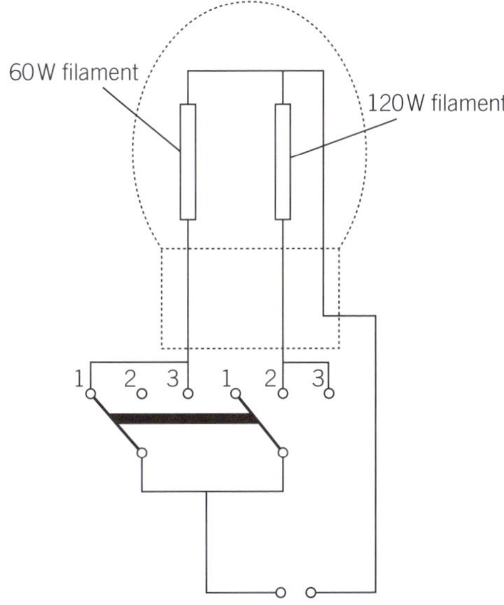

(i) Show that, when the switch is in position 1, the resistance of the 60 W filament is 240 Ω.
[1]

$$P = \frac{V^2}{R} \implies R = \frac{V^2}{P} = \frac{120^2}{60} = 240\,\Omega \checkmark$$

> Method and correct substitution.

(ii) The 60 W filament is a tungsten wire of length 580 mm and diameter 0.046 mm. Calculate the resistivity of tungsten.

✓ ✓

> Substitution and calculation.

$$\rho = \frac{RA}{L} = \frac{240 \times \pi \times (0.023 \times 10^{-3})^2}{580 \times 10^{-3}} = 6.9 \times 10^{-7}\,\Omega\,m^{-1}$$

> The units of resistivity are Ω m, not Ω m^{-1}.

> Method

resistivity of tungsten = $6.9 \times 10^{-7}\,\Omega\,m^{-1}$ ✗ .... [4]

(c) The switch is moved to position 2. Calculate the resistance of the 120 W filament.

✓

$$P = \frac{V^2}{R} \implies R = \frac{V^2}{P} = \frac{120^2}{120} = 120\,\Omega \checkmark$$

> Substitution and calculation.

> Method.

resistance of the 120 W filament = 120 .... Ω [2]

(d) The switch is moved to position 3. Calculate the total current drawn from the 120 V supply.

Total resistance = 240 + 120 = 360 Ω ✗

$$\text{Current I} = \frac{V}{R} = \frac{120}{360} = 0.33\,A \quad ✗$$

> The two filaments are connected in parallel, not in series. The combined resistance is 80 Ω, and the total current is $\frac{120}{80} = 1.5\,A$

current drawn from the supply = 0.33 .... A [2]

1 An electric toaster is labelled 230 V, 1 kW. It is switched on for 3 minutes. How many electrons pass through the toaster in this time?

A 8.2×10^{17} **B** 1.6×10^{19}

C 4.9×10^{21} **D** 8.0×10^{21} [1]

2 An electric lamp has a resistance of 720 Ω when switched on. If a charge of 25 C passes through the lamp in one minute, what is the power of the lamp?

A 30 W **B** 75 W **C** 125 W **D** 300 W [1]

3 A thin slice of germanium of thickness 0.2 mm and width 2.5 mm, carries a current of 30 µA.

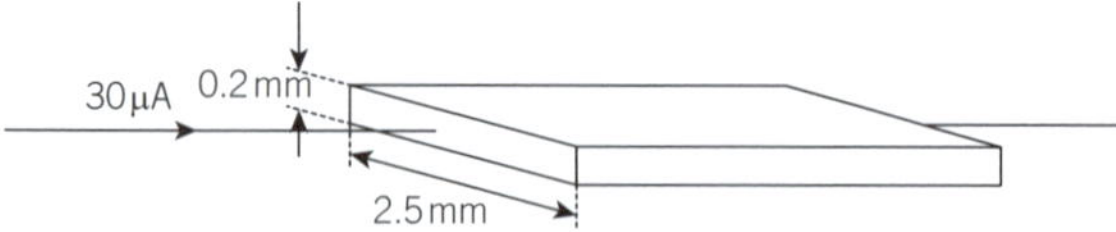

The density of charge carriers in germanium is $6.0 \times 10^{20}\,\text{m}^{-3}$. Assuming all the charge carriers are electrons, what is their drift velocity?

A $0.063\,\text{m s}^{-1}$ **B** $0.63\,\text{m s}^{-1}$

C $6.3\,\text{m s}^{-1}$ **D** $0.063\,\text{m s}^{-1}$ [1]

4 A student is investigating how the resistance of a metal wire varies with the diameter of the wire.

She measures the resistance R of wires made of the same material and the same length, but different diameters d.

Which graph should she plot to obtain a straight line?

A R against d^2 **B** R against $\sqrt{d}$

C R against $\dfrac{1}{d^2}$ **D** R against $\dfrac{1}{\sqrt{d}}$ [1]

5 A wire resistor has a resistance of R ohms. What is the resistance of a wire of the same material, but three times as long, with twice the diameter?

A $\dfrac{2}{3}R$ **B** $\dfrac{3}{4}R$ **C** $\dfrac{4}{3}R$ **D** $\dfrac{3}{2}R$ [1]

6 A 50 W heater is made from a metal wire of resistivity $4.9 \times 10^{-7}\,\Omega\,\text{m}$ and diameter 0.56 mm. The heater is connected to a 12 V car battery.

What is the length of the wire?

A 0.12 m **B** 0.48 m **C** 1.4 m **D** 5.8 m [1]

7 A power cable consists of six strands of copper wire, each of diameter 1.2 mm. The resistivity of copper is $1.7 \times 10^{-8}\,\Omega\,\text{m}$.

What is the resistance of 1.0 km of the cable?

A 0.10 m **B** 0.42 m **C** 2.5 m **D** 15 m [1]

8 A wire of resistance 5.0 Ω is to be made by extruding 1 cm³ of copper wire.

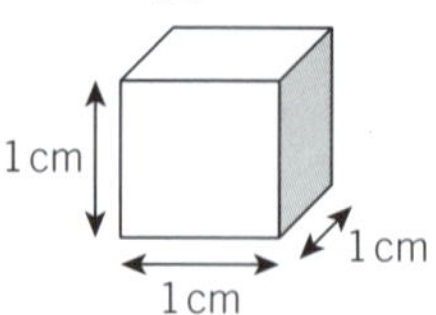

The resistivity of copper is $1.7 \times 10^{-8}\,\Omega\,\text{m}$.

Calculate:

(a) the length of the wire [2]

(b) the diameter of the wire. [2]

9 Two identical strain gauges, A and B, are connected to the upper and lower surfaces of a cantilever beam, as shown.

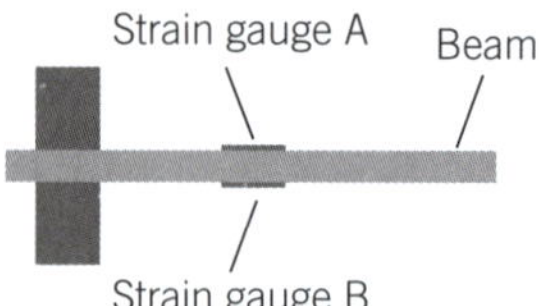

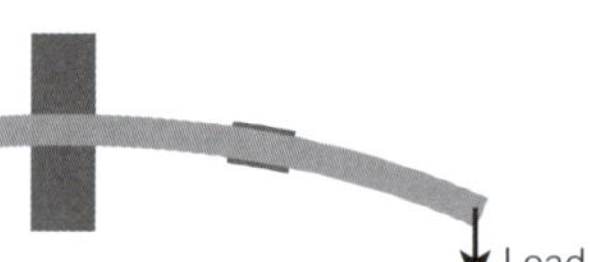

(a) Explain why the resistance of strain gauge A increases when the load is applied. [2]

(b) The two strain gauges are connected in a potential divider circuit, as shown.

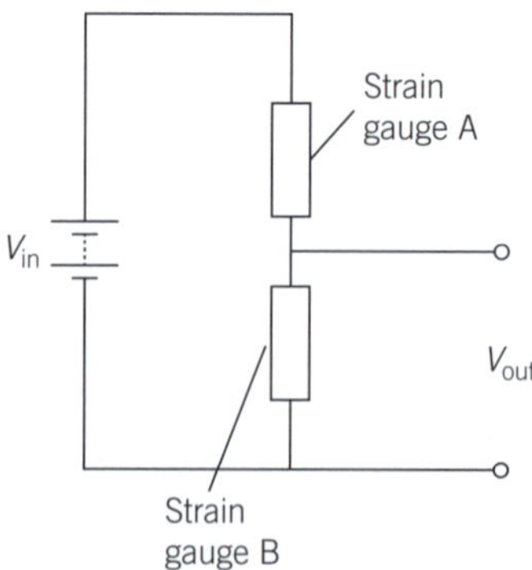

(i) The resistance of each strain gauge changes by 10%. Calculate the ratio $\dfrac{V_{out}}{V_{in}}$.

(ii) Suggest a reason why it is better to use two strain gauges rather than one. [4]

20 Direct current circuits (d.c.)

Key points

- ☐ Recall and use appropriate circuit symbols.
- ☐ Draw and interpret circuit diagrams containing sources, switches, resistors, ammeters, voltmeters, and other components referred to in the syllabus.
- ☐ Define electromotive force (e.m.f.) in terms of the energy transferred by a source in driving unit charge round a complete circuit.
- ☐ Distinguish between e.m.f. and potential difference (p.d.) in terms of energy considerations.
- ☐ Understand the effects of the internal resistance of a source of e.m.f. on the terminal potential difference.
- ☐ Recall Kirchhoff's first law, and appreciate the link to conservation of charge.
- ☐ Recall Kirchhoff's second law, and appreciate the link to conservation of energy.
- ☐ Derive, using Kirchhoff's laws, a formula for the combined resistance of two or more resistors in series, and solve problems using this formula.
- ☐ Derive, using Kirchhoff's laws, a formula for the combined resistance of two or more resistors in parallel, and solve problems using this formula.
- ☐ Apply Kirchhoff's laws to solve simple circuit problems.
- ☐ Understand the principle of a potential divider circuit as a source of variable p.d.
- ☐ Recall and solve problems using the potentiometer as a means of comparing potential differences.
- ☐ Understand that an electronic sensor consists of a sensing device and a circuit that provides an output voltage.
- ☐ Explain the use of thermistors, light-dependent resistors, and strain gauges in potential dividers to provide a potential difference that is dependent on temperature, illumination, and strain, respectively.

Practical circuits

Electromotive force (e.m.f.), potential difference (p.d.), and internal resistance

Some key terms:

- **Electromotive force**: The **electromotive force** (e.m.f.) E of a power supply, such as a cell or a laboratory power pack, is the energy given to each coulomb of charge as it passes through the supply. The name is slightly misleading as it is not a force at all!

- **Internal resistance**: Some energy is needed to 'push' the charge through the cell as the cell itself has some electrical resistance. The resistance of the cell or power supply is called the **internal resistance** r.

- **External potential difference**: As charge moves around a circuit it transfers energy to other forms (e.g., heat and light). The energy transferred by each coulomb as it passes through a component (such as a lamp or resistor) is the **potential difference** (p.d.) across the component.

In the circuit shown in Figure 20.1:

$$E = Ir + IR$$

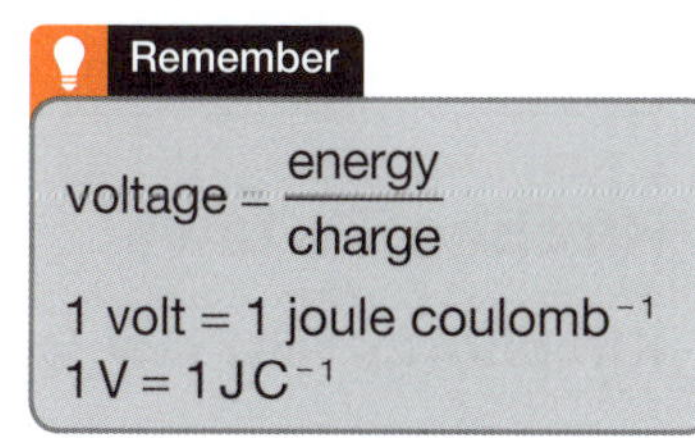

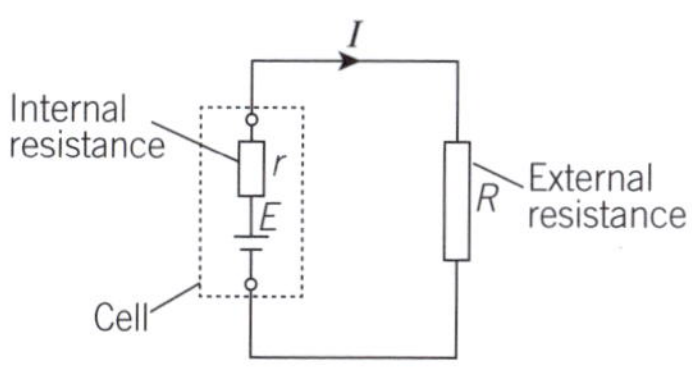

▲ **Figure 20.1** Electromotive force, internal resistance, and terminal potential difference

The precise meanings of terms such as e.m.f. and terminal p.d. are often confused. Table 20.1 gives exact descriptions of what each term means.

▼ **Table 20.1** Electromotive force, terminal potential difference, and 'lost volts'

Symbol	Name	Meaning
E	e.m.f.	energy supplied to each coulomb as it passes through the cell
Ir	the 'lost' volts	energy lost by each coulomb in passing through the cell
IR	external p.d.	potential difference across the external resistor
$E - Ir$	terminal p.d.	potential difference between the terminals of the cell or power supply

Worked example

A high-resistance voltmeter connected across a cell, as shown in Figure 20.2, reads 1.5 V. When the switch is closed, the reading on the ammeter is 270 mA.

a) For the cell calculate:

    i) the e.m.f.   **ii)** the internal resistance.

b) Determine the reading on the voltmeter when the switch is closed.

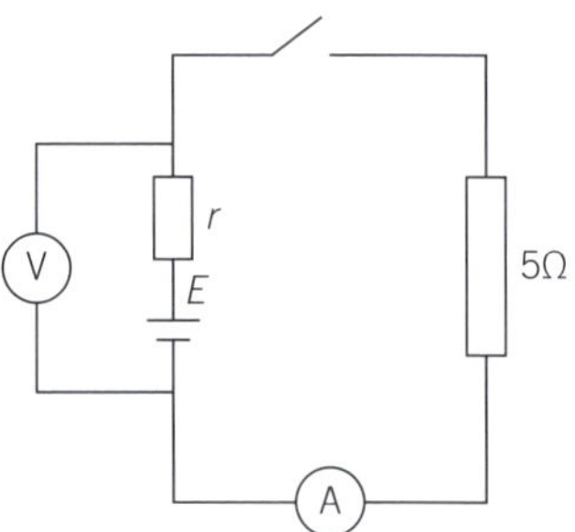

▲ **Figure 20.2**

Answer

a) i) The e.m.f. of the cell is 1.5 V – the switch is open so no current is flowing (there are no 'lost' volts inside the cell). In this instance the e.m.f. and the terminal p.d. are the same.

  ii) $1.5 = 0.27 (5 + r)$

$$r = \frac{1.5}{0.27} - 5 = 0.56\,\Omega$$

b) The reading on the voltmeter when the switch is closed is the terminal p.d.

$$\text{terminal p.d.} = \text{e.m.f.} - \text{'lost volts'}$$

$$= 1.5 - 0.27 \times 0.56$$

$$= 1.35\,\text{V}$$

Kirchhoff's laws

Kirchhoff's first law (conservation of charge)

Remember

Kirchhoff's first law: The total current flowing into a junction is equal to the total current flowing out of the junction.

It is effectively a statement of the conservation of electric charge (see Figure 20.3).

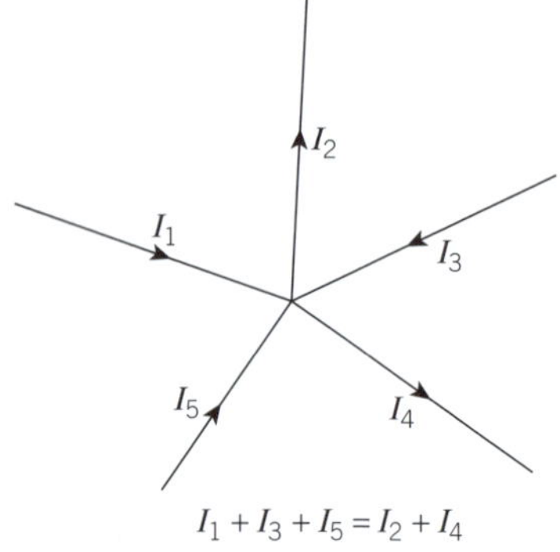

▲ **Figure 20.3** Kirchhoff's first law

Worked example

In the circuit shown in Figure 20.4, ammeter A_1 reads 38 mA, ammeter A_2 reads 15 mA, and ammeter A_4 reads 12 mA. Determine the readings on ammeters A_3 and A_5.

Answer

At junction P, the current divides with $38 - 12 = 26$ mA heading towards Q.

At junction Q, the current divides:

reading on ammeter $A_3 = 26 - 15 = 11$ mA

reading on ammeter $A_5 = 15 + 11 + 12 = 38$ mA.

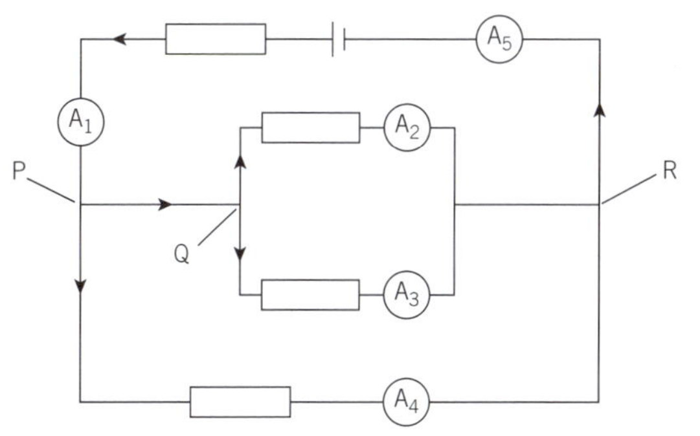

▲ **Figure 20.4**

Kirchhoff's second law (conservation of energy)

> **Remember**
>
> **Kirchhoff's second law**: For any complete loop in a circuit, the sum of the e.m.f.s round the loop is equal to the sum of the potential drops around the loop.
>
> It is a statement of the conservation of energy in electrical circuits (see Figure 20.5).

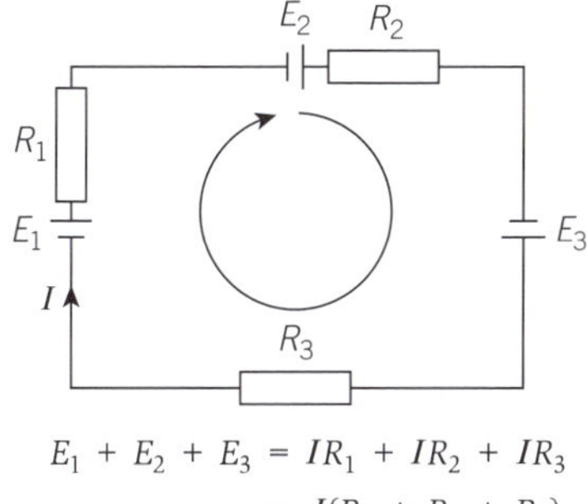

$$E_1 + E_2 + E_3 = IR_1 + IR_2 + IR_3$$
$$= I(R_1 + R_2 + R_3)$$

▶ **Figure 20.5** Kirchhoff's second law

Worked example

In the circuit shown in Figure 20.6, calculate the currents I_1 and I_2.

Answer

For the left-hand loop: $\quad 7 = 3I_1 + 2(I_1 + I_2)$ (eqn 1)

For the right-hand loop: $\quad 5 = I_2 + 2(I_1 + I_2)$ (eqn 2)

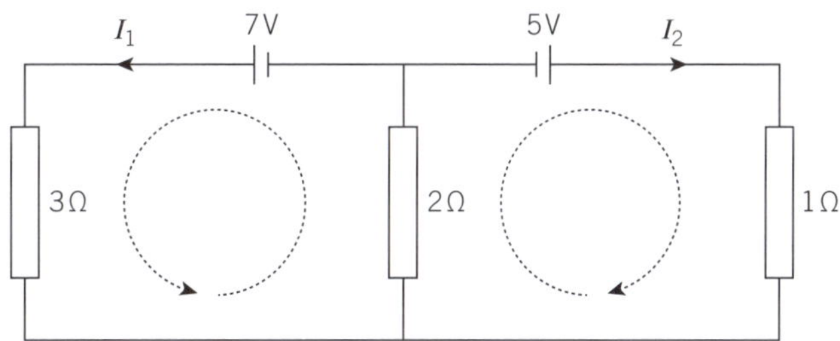

▲ **Figure 20.6**

Re-arranging eqn 1 and eqn 2:

$$7 = 5I_1 + 2I_2 \quad \text{and} \quad 5 = 3I_2 + 2I_1$$

Solving these equations gives $I_1 = I_2 = 1$ A.

A useful way of thinking about Kirchhoff's second law is to imagine a single coulomb of charge going around the circuit. It gains energy passing through a source of e.m.f (the voltage is the energy gained per coulomb) and loses energy in passing through resistors (the p.d. across a resistor is the energy lost per coulomb). By the time a coulomb completes a circuit, it will have lost as much energy as it has gained.

Resistors in series

The current passing through several resistors connected in series must be the same (see Figure 20.7).

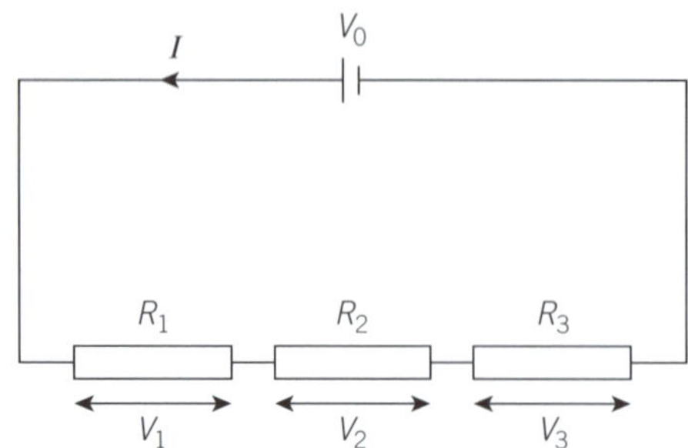

▲ **Figure 20.7** Resistors in series

From Kirchhoff's second law:

$$V_0 = V_1 + V_2 + V_3 = IR_1 + IR_2 + IR_3$$
$$V_0 = I(R_1 + R_2 + R_3)$$

The single resistor of resistance R that is equivalent to the three resistors in parallel is:

$$R = R_1 + R_2 + R_3$$

> ⭐ **Exam tip**
>
> For resistors in series:
>
> $$R = R_1 + R_2 + R_3 + ...$$
>
> This formula is provided in Exam Papers 1, 2, and 4, but you need to be able to derive it.

Resistors in parallel

The potential difference across a number of resistors connected in parallel is the same across each resistor (see Figure 20.8).

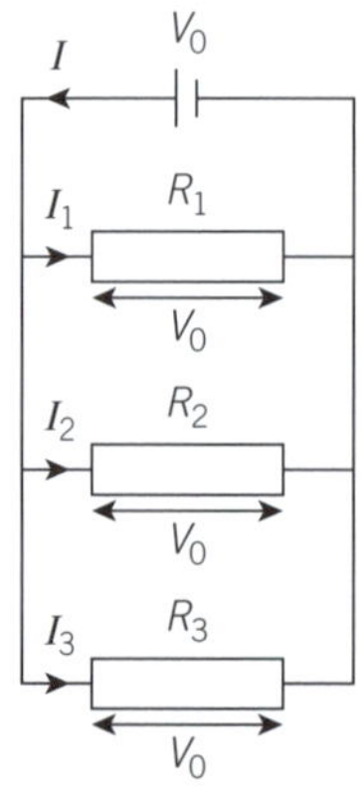

▲ **Figure 20.8** Resistors in parallel

> ⭐ **Exam tip**
>
> For resistors in parallel:
>
> $$\frac{1}{R} = \left(\frac{1}{R_1} + \frac{1}{R_2} + \frac{1}{R_3} ...... \right)$$
>
> When using this equation, don't forget to turn your answer 'upside-down' to find R.
>
> This formula is provided in Exam Papers 1, 2, and 4, but you need to be able to derive it.

From Kirchhoff's first law:

$$I = I_1 + I_2 + I_3 = \frac{V_0}{R_1} + \frac{V_0}{R_2} + \frac{V_0}{R_3}$$

$$I = V_0 \left(\frac{1}{R_1} + \frac{1}{R_2} + \frac{1}{R_3} \right)$$

The single resistor of resistance R that could replace the three resistors in parallel is given by:

$$\frac{1}{R} = \left(\frac{1}{R_1} + \frac{1}{R_2} + \frac{1}{R_3} \right)$$

Resistors connected in parallel have a smaller combined resistance.

Worked examples

1 Six equal resistors, each of resistance R, are connected as shown in Figure 20.9.

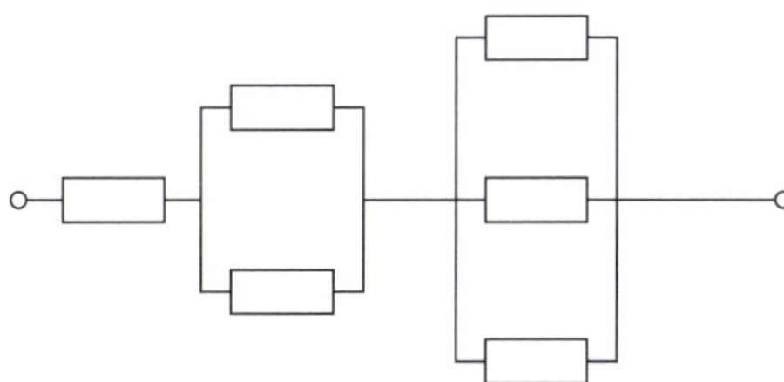

▲ **Figure 20.9**

Determine the total resistance of this combination of resistors.

Answer

The three resistors in parallel combine to give a total resistance of $\dfrac{R}{3}$. Similarly, the two resistors in parallel combine to give a resistance of $\dfrac{R}{2}$. The total resistance is then:

$$R + \frac{R}{2} + \frac{R}{3} = \frac{11R}{6}$$

2 Calculate the resistance of the resistor that must be placed in parallel with a resistor of resistance $15\,\Omega$ to have a combined resistance of $6\,\Omega$.

Answer

$$\frac{1}{6} = \frac{1}{15} + \frac{1}{R} \quad \text{so} \quad \frac{1}{R} = \frac{1}{6} - \frac{1}{15} = \frac{5-2}{30} = \frac{3}{30} = \frac{1}{10}$$

$$R = 10\,\Omega$$

Potential dividers

Two or more resistors can be connected in series to form a **potential divider** circuit. As the name suggests, a potential divider is a way of sharing, or dividing up, a potential difference between the different resistors. Figure 20.10 shows a simple potential divider circuit.

$$V_{\text{S}} = V_1 + V_2 = I(R_1 + R_2) \qquad \text{so } I = \frac{V_{\text{s}}}{R_1 + R_2}$$

Hence

$$V_1 = IR_1 = \left(\frac{R_1}{R_1 + R_2}\right)V_{\text{s}} \qquad \text{similarly} \qquad V_2 = IR_2 = \left(\frac{R_2}{R_1 + R_2}\right)V_{\text{s}}$$

If the resistors have the same resistance, half the supply voltage V_{s} is across each of the two resistors. If one resistor is twice as big as the other, the p.d. across the larger resistor is twice as much as the p.d. across the smaller resistor.

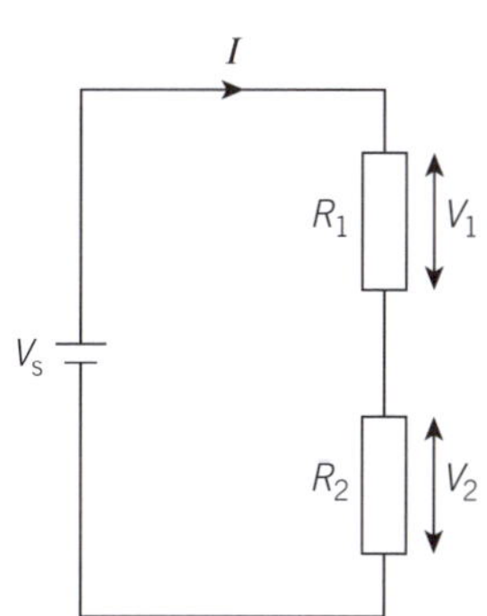

▲ **Figure 20.10** Potential divider

Worked examples

1 In the circuit shown in Figure 20.11, calculate the potential difference across the 5 kΩ resistor.

Answer

The 'fraction' of the supply p.d. across the 5 kΩ resistor is:

$\left(\dfrac{5}{5+15}\right) = \dfrac{1}{4}$, so the p.d. across the 5 kΩ resistor $= \dfrac{1}{4} \times 12 = 3\,V$

2 The resistance of a negative temperature coefficient thermistor decreases as its temperature increases. When its temperature is 20 °C its resistance is 90 kΩ; when its temperature is 50 °C its resistance is 30 kΩ.

For the circuit shown in Figure 20.12:

a) Determine the output p.d. V_{out} when the temperature is:

    i) 20 °C   **ii)** 50 °C.

b) State how you would alter the circuit so that the output p.d. decreased as the temperature increased.

Answer

a) **i)** $V_{out} = \left(\dfrac{10}{10+90}\right) \times 10 = 1.0\,V$ **ii)** $V_{out} = \left(\dfrac{10}{10+30}\right) \times 10 = 2.5\,V$

b) Swapping the thermistor and the fixed resistor around would mean that V_{out} would decrease as the temperature increased.

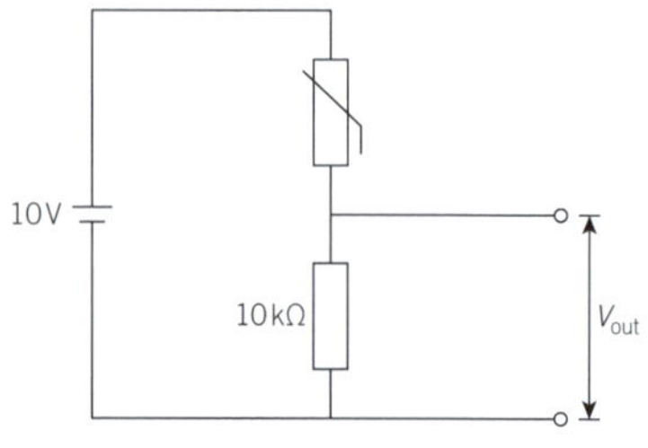

▲ **Figure 20.11**

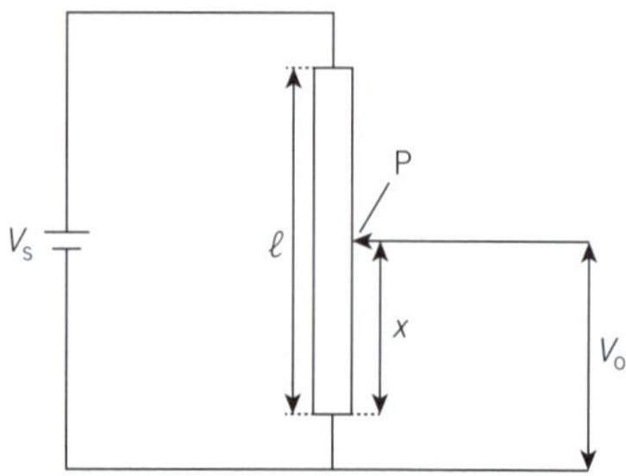

▲ **Figure 20.12**

A continuously variable potential divider (Figure 20.13) can be made by replacing the fixed resistors with a variable resistor (e.g., a length of resistance wire or a rheostat).

The output p.d. V_0 can have any value from 0 V to V_s, by moving the sliding connector P up or down.

$$V_0 = \left(\frac{x}{l}\right)V_s$$

Potentiometers

A potential divider is also known as a **potentiometer**, particularly when used to compare potential differences. Figure 20.14 is an example of a simple potentiometer circuit used to measure the e.m.f. E of a **dry cell** (the correct name for a torch 'battery').

The e.m.f. of the **driver cell** is V_0 and is known to a high degree of accuracy. The slider is moved up or down until the reading on the centre-zero galvanometer (a sensitive ammeter) reads zero – the potentiometer is then balanced. The e.m.f. of the dry cell can be found from:

$$E = \left(\frac{x}{l}\right)V_0$$

The advantages of this method of measuring potential differences are:

- it is a *null* method (the galvanometer only has to read '0' accurately)

- no current is drawn from the unknown p.d. so there are no 'lost' volts across the internal resistance of the dry cell

- the accuracy can be increased by using a longer resistance wire.

A disadvantage of this method is the apparatus is bulky and slow to use compared to a digital voltmeter.

▲ **Figure 20.13** Continuously variable potential divider

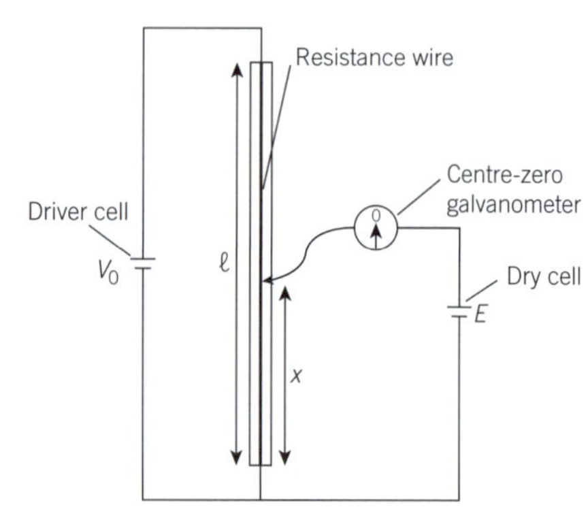

▲ **Figure 20.14** Potentiometer circuit

Worked example

A potentiometer is connected to a dry cell of unknown e.m.f. E as shown in Figure 20.15. The balance length (when the galvanometer reads zero) is 54.7 cm.

When the dry cell is replaced with another cell of e.m.f. 1.02 V the balance length changes to 37.4 cm. Calculate the e.m.f. of the dry cell.

Answer

$$\frac{E}{1.02} = \frac{54.7}{37.4} = 1.463 \qquad \text{so } E = 1.49\,\text{V}$$

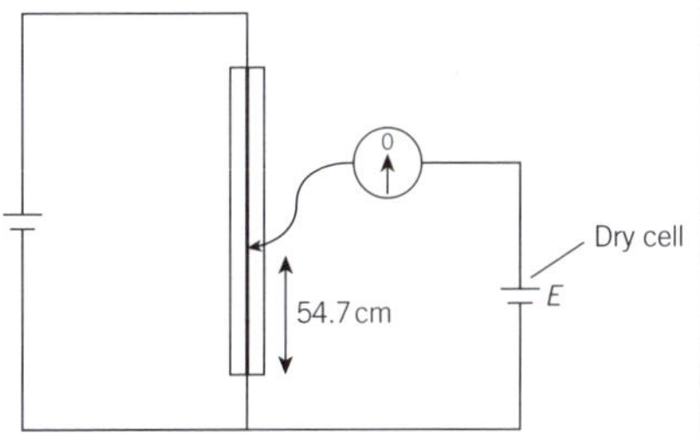

▲ Figure 20.15

Note that neither the e.m.f. of the driver cell, nor the length of the resistance wire, are needed to answer this question.

Electronic sensors

Circuits designed to detect changes in the environment, such as temperature or light intensity, depend on producing an output p.d. that changes with the variable being monitored. The sensing device is a resistor which changes resistance with changes in the variable, and is part of a potential divider circuit.

Temperature sensor

The resistance of a negative temperature coefficient thermistor decreases as the temperature increases. In the potential-divider circuit shown in Figure 20.16, a temperature increase will reduce the resistance of a thermistor and increase the output p.d. V_{out}. A decrease can be achieved by either swapping the fixed resistor and the thermistor, or using a positive temperature coefficient thermistor.

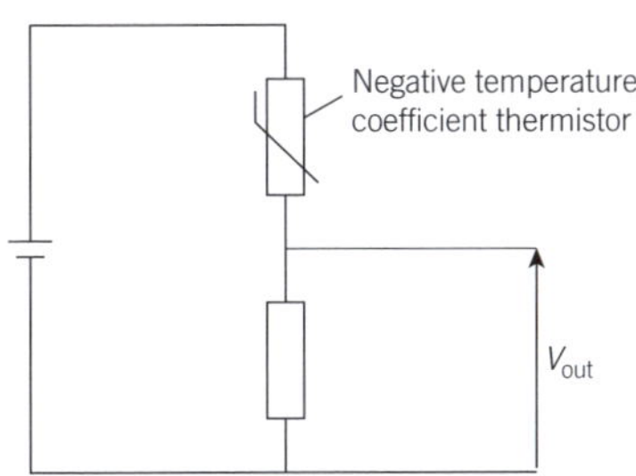

▲ Figure 20.16 Temperature sensor

Light sensor

A light sensor (Figure 20.17) can be made in a similar way, replacing the thermistor with a light-dependent resistor (LDR). As the incident light intensity increases, the resistance of the LDR decreases and the p.d. across the LDR decreases. The output V_{out} increases.

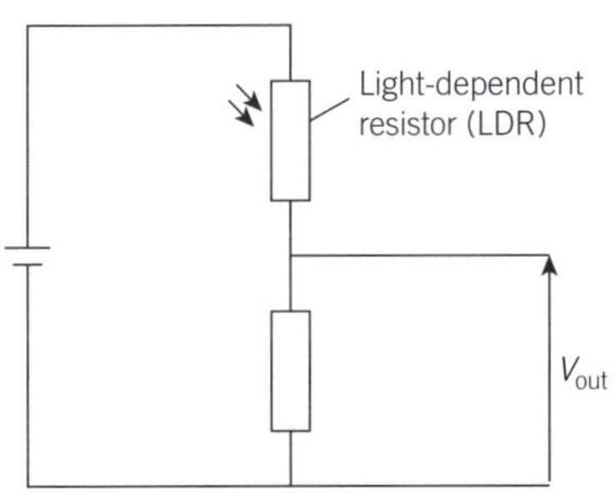

▲ Figure 20.17 Light sensor

Strain gauge

A strain gauge is a long length of resistance wire embedded in plastic as shown in Figure 20.18. The strain gauge is glued to the structure being tested. As loads are applied to the structure the extension of the wire changes the resistance of the strain gauge. From $R = \rho L/A$, if there is no change in the cross-sectional area of the wire, the change in resistance of the wire is given by:

$$\frac{\Delta R}{R} = \frac{\Delta L}{L}$$

where L is the original length of the wire and R its resistance. When a wire is stretched it also becomes thinner (the cross-sectional area of the wire decreases) which also increases the resistance of the wire. If the volume of the wire stays roughly constant, it can be shown that $\frac{\Delta R}{R} = \frac{2\Delta L}{L}$. In either case, the change in resistance is directly proportional to the change in length, which makes it ideal as a sensor.

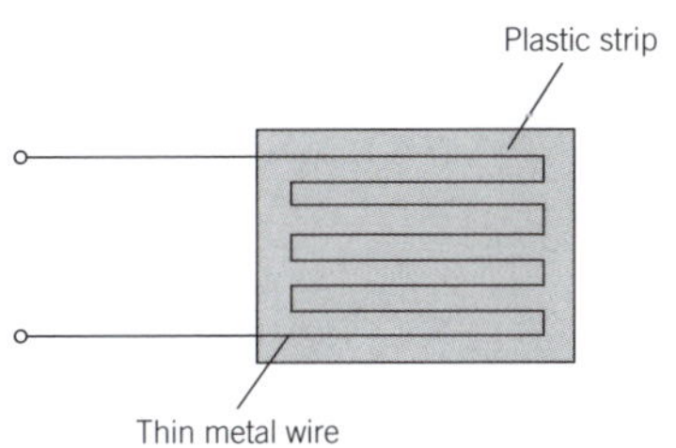

▲ Figure 20.18 Metal-wire strain gauge

 **Link**

The change in length of the wire can often be very small, producing a correspondingly small change in V_{out}. The sensor can be improved by connecting V_{out} to an operational amplifier to magnify the change. See Unit 21 *Electronics* for more details.

↑ Raise your grade

(a) In relation to a cell, explain the meaning of:

(i) electromotive force (e.m.f.)

The force from the cell pushing the current ✗

> e.m.f. is **not** a force. The e.m.f. is the electrical energy per unit charge produced inside the cell.

(ii) terminal potential difference (p.d.).

The voltage from the cell ✗

> As the name suggests, it is the potential difference between the two terminals of the cell. ..... [2]

(b) State Kirchhoff's second law.

The sum of the e.m.f.s is equal to the sum of the potential drops ✓ ✗

> For the second mark, the candidate should have added 'around any complete loop of a circuit'. .. [2]

(c) A dry cell, of e.m.f. E and internal resistance r is connected to an external resistor of resistance R. The current I is recorded for two different values of R, as shown

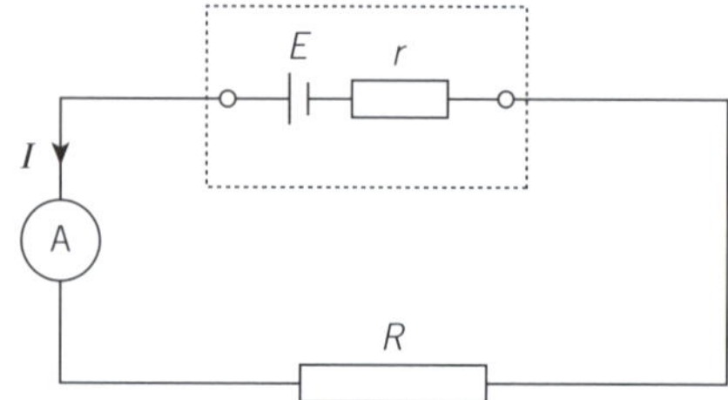

R/Ω	I/A
10	0.5
18	0.3

(i) Show that the internal resistance r is 2 Ω.

Using Kirchhoff's 2nd law: $E = 0.5\,(r + 10) = 0.3(r + 18)$ ✓

$$0.5r + 5 = 0.3r + 5.4 \quad ✓$$

$$r = 2\Omega$$

> Method.

> Calculation.

(ii) Calculate the e.m.f. E.

$$E = 0.5(r + 10) = 0.5 \times 12 = 6.0V \quad ✓$$

> Calculation.

$E = \underline{6.0}$ V [3]

(d) A different cell, of e.m.f. 9.0 V and internal resistance 0.5 Ω, is connected to a 4 Ω resistor. Calculate:

(i) the terminal p.d.

$$\text{current } i = \frac{9.0}{4.5} = 2.0A$$

> ✓ Method. ✓ Calculation.

terminal p.d. = e.m.f. − 'lost volts' = 9.0 − (2.0 × 0.5) = 8.0 V

terminal p.d. = $\underline{8.0}$ V

(ii) the energy per second dissipated in the 4 Ω resistor.

$$P = i^2R = 2.0^2 \times 4 = 16\,W$$

> ✓ Method. ✓ Calculation.

energy dissipated per second = $\underline{16}$ W [4]

? Exam-style questions

1 The diagram shows different currents entering or leaving a circuit junction.

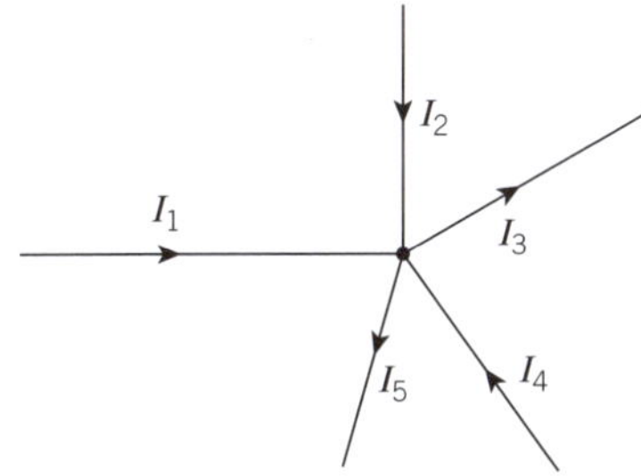

Which statement is correct?

A $I_1 + I_2 = I_3 + I_4 + I_5$

B $I_1 + I_2 = I_3 + I_4 - I_5$

C $I_1 + I_2 = I_3 - I_4 - I_5$

D $I_1 + I_2 = I_3 - I_4 + I_5$ [1]

2 For the circuit shown, which statement is correct?

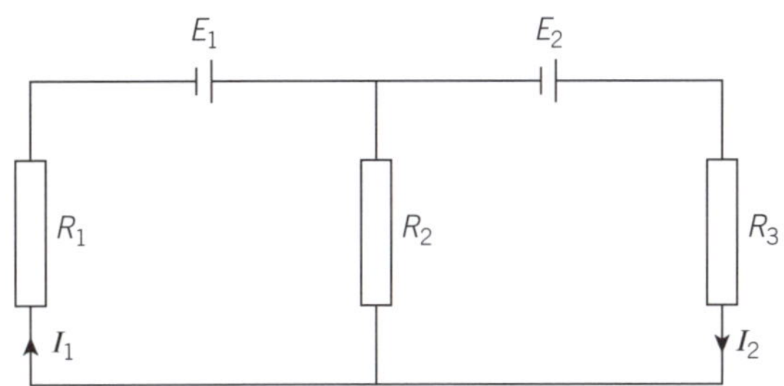

A $E_1 = (I_1 - I_2)R_2 + I_1R_1$

B $E_1 = (I_1 + I_2)R_2 + I_1R_1$

C $E_2 = I_2R_3 + (I_1 + I_2)R_2$

D $E_2 = I_2R_3 - (I_1 + I_2)R_2$ [1]

3 A cell, of e.m.f. 24 V and internal resistance 4 Ω, is connected to two resistors, each of resistance 8 Ω, as shown.

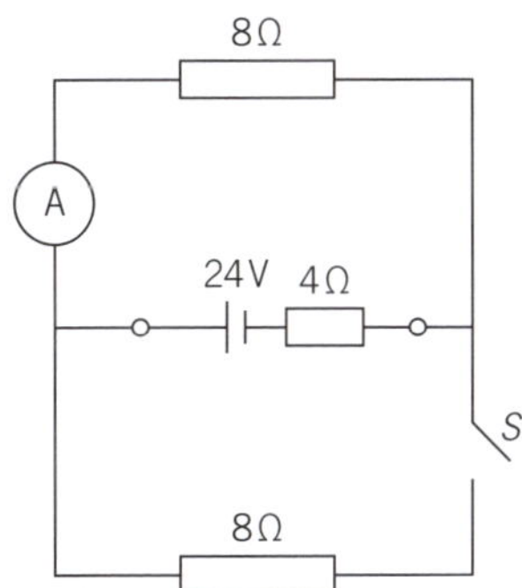

What is the **change** in the ammeter reading when switch S is closed?

A 0.5 A **B** 1.0 A **C** 1.5 A **D** 2.0 A [1]

4 A battery, of e.m.f. E and internal resistance r, is connected to a high-resistance voltmeter and a variable resistor of resistance R as shown.

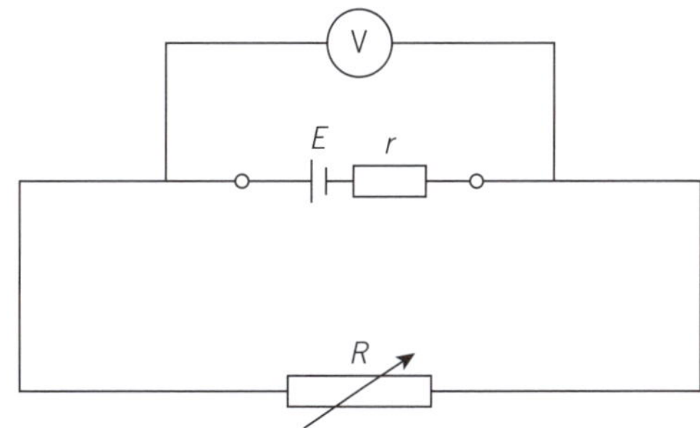

When $R = 1\,\Omega$ the voltmeter reading was 3 V.

When $R = 3\,\Omega$ the voltmeter reading was 6 V.

Which line in the table gives the correct values of E and r?

	E/V	r/Ω
A	6.0	1.5
B	6.0	3.0
C	12.0	1.5
D	12.0	3.0

[1]

5 A potentiometer circuit is used to measure the output p.d. of a thermocouple using a uniform resistance wire of length 99.6 cm and resistance 12 Ω, as shown.

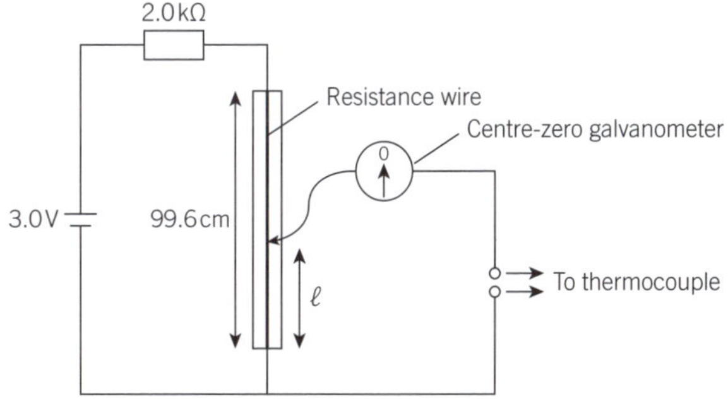

The output p.d. of the thermocouple is known to be between 0 V and 15 mV.

(a) Explain why the 2.0 kΩ resistor is needed. [1]

(b) The centre-zero galvanometer reads zero when $\ell = 68.3$ cm.

Calculate:

(i) the current in the 2.0 kΩ resistor

(ii) the thermocouple p.d. [4]

Key points

- ☐ Recall the main properties of the ideal operational amplifier (op-amp).
- ☐ Deduce, from the properties of an operational amplifier, its use as a comparator.
- ☐ Understand the effects of negative feedback on the gain of an operational amplifier.
- ☐ Recall the circuit diagrams for both the inverting and the non-inverting amplifier.
- ☐ Understand the virtual earth approximation and derive an expression for the gain of inverting amplifiers.
- ☐ Recall and use expressions for the voltage gain of inverting and of non-inverting amplifiers.
- ☐ Understand that an output device may be required to monitor the output of an op-amp circuit.
- ☐ Understand the use of relays in electronic circuits.
- ☐ Understand the use of LEDs as devices to indicate the state of the output of electronic circuits.
- ☐ Understand the need for calibration where digital or analogue meters are used as output devices.

Ideal operational amplifiers

An **operational amplifier** (op-amp) is one of the most useful electronic devices. As the name suggests, it can amplify (enlarge) a voltage from a sensor circuit, or carry out mathematical operations such as multiplication and integration. In circuit diagrams it is represented by a triangle with two input terminals (see Figures 21.1 and 21.2) called the **inverting** (V_-) and **non-inverting** (V_+) **inputs**, and one output terminal.

To work, the op-amp needs a three-terminal power supply, typically +15 V, 0 V and +15 V, though the connections to the power supply are usually omitted from circuit diagrams for greater clarity. The output voltage can be positive or negative, but cannot be greater than the supply voltage.

The key properties of an **ideal operational amplifier** are:

- infinite open-loop gain
- infinite input resistance
- zero output resistance
- infinite bandwidth
- infinite slew rate.

Used on its own (see Figure 21.2), the output of the op-amp is proportional to the difference between the two input voltages:

$$V_{out} = A(V_+ - V_-)$$

where A is the **open-loop gain** of the amplifier, typically equal to 10^5.

Bandwidth: An infinite bandwidth means the op-amp amplifies all alternating voltages by the same factor, regardless of the frequency of the voltage.

Slew rate: An infinite slew rate means there is no delay between the input voltage changing and the output voltage changing.

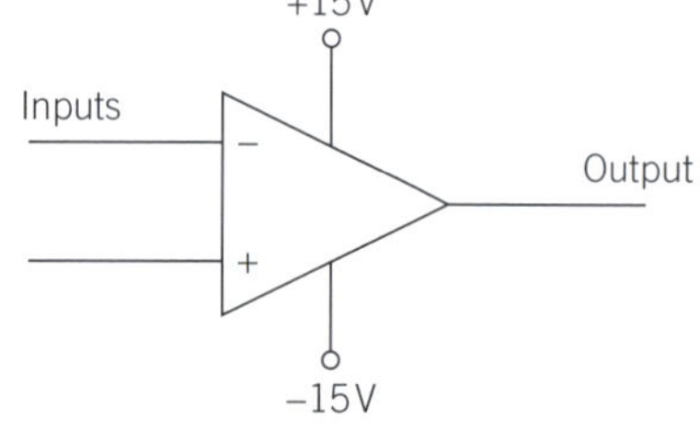

▲ **Figure 21.1** Operational amplifier

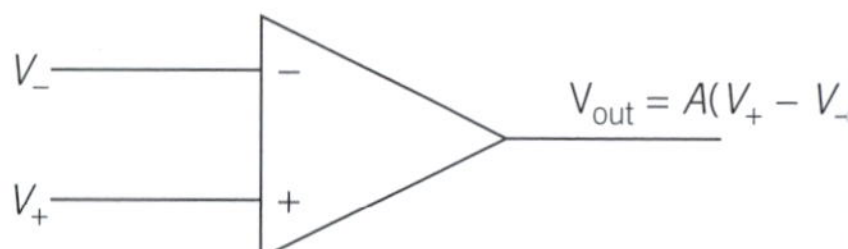

▲ **Figure 21.2** Open-loop gain

💡 **Remember**

The input current to the op-amp is so small it can be considered as zero.

Operational amplifier circuits

Op-amp as a comparator

A very small difference between V_+ and V_- is amplified up to the maximum output voltage called the **saturation voltage**. An op-amp used in this way is a **comparator** – it compares the two input voltages. If V_+ is greater than V_- then the output is the positive saturation voltage; if V_+ is less than V_- then the output is the negative saturation voltage. In this way the op-amp can be used as a switch or indicator.

In Figure 21.3, the LED comes on if V_+ is greater than V_-, but goes off if the reverse is true.

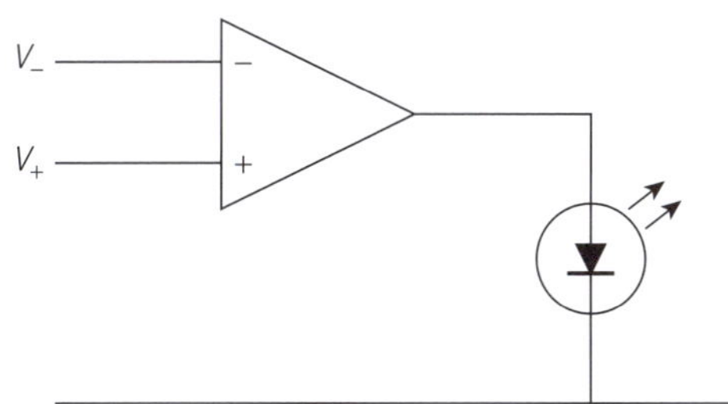

▲ **Figure 21.3** Comparator

Inverting amplifier

For most applications of op-amps some **negative feedback** is introduced – some of the output voltage is fed back and subtracted from the input voltage – which reduces the size of the amplification but greatly increases **stability** (reduces sudden fluctuations in the output voltage). Figure 21.4 shows an op-amp with negative feedback.

As well as feedback from output to input through a resistor R_f, most op-amp circuits also have an input resistor R_{in} (see Figure 21.5).

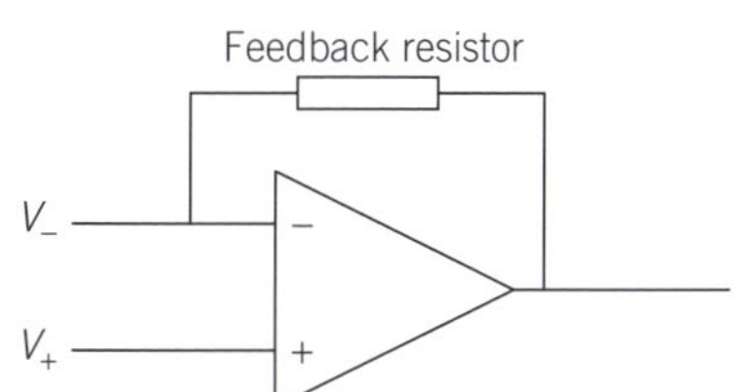

▲ **Figure 21.4** Negative feedback

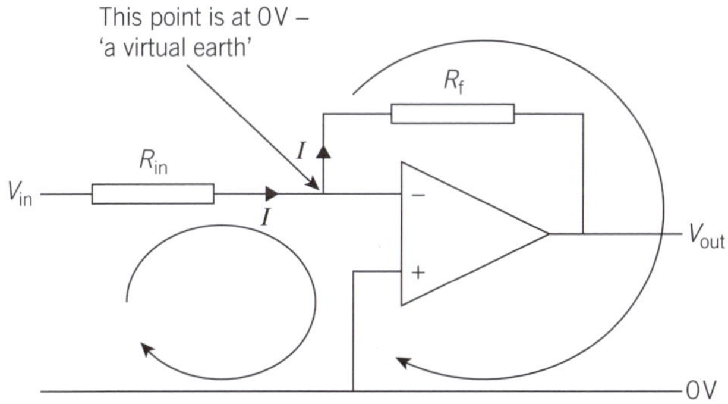

▲ **Figure 21.5** Inverting amplifier

The non-inverting input is connected to earth (0 V) and $V_+ \approx V_-$ so:

$$V_+ = V_- = 0\,\text{V}$$

This means the point indicated with an arrow in the circuit in Figure 21.5 is also at 0 V – a 'virtual earth'. The current going into the op-amp is very small ($\approx 15\,\text{pA}$) and can be treated as zero. The current in the input resistor R_{in} is the same as the current in resistor R_f.

Applying Kirchhoff's second law to the two loops:

$$V_{in} = IR_{in} \quad \text{and} \quad IR_f + V_{out} = 0 \quad \text{so} \quad V_{out} = -IR_f$$

combining these two equations: $\qquad \dfrac{V_{out}}{V_{in}} = -\dfrac{R_f}{R_{in}}$

$\dfrac{V_{out}}{V_{in}}$ is the **voltage gain** (the amplification) of the amplifier.

For the inverting amplifier the voltage gain is $-\dfrac{R_f}{R_{in}}$.

> 💡 **Remember**
>
> The maximum output voltage is called the **saturation voltage** (usually slightly less than the supply voltage).

> 💡 **Remember**
>
> For an **inverting amplifier**, the voltage gain:
> $$\frac{V_{out}}{V_{in}} = -\frac{R_f}{R_{in}}$$

The voltage gain depends only on the relative sizes of the two resistors and not on the gain or input resistance of the op-amp itself. The amplifier is an inverting amplifier because V_{out} is the opposite sign to V_{in} (the output will be in **antiphase** with the input) as shown in Figure 21.6 where an a.c. signal is the input to an inverting amplifier.

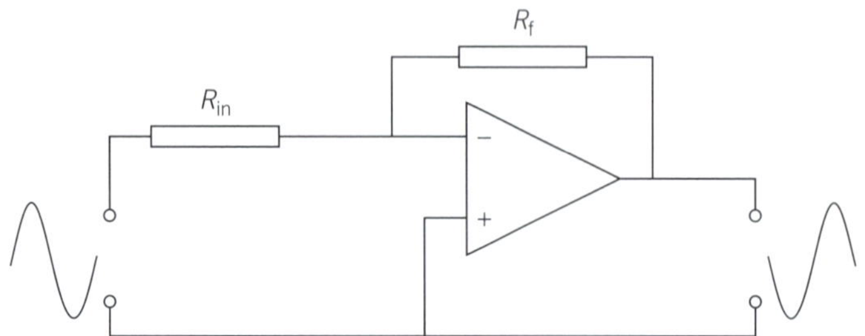

▲ **Figure 21.6** Inverting amplifier with a.c.

Worked example

a) For the circuit shown in Figure 21.7, determine the potential at point P.

b) Calculate:

   i) the current in the $10\,k\Omega$ resistor

   ii) the current in the $200\,k\Omega$ resistor.

c) Show that $V_{out} = -10\,V$.

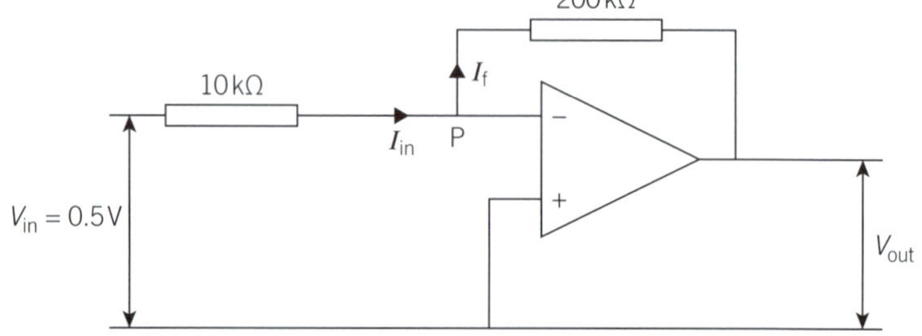

▲ **Figure 21.7** Inverting amplifier circuit

Answer

a) $V_P = V_- \approx V_+ = 0\,V$ (a 'virtual earth')

b) i) $I_{in} = \dfrac{V_{in}}{R_{in}} = \dfrac{0.5}{10 \times 10^3} = 5 \times 10^{-5}\,A$ ii) $I_f = I_{in} = 5 \times 10^{-5}\,A$

c) $(5 \times 10^{-5}) \times (200 \times 10^3) + V_{out} = 0$ so $V_{out} = -10\,V$

Non-inverting amplifier

The non-inverting amplifier (see Figure 21.8) produces an amplified output p.d. which is **in phase** with the input p.d.

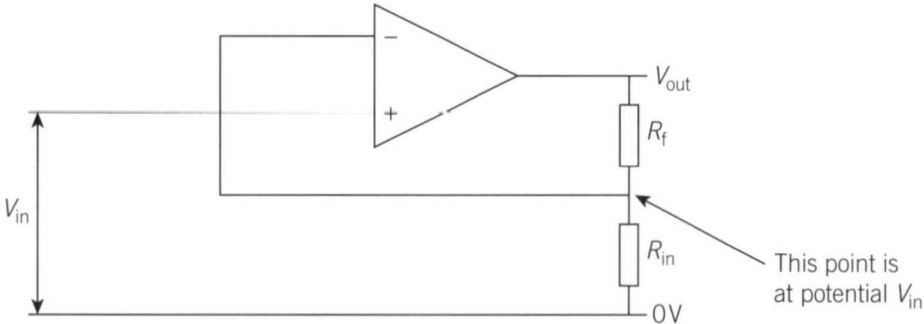

▲ **Figure 21.8** Non-inverting amplifier

$V_{in} = V_+ \approx V_-$, so the potential between the two resistors R_f and R_{in} is V_{in}.

The output is a potential divider circuit with:

$$\frac{V_{out}}{V_{in}} = \frac{R_f + R_{in}}{R_{in}} = 1 + \frac{R_f}{R_{in}}$$

For the non-inverting input the voltage gain is $1 + \dfrac{R_f}{R_{in}}$.

Output devices

Operational amplifier circuits are used in measurement, signal processing, and control circuits. Light-emitting diodes (LEDs) can be used to monitor the state of the voltage output in, for example, a comparator circuit.

Operating relays

Op-amp circuits cannot switch on appliances such as heaters and lights directly as they use only very small currents; instead the output from an op-amp circuit is usually connected to a **relay** (see Figure 21.9), which uses an electromagnetic switch that turns on the power to the appliance.

When a small current from the op-amp output passes through the coil (the rectangle in the symbol for a relay) it becomes an electromagnet (the core of the coil is usually soft iron). The electromagnet closes a spring-loaded switch which completes the secondary (high-current) circuit. When the current from the op-amp stops, the switch is pulled open again by the spring.

When using a relay, large e.m.f.s can be induced across the coil of the relay when the switch is opening or closing. To avoid this, diodes are used (see Figure 21.10).

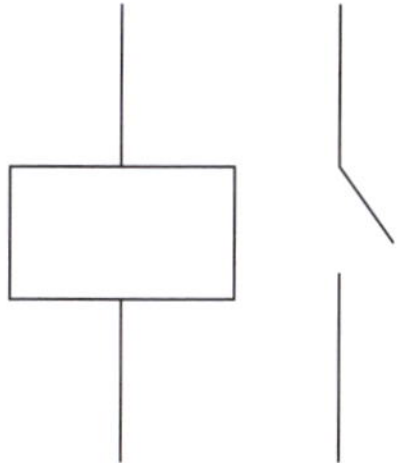

▲ **Figure 21.9** Symbol for a relay

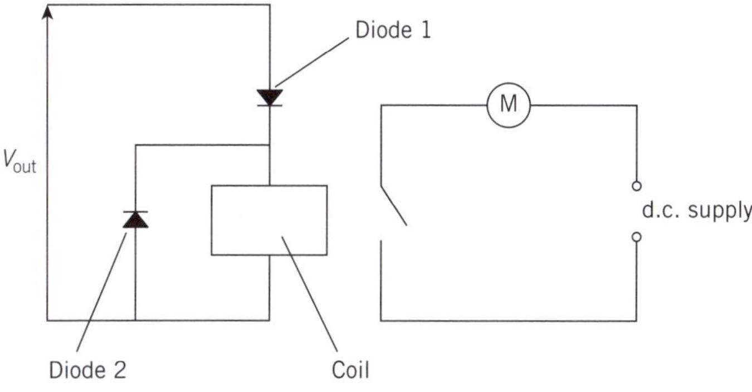

▲ **Figure 21.10** Diode protection

Suppose, for example, the output from an op-amp is used to control a large electric motor. When the output from the op-amp is positive, a small current passes through diode 1 and the relay coil, closing the switch and completing the motor circuit.

When the output from the op-amp is negative, the current stops flowing through the coil and the relay is switched off. The sudden drop in current causes the magnetic field in the coil to collapse which can cause a large back e.m.f. which could destroy the op-amp; diode 2 prevents this by effectively short-circuiting the coil.

Calibrating output devices

Op-amp circuits can be used to monitor changes in a range of properties. This is achieved by using an electrical component whose resistance varies with the property being investigated. The resistance does not usually vary linearly with the property being monitored, and so a calibration curve is needed.

Figure 21.11 shows how a thermistor in an op-amp circuit can be used to monitor temperature.

This is an inverting amplifier, so:

$$V_{\text{out}} = -\frac{R_{\text{f}}}{(R_{\text{th}} + R_{1})}V_{\text{in}}$$

Link

For more about Faraday's law and Lenz's law see Unit 23 *Electromagnetic induction*.

Link

See page 11 for more details about calibration curves.

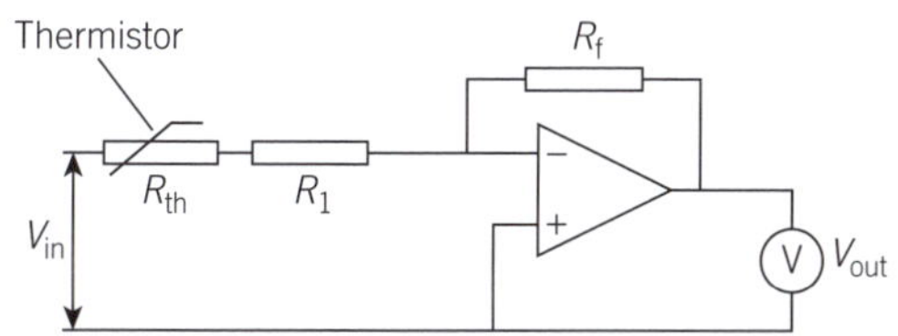

▲ **Figure 21.11** Op-amp used as a thermometer

 **Raise your grade**

(a) State three properties of *an ideal* operational amplifier.

Ideal op-amps have – infinite resistance ✗

      – infinite gain ✗

      – infinite slew rate ✔

   [3]

> The **input** resistance is infinite.
> The **open-loop** gain is very high.

(b) An ideal operational amplifier is connected as shown. The saturation voltage is 8.0 V.

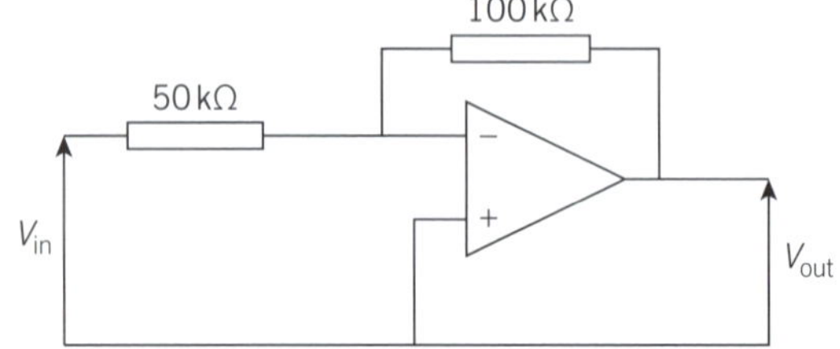

Calculate:

(i) the voltage gain of the circuit

$$\frac{V_{out}}{V_{in}} = -\frac{R_f}{R_{in}} = -\frac{100\,k\Omega}{50\,k\Omega} = -2 \quad ✔$$

> One mark for the correct calculation. One mark for remembering the minus sign.

voltage gain = ...−2..... ✔ [2]

(ii) the output voltage when $V_{in} = 4.5$ V.

$$\frac{V_{out}}{V_{in}} = -2 \rightarrow V_{out} = -2 \times 4.5 = 9\,V$$

> The value obtained is greater than the saturation voltage. The correct answer is 8.0 V.

$V_{out} = $ .....9..... V ✗ [1]

(c) The input voltage V_{in} is an alternating voltage of amplitude 5.0 V as shown by the dotted line on the graph.

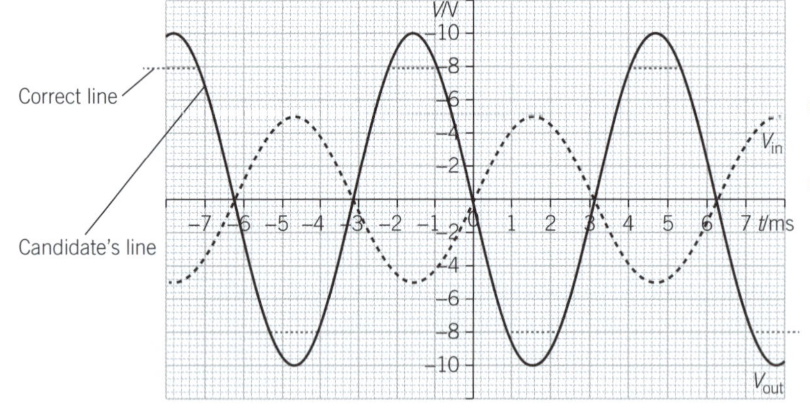

> ✔ V_{out} is twice the input for $V_{in} \leq \pm 4.0$ V.
> ✔ V_{out} is in antiphase (π out of phase) with V_{in}.
> ✗ The output voltage should not exceed 8.0 V (the saturation voltage).

Draw the output voltage V_{out} on the same graph. [3]

? **Exam-style questions**

1 In the context of operational amplifier circuits, explain the meaning of the following terms:

(a) *open-loop gain*

(b) *non-inverting*

(c) *negative feedback*

(d) *calibration curve.* [4]

2 An operational amplifier can be used as a *comparator.*

(a) Explain what is meant by the term *comparator.* [2]

(b) In the circuit shown, calculate the potential at point P. [1]

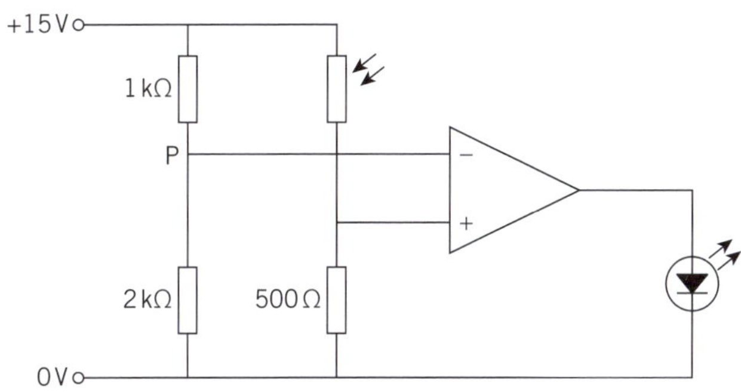

(c) In daylight the LDR has a resistance of 500 Ω. State whether the LED is on or off. Explain your answer. [2]

3 (a) State three properties of an ideal operational amplifier. [3]

(b) (i) State the name of the circuit shown below.

(ii) Point Q is referred to as a 'virtual earth'. Explain what is meant by the term *virtual earth.* [2]

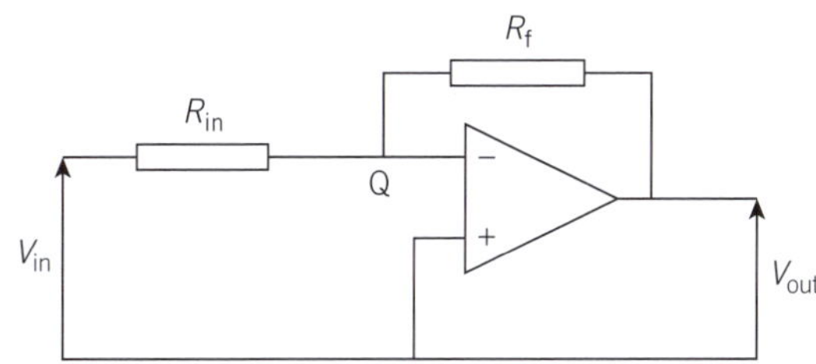

(c) Show that $\dfrac{V_{out}}{V_{in}} = -\dfrac{R_f}{R_{in}}$. Explain your working. [2]

4 (a) State what is meant by *voltage gain*. [1]

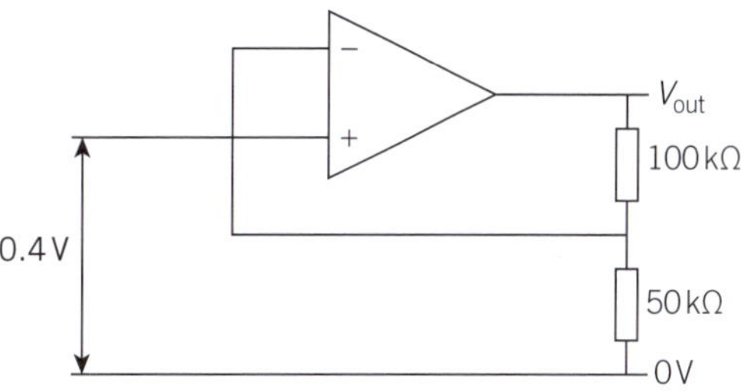

(b) Calculate:

(i) the voltage gain of the circuit shown

(ii) the output voltage V_{out}. [2]

5 An electronic system can be divided into three stages:

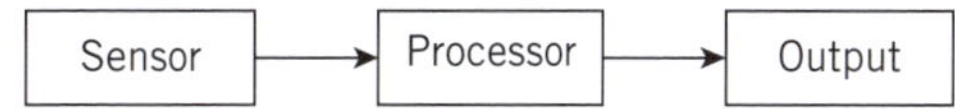

An engineer is designing a system to make a heater come on when the temperature falls below a certain value.

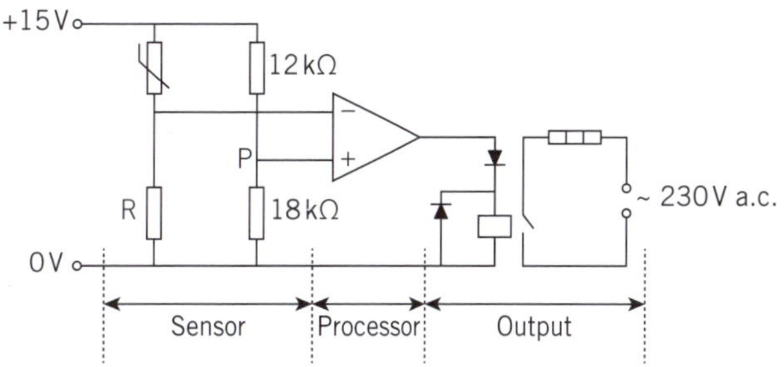

(a) (i) Name the components in the output.

(ii) Explain the purpose of the processor. [3]

(b) Calculate the potential at point P. [1]

(c) The graph shows the resistance of the thermistor at different temperatures.

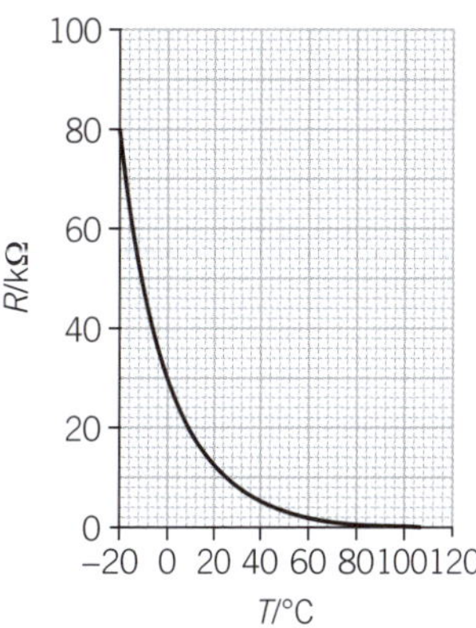

The heater must switch on when the temperature falls below 0 °C. Calculate the value of R needed. [1]

(d) State the adjustment needed for the heater to switch on at a higher temperature. [1]

Key points

- ☐ Understand that a magnetic field is an example of a field of force produced either by current-carrying conductors or by permanent magnets.
- ☐ Represent a magnetic field by field lines.
- ☐ Appreciate that a force might act on a current-carrying conductor placed in a magnetic field.
- ☐ Recall and solve problems using the equation $F = BIl\sin\theta$, with directions found by using Fleming's left-hand rule.
- ☐ Define magnetic flux density and the tesla.
- ☐ Understand how the force on a current-carrying conductor can be used to measure the flux density of a magnetic field using a current balance.
- ☐ Predict the direction of the force on a charge moving in a magnetic field.
- ☐ Recall and solve problems using $F = Bqv\sin\theta$.
- ☐ Derive the expression $V_{\mathrm{H}} = \dfrac{BI}{ntq}$ for the Hall voltage, where t is the thickness and n is the density of charge carriers.
- ☐ Describe and analyse qualitatively the deflection of beams of charged particles by uniform electric and uniform magnetic fields.
- ☐ Explain how electric and magnetic fields can be used in velocity selection.
- ☐ Explain the main principles of one method for the determination of v and e/m_{e} for electrons.
- ☐ Sketch flux patterns due to a long straight wire, a flat circular coil, and a long solenoid.
- ☐ Understand that the field due to a solenoid is influenced by the presence of a ferrous core.
- ☐ Explain the forces between current-carrying conductors and predict the direction of the forces.
- ☐ Describe and compare the forces on mass, charge, and current in gravitational, electric, and magnetic fields, as appropriate.
- ☐ Explain the main principles behind the use of nuclear magnetic resonance imaging (NMRI) to obtain diagnostic information about internal structures.
- ☐ Understand the function of the non-uniform magnetic field, superimposed on the large constant magnetic field, in diagnosis using NMRI.

Magnetic fields

A **magnetic field** is the space around a magnet or a current-carrying wire in which magnetic forces (on a ferrous metal, another magnet or another wire carrying a current, for example) can act. The shape, size and direction of a magnetic field can be illustrated using magnetic field lines (see Figure 22.1).

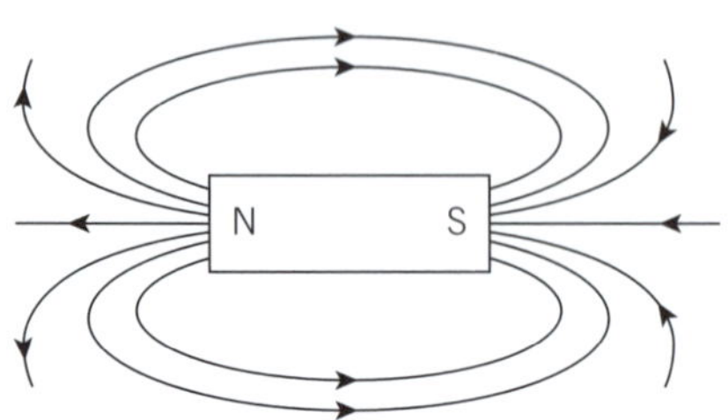

a Bar magnet

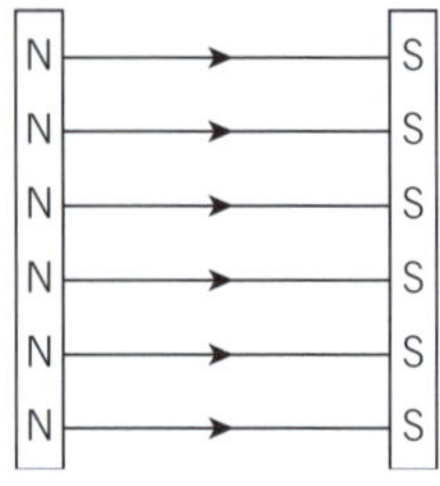

b Uniform magnetic field

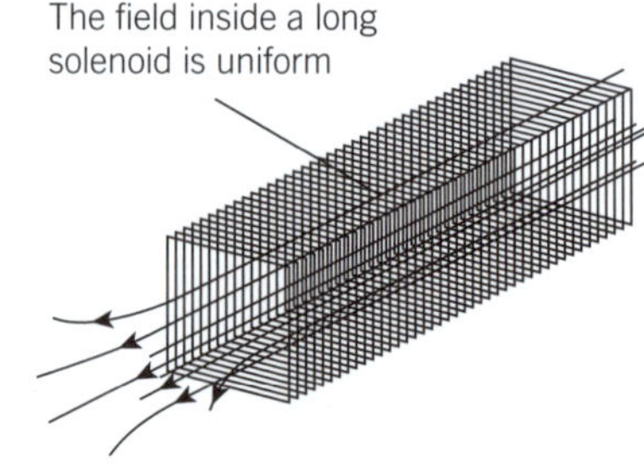

c Long coil (solenoid)

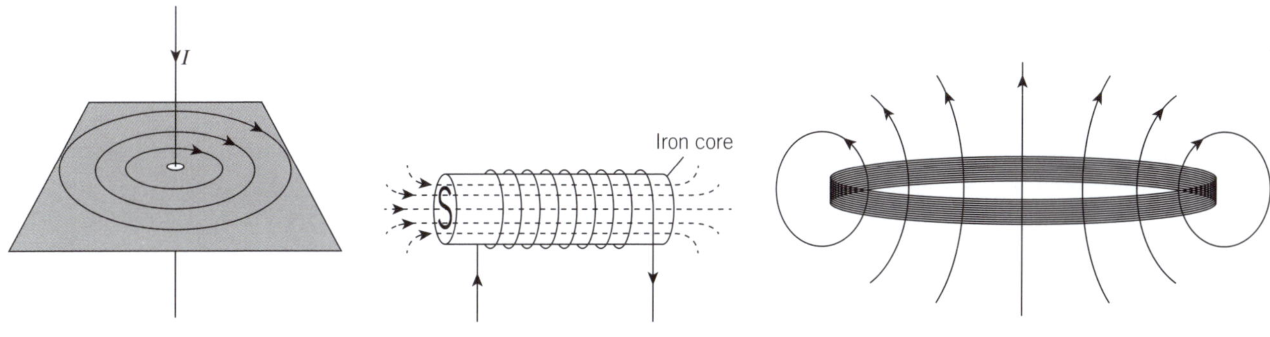
d Current carrying wire   **e** Solenoid with iron core   **f** Flat coil

▲ **Figure 22.1** Magnetic field patterns

> 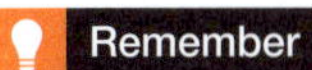
> **Remember**
>
> The **corkscrew rule** can be used to find the direction of the magnetic field lines due to a current in a long straight wire. To make the corkscrew (current) move down, turn the corkscrew (field lines) clockwise (see Figure 22.2)

Key points to remember about magnetic fields:

- the closer together the field lines, the stronger the field

- the direction of a field line is the direction a small plotting compass would point (its north pole would point to the south pole of another magnet, so field lines always point from N to S)

- inserting a ferrous metal core into a coil or solenoid (e.g., iron) can increase the strength of the field many thousands of times.

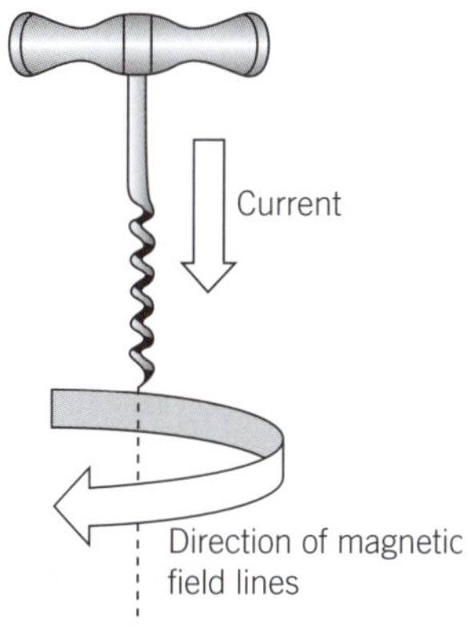

▲ **Figure 22.2** Corkscrew rule

Force on a current-carrying conductor in a magnetic field

A wire of length l carrying a current I at right angles to a magnetic field B experiences a force F at right angles to both the direction of the current and the magnetic field, as shown in Figure 22.3.

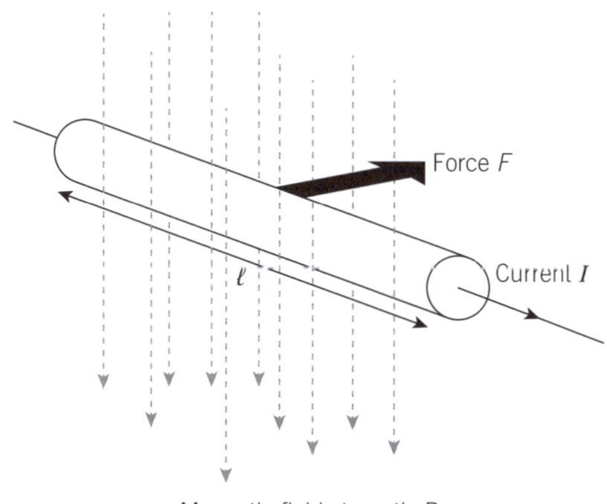

▲ **Figure 22.3** Force on a current-carrying wire

The force F is given by the equation: $F = BIl$

Re-arranging this equation gives: $B = \dfrac{F}{Il}$

> **Remember**
>
> $$F = BIl$$
>
> The SI unit of magnetic field strength is the **tesla** (T). A magnetic field strength of 1 T is the strength of the magnetic field normal to a long straight wire carrying a current of 1 A that would produce a force of 1 N m^{-1} on the wire.
>
> $$1\,T = 1\,N\,A^{-1}\,m^{-1}$$
>
> Magnetic field strength is also called **magnetic flux density** (see Unit 23 *Electromagnetic induction* to see why), and so can also have units of Wb m^{-2}.

This equation defines the **magnetic field strength** B (also called the **magnetic flux density**) as **the force per unit current-length on a wire placed at right angles to the magnetic field**. The SI unit of magnetic field strength is the **tesla** (T). A magnetic field strength of $1\,T$ is the strength of the magnetic field normal to a long straight wire carrying a current of $1\,A$ that would produce a force of $1\,N\,m^{-1}$ on the wire.

The direction of the force can be found using **Fleming's left-hand rule** (see Figure 22.4). The thumb, first finger and second finger are first held at right angles to each other. The **F**irst finger is then aligned with the direction of the magnetic **F**ield and the se**C**ond finger made to point in the direction of the **C**urrent. The **Th**umb then automatically shows the force (**Th**rust) on the wire.

The equation $F = BIl$ only applies if the current and the magnetic field are at right angles to each other. If the magnetic field and the current are at an angle θ to each other, as shown in Figure 22.5, the magnetic field B can be resolved into two perpendicular components $B\cos\theta$ and $B\sin\theta$.

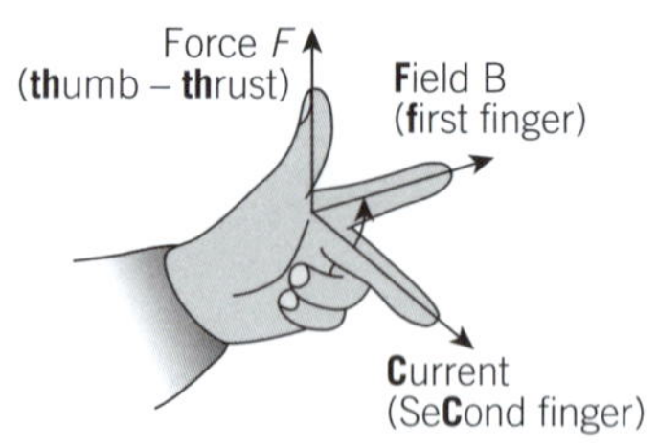

▲ **Figure 22.4** Fleming's left-hand rule

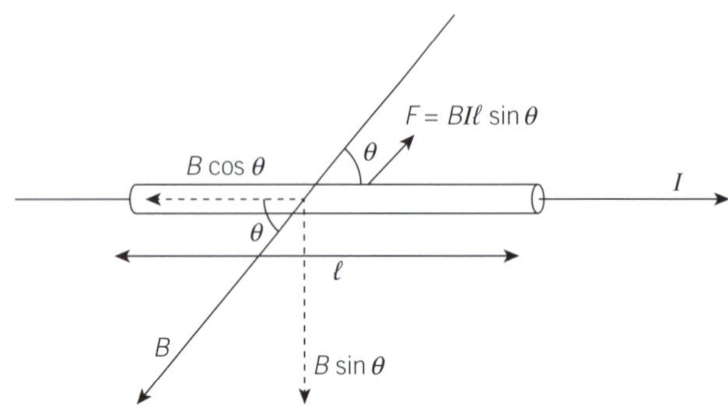

▲ **Figure 22.5** Force on a wire at an angle θ to the B-field

The component of the magnetic field parallel to the current has no effect. Only the component of the magnetic field perpendicular to the wire exerts a force. In this case:

$$F = BIl\sin\theta$$

Worked examples

1. In the UK, the magnetic field strength of the Earth is $1.7 \times 10^{-4}\,T$ in a direction making $25°$ to the vertical (Figure 22.6).

   Calculate the current needed in a copper cable of diameter $1\,mm$ for the cable to be self-supporting in the Earth's magnetic field.

   [The density of copper is $8900\,kg\,m^{-3}$.].

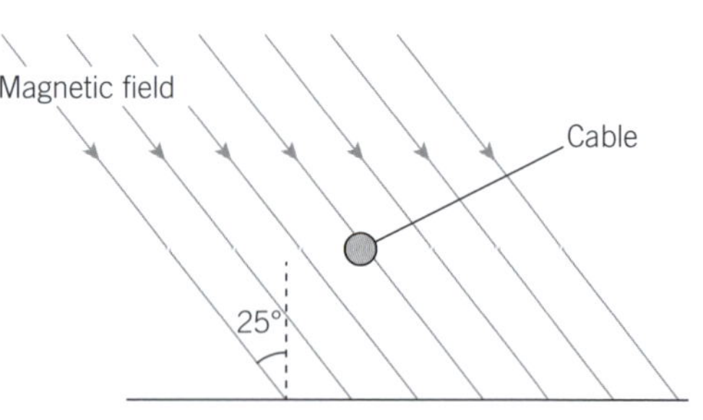

▲ **Figure 22.6**

Answer

Consider a $1.0\,m$ length of the copper cable. Its weight would be:

$$mg = \pi \times (0.5 \times 10^{-3})^2 \times 1.0 \times 8900 \times 9.81 = 0.069\,N$$

For the cable to be self-supporting:

$$BIl\sin\theta = 0.069$$

$$I = \frac{0.069}{1.7 \times 10^{-4} \times 1.0 \times \sin 25} = 960\,A\,!$$

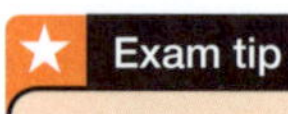

> ★ **Exam tip**
>
> Notice it is the **horizontal** component of the magnetic field which produces a vertical force on the cable.

2 Two thin aluminium strips are suspended vertically and held close to each other. They are connected in series to a d.c. supply as shown in Figure 22.7.

Describe and explain what happens when the switch is closed.

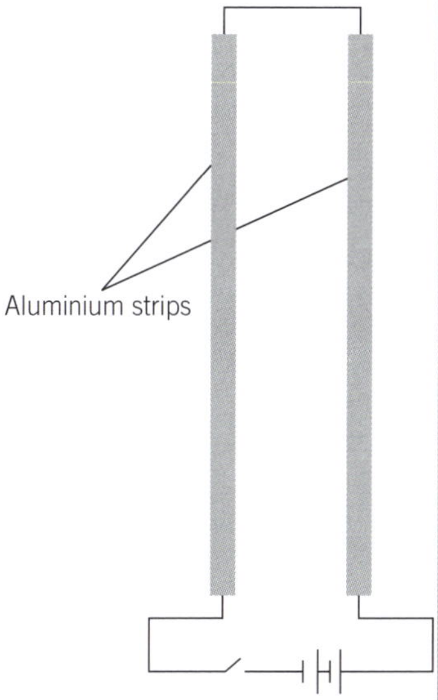

Answer

When viewed from above, the current is descending ($\otimes$) in the left-hand strip of aluminium and ascending ($\odot$) in the right-hand strip. The magnetic field around the left-hand strip is circular, and (using the corkscrew rule) the field lines rotate clockwise (Figure 22.8).

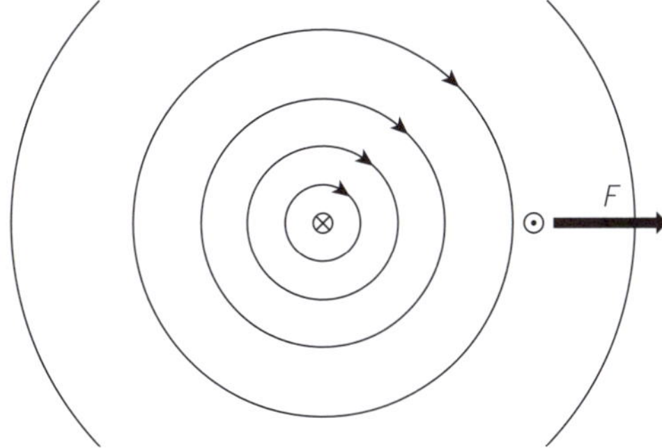

▲ **Figure 22.8**

Using the left-hand rule, the force on the right-hand strip, due to the magnetic field of the left-hand strip, is outwards, towards the right. Applying the same analysis to the left-hand strip shows it will be pushed out to the left – the two strips appear to repel each other. If the currents in the two strips are in the same direction, the strips will appear to attract each other.

▲ **Figure 22.7**

Measuring magnetic field strength *B* using a current balance

Figure 22.9 shows a simple **current balance**.

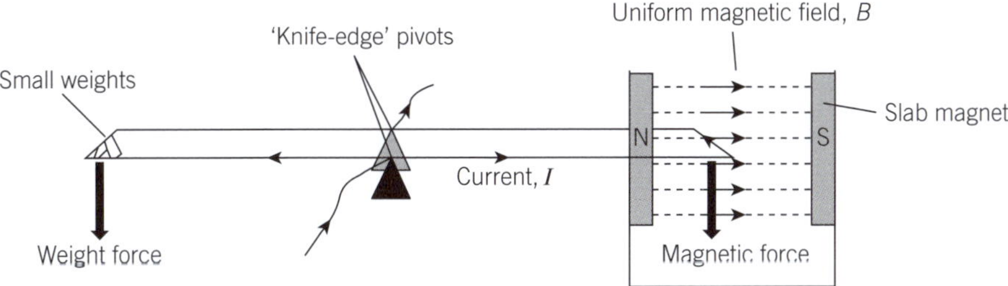

▲ **Figure 22.9** Current balance

A pair of slab magnets provide a uniform magnetic field. When a current is passed through the rectangular wire frame (via the knife edges) the balance is deflected down. The deflecting force can be found by balancing the wire frame using small weights (e.g., small pieces of graph paper) and the strength of the magnetic field calculated using $B = \dfrac{F}{Il}$ where l is the length of the conductor in the magnetic field and I the current in the wire.

Force on a charged particle moving in a magnetic field

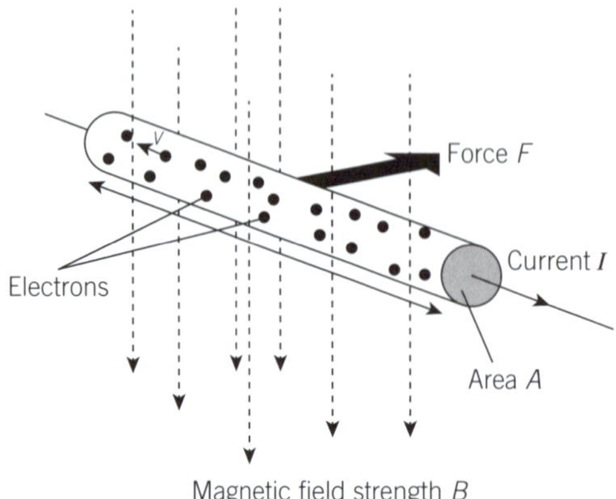

▲ **Figure 22.10** Force on a charged particle in a magnetic field ($F = Bqv$)

The magnetic force on a wire of length l carrying a current I perpendicular to a magnetic field B (Figure 22.10) is given by $F = BIl$. The current I is:

$$I = nAqv$$

where n is the density of charge carriers, A the cross-sectional area of the wire, q the charge on an individual charge carrier and v the drift velocity.

Combining these two equations:

$$F = B(nAqv)l = Bq(nAl)v$$

nAl is the total number of charge carriers, N, in the conductor of length l. The force on N charge carriers is:

$$F = BqNv$$

So the force on a single charge carrier is:

$$F = Bqv$$

If the angle between the field and the velocity is θ the force F is given by:

$$F = Bqv \sin \theta$$

> **Remember**
>
> Current flow and electron flow.
>
> In Figure 22.10 the direction of the conventional current is from left to right. If the charge carriers are electrons (e.g., in metal wires), they will be flowing from right to left.

> **Link**
>
> See Unit 19 *Current of electricity* for more information about $I = nAqv$

> **Remember**
>
> If a particle of charge q is moving at a speed v in a direction at an angle θ to a magnetic field B, the force on the particle is:
>
> $$F = Bqv \sin \theta$$
>
> You must be able to recall and use this equation.

Worked example

A proton, travelling horizontally at a speed of $3.0 \times 10^7\,\text{m s}^{-1}$, enters a uniform magnetic field of strength $300\,\text{mT}$, at an angle of $40°$ to the field, as shown in Figure 22.11.

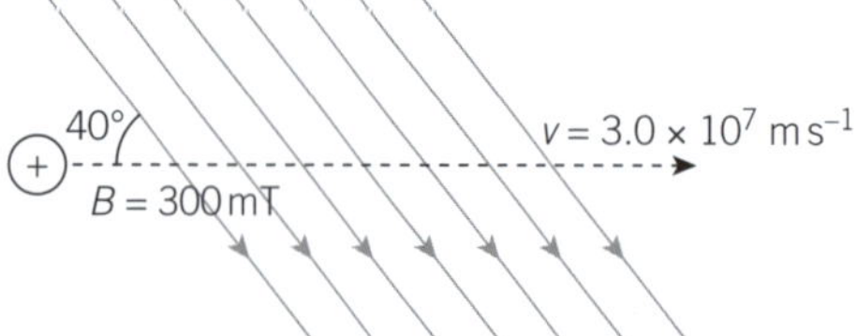

▲ **Figure 22.11** Force on a charged particle moving in a magnetic field

a) Calculate the force on the proton.

b) State the direction of the force.

Answer

a) $F = Bqv = (300 \times 10^{-3} \times \sin 40°) \times 1.6 \times 10^{-19} \times 3.0 \times 10^7 = 9.3 \times 10^{-13}\,\text{N}$

b) Using Fleming's left-hand rule, the first finger (the magnetic field) points downwards, the second finger (the current) goes from left to right, leaving the thumb (the thrust or motion) pointing towards the paper.

> Notice that only the vertical component of the magnetic field ($B \sin \theta$) is used in the calculation.

Deflecting charged particles in a magnetic field

A beam of electrons passing through a magnetic field, with a component of the field in a direction perpendicular to the path of the electrons, will experience a force at right angles to the direction in which they are travelling.

In Figure 22.12 the symbol $\times$ indicates a magnetic field into the paper (we are seeing the back of an arrow). The electron beam is moving left to right which means the conventional current flow is right to left. Using the left-hand rule, the force on the electron beam is downwards. As the beam deflects downwards the force alters direction so that it is always at right angles to the path of the electrons. The speed of the electrons does not change, only the direction.

For a beam of electrons, with each electron moving at speed v, and a uniform magnetic field B, the force on each electron will be Bev in a direction at right angles to the path of the beam. While the beam is in the magnetic field the electrons will follow a circular path.

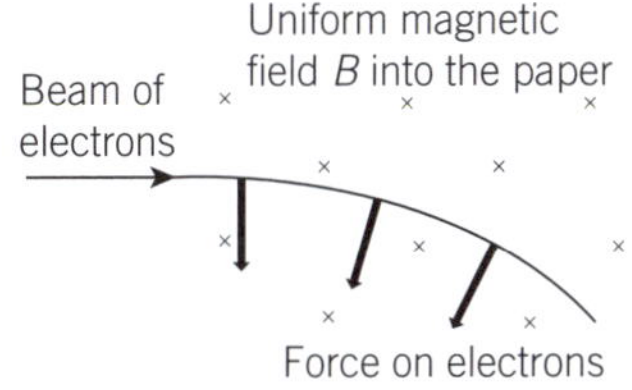

▲ **Figure 22.12** Force on a beam of electrons in a magnetic field

---

Worked example

Electrons are accelerated from rest across a p.d. of 5 kV in an electron 'gun' (Figure 22.13a). They then enter a region with a uniform magnetic field of magnetic flux density 4.0 mT and travel in a circular path. The direction of the magnetic field is perpendicular to the path of the electrons, as shown in Figure 22.13b.

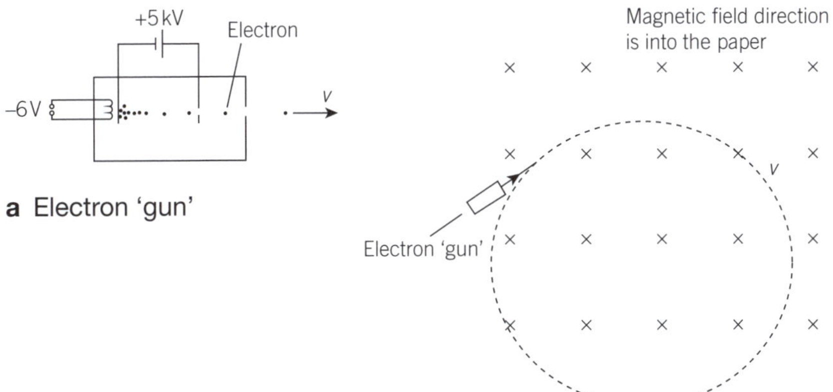

Electrons are produced in an electron 'gun' by heating a metal plate to release electrons – a process called **thermionic emission**. They then accelerate towards the positive anode.

▲ **Figure 22.13** Electrons moving in circles

a) Calculate the speed of the electrons leaving the electron 'gun'. [Charge on an electron is 1.6×10^{-19} C; mass of an electron is 9.11×10^{-31} kg.]

b) Explain why the electrons travel in circles. **c)** Calculate the radius of the circle.

Answer

a) $\frac{1}{2}mv^2 = eV$

$$v = \sqrt{\frac{2eV}{m}} = \sqrt{\frac{2 \times 1.6 \times 10^{-19} \times 5 \times 10^{3}}{9.11 \times 10^{-31}}} = 4.2 \times 10^{7}\,\text{ms}^{-1}$$

b) The force on an electron is always at right angles to its direction of motion.

c) Using $F = ma$, where a is the **centripetal acceleration** $\dfrac{v^2}{r}$:

$$Bev = \frac{mv^2}{r}$$

$$r = \frac{mv}{Be} = \frac{9.11 \times 10^{-31} \times 4.2 \times 10^{7}}{4.0 \times 10^{-3} \times 1.6 \times 10^{-19}} = 0.060\,\text{m}\,(6.0\,\text{cm})$$

See Unit 7 *Motion in a circle* for more details about centripetal acceleration.

Measuring e/m_e, the charge-to-mass ratio of an electron

The charge-to-mass ratio for electrons e/m_e can be found by making electrons move in circles in a fine beam tube filled with a gas such as hydrogen at low pressure.

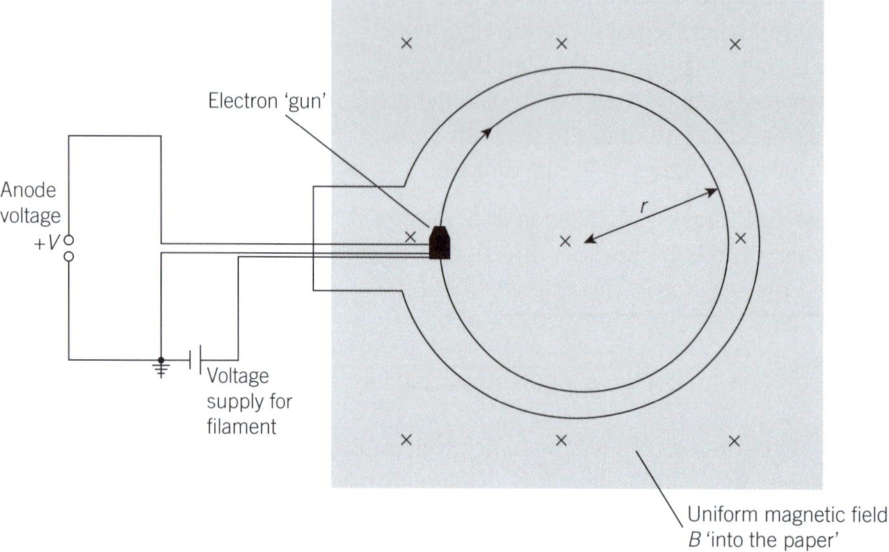

▲ **Figure 22.14** Measuring e/m_e

Electrons are accelerated by the anode voltage V and emerge from the electron 'gun' with a velocity v, as shown in Figure 22.14. Using the principle of the conservation of energy:

$$\frac{1}{2}m_e v^2 = eV \qquad \text{(eqn 1)}$$

where m_e is the mass of an electron. The electrons then experience a force Bev at right angles to the direction of travel, causing them to move in a circle of radius r. Using $F = ma$:

$$Bev = m_e \frac{v^2}{r} \qquad \text{(eqn 2)}$$

Combining eqn 1 and eqn 2 to eliminate v:

$$\frac{e}{m_e} = \frac{2V}{B^2 r^2}$$

The path of the electrons can be seen because they collide with the gas atoms in the tube, causing them to emit light.

Worked example

Find the ratio of e/m_e from the following results:

$V = 200\,\text{V}$, $B = 1.2\,\text{mT}$, $r = 4.0\,\text{cm}$.

Answer

$$\frac{e}{m_e} = \frac{2 \times 200}{(1.2 \times 10^{-3})^2 \times (4.0 \times 10^{-2})^2} = 1.74 \times 10^{11}\,\text{C kg}^{-1}$$

The true value is $1.76 \times 10^{11}\,\text{C kg}^{-1}$.

Separating particles with different velocities

A **uniform magnetic field** set at right angles to a **uniform electric field** can be used as a way of separating charged particles with different energies and velocities.

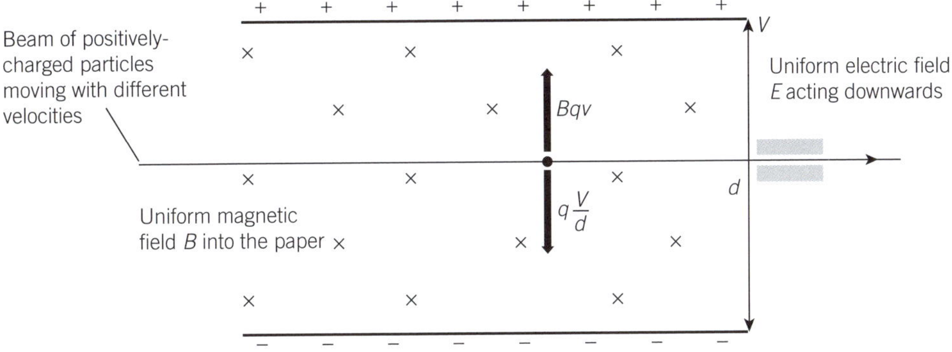

▲ **Figure 22.15** Separating charged particles moving with different velocities

Positively charged particles, moving with different velocities, enter a region with a uniform E-field and a uniform E-field, as shown in Figure 22.15. The magnetic field exerts a force Bqv upwards and the electric field exerts a force qV/d downwards. These forces will cancel out and the particle will travel in a straight line, undeflected by either field, if:

$$Bqv = q\frac{V}{d}$$

$$v = \frac{V}{Bd}$$

Particles travelling faster than this will be deflected upwards; particles slower than this will be deflected downwards. The experiment illustrates how particles moving with different speeds can be selected.

Hall effect

A thin slice of a semiconductor such as germanium is placed at right angles to a magnetic field of flux density B as shown in Figure 22.16.

The charge carriers in the thin slice are positively charged. If a current I passes through the slice the charge carriers will feel a force pushing them towards the back of the slice.

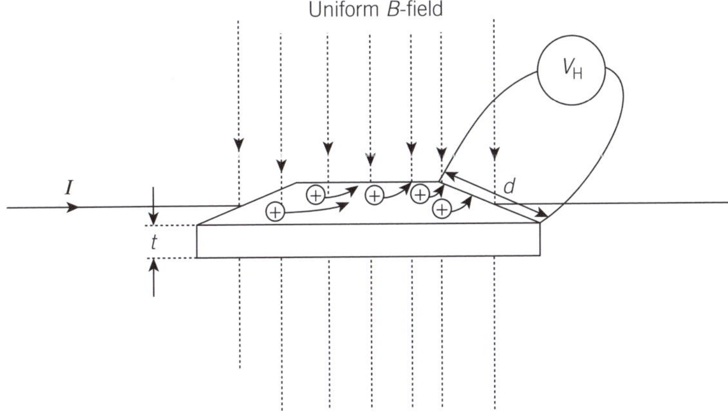

▲ **Figure 22.16** Hall effect

A potential difference V_H (the **Hall voltage**) builds up between the front and the back of the slice, creating an electric field similar to the electric field between two oppositely-charged parallel plates. This is known as the **Hall effect**.

The electric field tries to push the positively-charged particles back towards the front of the slice while the magnetic field continues to push the positively charged particles towards the back of the slice. A balance is achieved when the magnetic force on each particle (Bqv) is equal to the electric force (Eq), as shown in Figure 22.17.

$$Eq = Bvq$$

But $E = \dfrac{V_H}{d}$, so:
$$V_H = Bvd$$

For a current-carrying conductor, $I = nAqv$. Combining these equations to eliminate v:

$$V_H = \frac{BId}{nAq} = \frac{BI}{ntq}$$

as the cross-sectional area of the slice, $A = dt$, where t is the thickness of the slice and d the depth.

This equation shows why the Hall effect is most easily observed using:

- a **thin** slice of material (so that the value of t is small)

- a semiconductor material such as germanium or silicon, as the value of n, the density of charge carriers, is much smaller for a semiconductor than a conductor like copper.

Using a Hall probe

A **Hall probe** uses the Hall effect to measure the strength of magnetic fields, the Hall voltage being directly proportional to the magnetic field strength B. Figure 22.18 shows a Hall probe being used to investigate how the magnetic field strength inside a solenoid varies with distance from the centre of the solenoid.

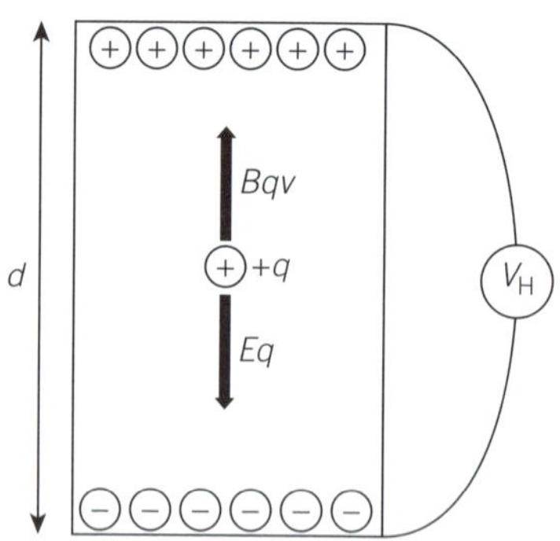

▲ **Figure 22.17** Hall voltage

> ★ **Exam tip**
>
> You may be asked to derive this equation.

> The Hall effect can be used to measure the strength of magnetic fields and as magnetic field sensors; for example, some computer printers use Hall effect sensors to detect whether there is missing paper or the cover is open.

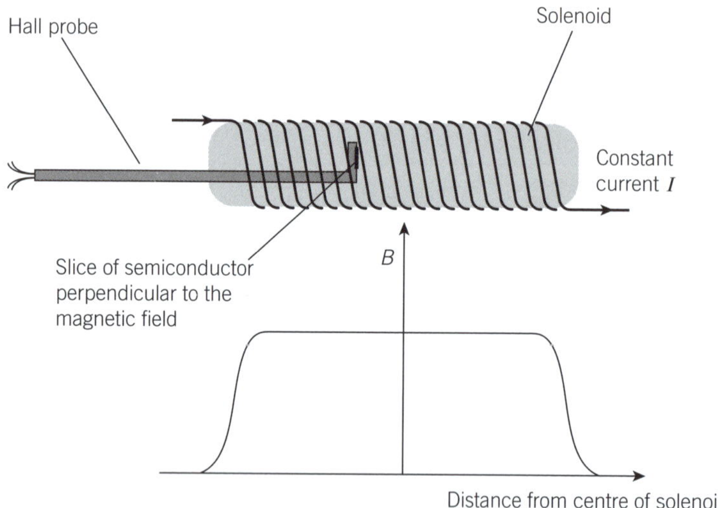

▲ **Figure 22.18** Using a Hall probe

> 💡 **Remember**
>
> Key points to remember when using a Hall probe:
>
> - The semiconductor slice must be perpendicular to the magnetic field being investigated.
>
> - The probe can only measure magnetic fields which do not vary with time.

In order to measure a magnetic flux density the Hall probe must first be calibrated using a known magnetic field. The probe is placed with the semiconductor slice perpendicular to a uniform magnetic field of known magnitude. A current I is passed through the slice and the Hall voltage V_H recorded. For a constant current I:

$$V_H = kB$$

The constant k can be found from the measurements taken.

Comparing the forces in different types of field

Figure 22.19 compares the forces on mass, charge, and current in gravitational, electrical, and magnetic fields:

- All three forces are examples of 'action at a distance'.

- All three fields can be represented by field lines which show the direction of the force at points along the line.

- The density of the field lines indicates the strength of the field

- The field strength is defined as the *force per unit … mass* (for gravitational fields), **charge** (electric fields), or **current-length** (for magnetic fields).

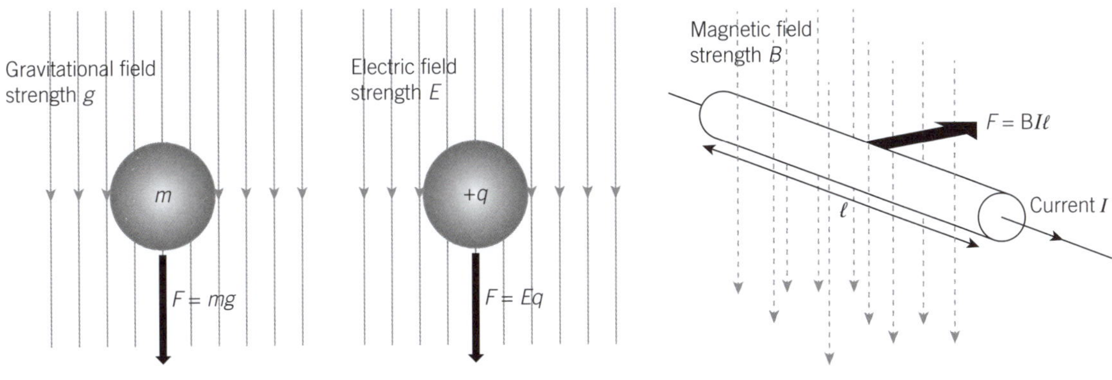

a Gravitational field   **b** Electric field   **c** Magnetic field

▲ **Figure 22.19** Forces in gravitational fields, electric fields, and magnetic fields

---

Worked example

a) Calculate:

  i) the electric force

  ii) the gravitational force

between an electron and a proton in a hydrogen atom (Figure 22.20). Assume the radius of a hydrogen atom is 0.5×10^{-10} m.

b) Calculate the ratio of these two forces.

c) State what this ratio would be if the proton and electron were:

  i) 1 m apart   **ii)** 1 light-year apart.

[Charge on an electron $= 1.6 \times 10^{-19}$ C; mass of an electron $= 9.1 \times 10^{-31}$ kg; mass of a proton $= 1.7 \times 10^{-27}$ kg.]

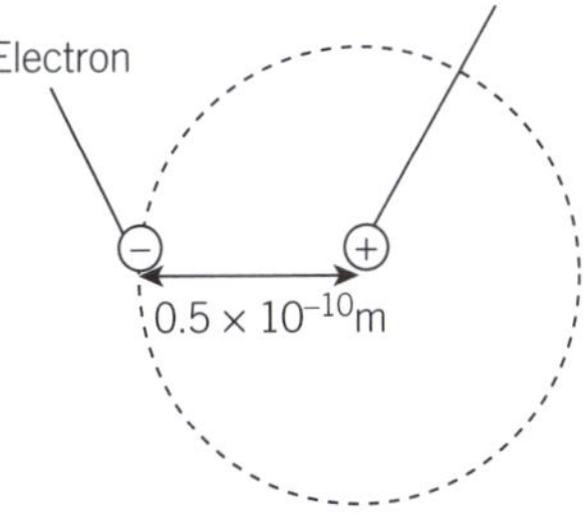

▲ **Figure 22.20** Hydrogen atom

Answer

a) **i)** Using Coulomb's law: $F_e = k\dfrac{q_1 q_2}{r^2} = 9 \times 10^9 \times \dfrac{\left(1.6 \times 10^{-19}\right)^2}{\left(0.5 \times 10^{-10}\right)^2} = 9.2 \times 10^{-8}$ N

    ii) Using Newton's law: $F_g = G\dfrac{m_1 m_2}{r^2} = 6.67 \times 10^{-11} \times \dfrac{9.1 \times 10^{-31} \times 1.7 \times 10^{-27}}{(0.5 \times 10^{-10})^2} = 4.1 \times 10^{-47}$ N

b) $\dfrac{F_e}{F_g} = \dfrac{9.2 \times 10^{-8}}{4.1 \times 10^{-47}} = 2.2 \times 10^{39}$

c) **i)** and **ii)** In both cases the ratio would be the same. Both Coulomb's law and Newton's law are inverse square laws; the ratio of the two is independent of the distance between the two masses/charges.

Nuclear magnetic resonance imaging

Nuclear magnetic resonance imaging (NMRI, or sometimes simply MRI) is a method of obtaining detailed medical images of the internal organs and soft tissue of a patient using radio-frequency electromagnetic waves (RF). The human body is mostly water, and so contains many hydrogen atoms. NMRI relies on a property of the nuclei of hydrogen atoms called **spin**. The nucleus of a hydrogen atom is a proton, and because it spins, it behaves like a small magnet, with north and south poles.

When the protons are subjected to a strong magnetic field, most of the protons align themselves along the magnetic field lines (in the same way that plotting compasses line up along magnetic field lines) and are in a low energy state, though a few line up the other way, which is an (unstable) higher energy state (see Figure 22.21).

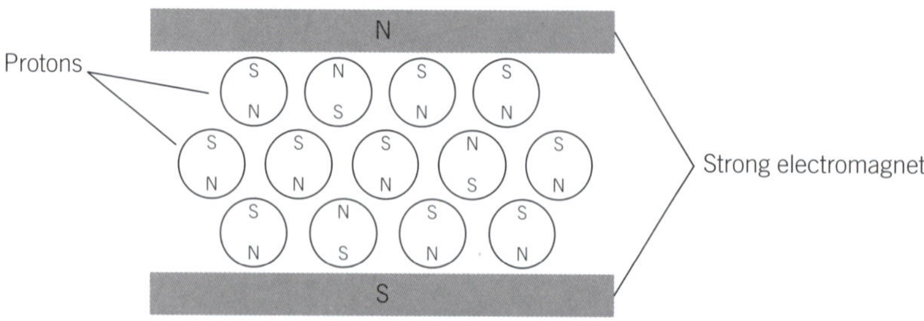

▲ **Figure 22.21** Hydrogen nuclei in a strong magnetic field

The protons do not line up exactly with the magnetic field lines. Instead, each proton spins about its own axis, and its axis turns around the direction of the magnetic field (**precesses**), as shown in Figure 22.22. The precession frequency depends on the strength of the magnetic field. In an NMRI scanner the magnetic field strength of the superconducting magnet is approximately 1.5 T. Superimposed on this field is a gradually increasing magnetic field (called the gradient field) produced by two perpendicular coils which 'ripples' through the patient.

The protons are exposed to a burst of radio waves. Where the precession frequency (determined by the strength of the magnetic field) matches the frequency of the radio waves, the protons resonate; that is they absorb an RF photon, which flips them into a higher energy state – hence the term 'nuclear magnetic resonance'. The gradient field ensures that only a small region of the patient has exactly the correct magnetic field value for resonance to occur.

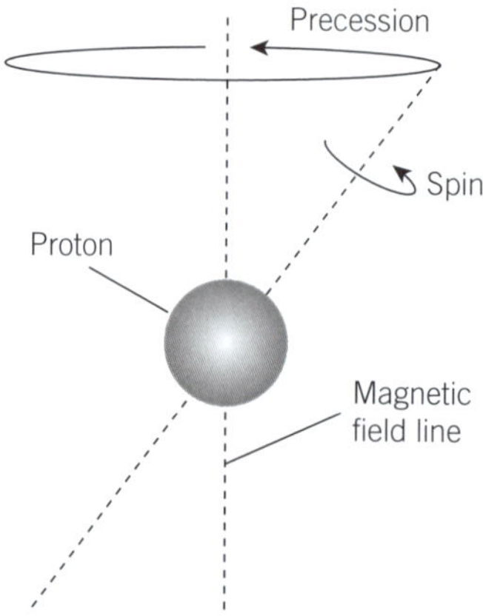

▲ **Figure 22.22** Precession

The protons that have been excited into a higher energy state gradually 'relax', returning to the lower energy state by emitting a radio frequency photon. The time taken to do this depends on the type of molecules around the proton:

- watery tissues have relaxation times of several seconds

- fatty tissues have relaxation times which are less than a second (e.g., grey matter in the brain has a relaxation time of 0.37 s compared to 0.30 s for white matter)

- cancerous tissues have relaxation times in between these two extremes.

A radio-frequency receiving coil (the same coil that produces the pulses of radio waves) detects the emitted radio waves and hence the relaxation times. All this information is passed to a computer which builds an image of the 'slice' of the patient. The patient moves slowly through the coils, enabling the computer to produce a complete body or brain scan of the patient (see Figure 22.23).

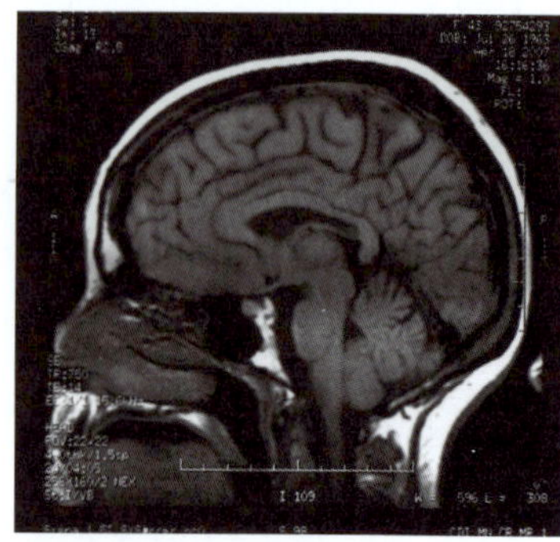

▲ **Figure 22.23** NMRI scan of the brain

⬆ Raise your grade

(a) Define the *tesla*.

The tesla is the SI unit of magnetic flux density (magnetic field strength). ✔

A field of strength 1 tesla exerts a force of 1N on a wire carrying a current of 1A ✗ ✗

> A conductor carrying a current of 1A **in a direction normal to the magnetic field** experiences a force of **1N per metre length** of the wire. [3]

An experiment is carried out to measure the strength of a uniform magnetic field between two magnets by measuring the force on a current-carrying wire. The magnets are repelling each other. The wire is supported by two clamps and connected to a d.c. power supply as shown.

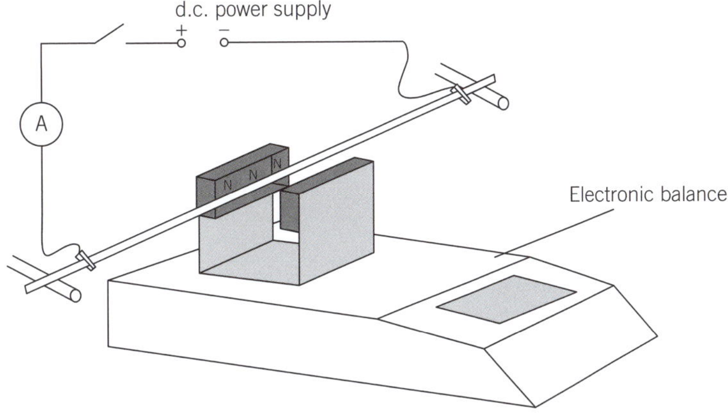

(b) (i) When the switch is closed a force acts on the wire. State the direction of this force.

downwards ✔   (Using Fleming's left-hand rule.)

(ii) The reading on the electronic balance decreases. Explain why.

From Newton's 3rd law, if the magnets exert a downward force on the wire, the wire ✔✔ exerts an equal and opposite (upwards) force on the magnets (which tries to 'lift' them). [3]

(c) The results of the experiment are shown in the table.

Length of wire in magnetic field	8.0 cm
Electric current	2.7 A
Initial reading on balance	95.4 g
Reading on balance when switch is closed	89.7 g

(i) Show that the force acting on the wire due to the magnetic field is approximately 5.6×10^{-2} N.

Change in force on balance = $(95.4 - 89.7) \times 10^{-3} \times 9.81 = 5.59 \times 10^{-2}$ N

✔   ✔

(ii) Calculate the strength of the magnetic field between the two magnets.

$$F = Bil \Rightarrow B = \frac{F}{il} = \frac{5.6 \times 10^{-2}}{2.7 \times 8.0} = 2.6 \times 10^{-3} \text{ T}$$

✔ ✗

> The correct equation, re-arranged correctly to find B, but the value of ℓ should be converted to m. Correct answer is 0.26 T.

strength of magnetic field = 2.6×10^{-3} T [4]

[Charge on an electron is 1.6×10^{-19} C; mass of electron is 9.11×10^{-31} kg.]

1 A rectangular wire loop ABCD, with dimensions 10.0 cm × 8.0 cm, lies with its plane parallel to a uniform magnetic field of strength 0.50 T. The loop carries a current of 2.4 A, as shown.

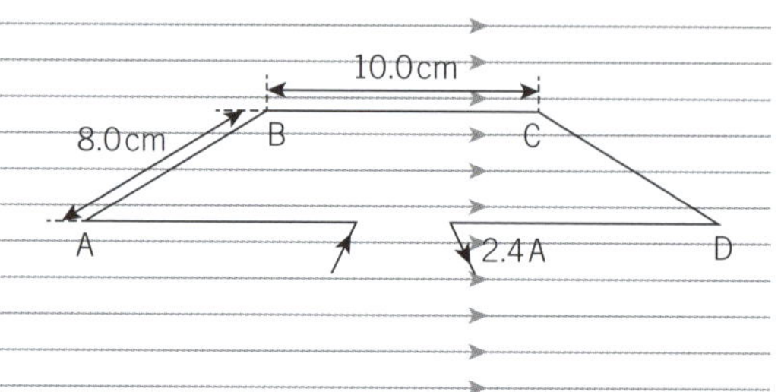

(a) State the direction of the force on:
  (i) AB   **(ii)** CD.   [2]

(b) Calculate the magnitude of the force on:
  (i) AB   **(ii)** BC.   [2]

(c) Calculate the total torque acting on the loop in this position.   [2]

(d) Describe and explain what would happen if the loop was in a vertical plane.   [2]

2 A β-particle, travelling at a speed v, enters a region with a uniform magnetic field of flux density 0.25 mT and a uniform electric field. The β-particle is undeflected, travelling in a straight, horizontal line, as shown.

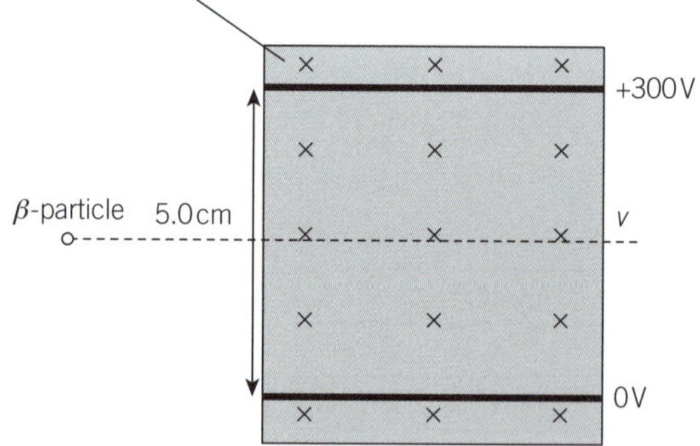

(a) Using information in the diagram, state the direction of:
  (i) the electric field
  (ii) the magnetic field.   [2]

(b) Show that the electric field strength is 6 kV/m.   [2]

(c) Calculate the velocity of the β-particle.   [3]

(d) Describe and explain what would happen to the path of the β-particle if it was moving faster.   [2]

3 A charged particle of mass m and charge $-e$ is travelling in a vacuum with velocity v. It enters a uniform field of flux density B as shown.

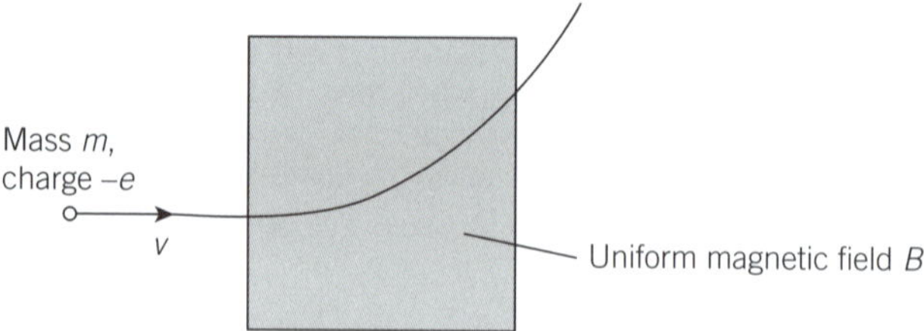

(a) State the direction of the magnetic field.   [1]

(b) (i) Calculate the magnetic force on the charged particle as it enters the uniform magnetic field.
  (ii) State the direction of this force.   [3]

(c) (i) Explain why the speed of the charged particle does not change.
  (ii) The charged particle travels along the arc of a circle. Show that the radius r of the circle is $\dfrac{mv}{Be}$.
  (iii) Describe how the path of the charged particle would change if its velocity decreased.   [4]

4 The diagram shows the track of a negatively charged particle, of mass m and charge $-q$, in a *cloud chamber*. A cloud chamber is a device containing air saturated with a vapour at low temperatures. As the particle travels through the chamber it creates ions on which droplets can form.

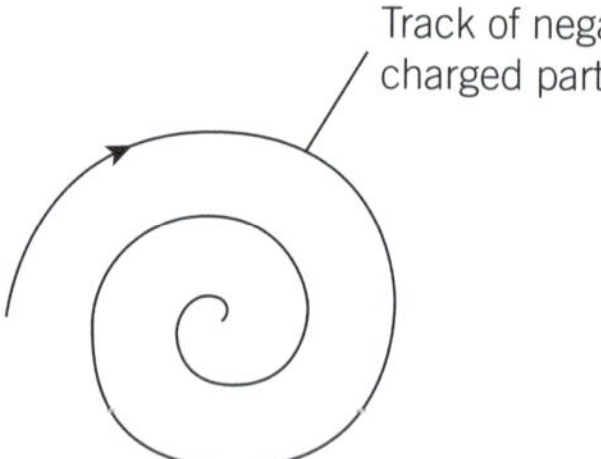

The track is curved because there is a uniform magnetic field of flux density B perpendicular to the plane of rotation of the charged particle. The particle travels clockwise and the radius decreases.

(a) State whether the direction of the magnetic field is 'down' (into the paper) or 'up' (out of the paper).   [1]

(b) Suggest a reason why the track spirals inwards (i.e., why the radius decreases).   [1]

(c) Calculate the momentum of the particle when it is moving with speed v and the radius of the track is r.   [2]

23 Electromagnetic induction

Key points

- [ ] Define magnetic flux and the weber.
- [ ] Recall and use $\phi = BA$.
- [ ] Define magnetic flux linkage.
- [ ] Infer from appropriate experiments on electromagnetic induction:
    - that a changing magnetic flux can induce an e.m.f. in a circuit
    - that the direction of the induced e.m.f. opposes the change producing it
    - the factors affecting the magnitude of the induced e.m.f.
- [ ] Recall and solve problems using Faraday's law of electromagnetic induction and Lenz's law.
- [ ] Explain simple applications of electromagnetic induction.

Electromagnetic induction

When a wire 'cuts through' a magnetic field, an e.m.f. is induced across the ends of the wire. This is an example of **electromagnetic induction**.

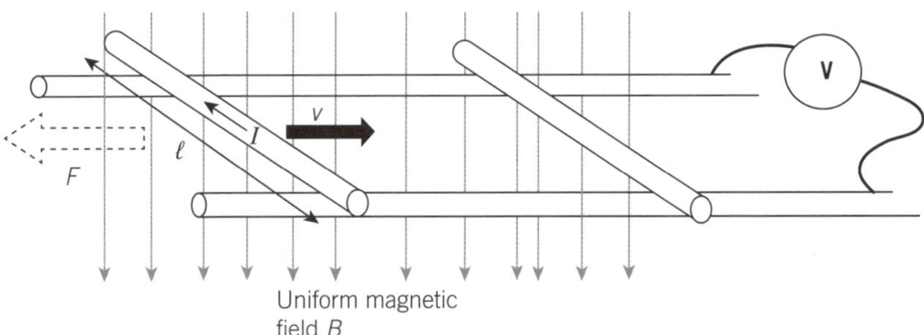

▲ **Figure 23.1** Electromagnetic induction

A wire of length l moving perpendicular to a magnetic field B with velocity v (See Figure 23.1.) will have an e.m.f. E induced between the ends of the wire. Experiments show that the e.m.f. induced can be increased by:

- moving the wire faster
- increasing the magnetic field strength B
- increasing the length of wire in the magnetic field.

If a second wire is connected between the two rails to complete a circuit, a current I will flow and a force will act on the wire which is perpendicular to both the direction of the current and the magnetic field. The force will act in the opposite direction to the direction the wire is moving (i.e., it opposes the motion which produces it).

External work must be done against this force for the wire to move. From $F = BIl$, and using the principle of the conservation of energy,

$$\text{electrical power generated} = \text{work done/second}$$

$$IE = (BIl) \times v$$

$$E = Blv$$

> **Remember**
>
> The force F always acts in a direction which opposes the motion producing the force.

> **Remember**
>
> The **magnetic field strength** B (T) is also called the **magnetic flux density** ϕ (Wb m^{-2})

> lv is the 'area swept out in one second'. Think of the wire as a brush – lv is the area of floor that would be swept in one second.

Worked examples

1. A car radio has a vertical aerial 0.5 m long. The car is travelling at a constant speed of $25\,\text{m s}^{-1}$, and the horizontal component of the Earth's magnetic field in the direction of the road is $4.0 \times 10^{-6}\,\text{T}$.

   a) Calculate the induced e.m.f. between the ends of the aerial.

   b) State and explain what the reading on a voltmeter connected to the ends of the aerial would be.

Answer

   a) Induced e.m.f. $E = Blv = 4.0 \times 10^{-6} \times 0.5 \times 25 = 5.0 \times 10^{-5}\,\text{V}$

   b) The reading on the voltmeter would be zero – the wires connecting the meter to the aerial would be cutting the same field. The number of magnetic field lines 'enclosed' by the electric circuit does not change (see Figure 23.2).

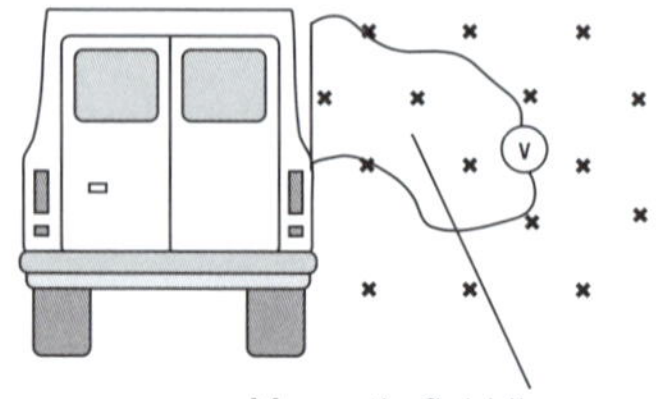

Magnetic field lines enclosed by area is constant so no induced e.m.f.

▲ **Figure 23.2**

2. A rectangular loop ABCD of metal wire, of dimensions $20.0\,\text{cm} \times 40.0\,\text{cm}$ and electrical resistance $50\,\Omega$, is held perpendicular to a uniform magnetic field of strength 150 mT, as shown in Figure 23.3. The loop is pulled out of the field at a steady speed of $0.10\,\text{m s}^{-1}$.

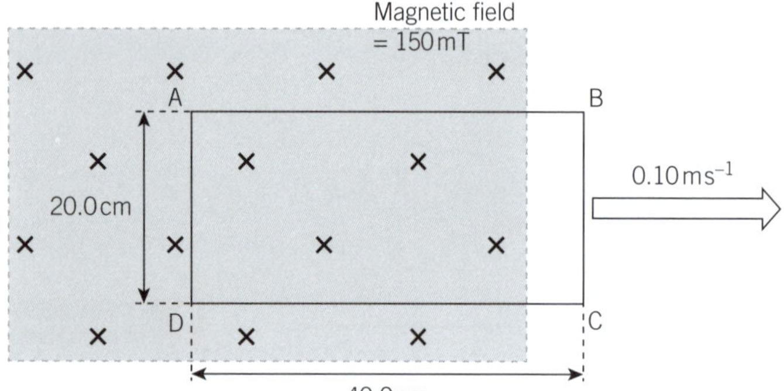

▲ **Figure 23.3** Induced e.m.f.s

> $\times$ This symbol indicates a magnetic field going into the paper (we see the back of an arrow).
>
> $\odot$ This symbol indicates a magnetic field coming up out of the paper (the front of an arrow).

Determine:

   a) the induced e.m.f.

   b) the current in the loop

   c) the direction of the current in the loop

   d) the force needed to pull the loop.

Answer

   a) Using $E = Blv$: $E = 150 \times 10^{-3} \times 20.0 \times 10^{-2} \times 0.10 = 3.0\,\text{mV}$

   b) $I = \dfrac{V}{R} = \dfrac{3.0 \times 10^{-3}}{50} = 6.0 \times 10^{-5}\,\text{A}\ (60\ \text{A})$

   c) The force on the wire must oppose the motion (from the conservation of energy). The force acts to the left on the vertical wire AD (as this is the only wire which is perpendicular to the magnetic field). Using the left-hand rule (the motor effect) the current must be from D to A (i.e., clockwise around the loop).

   d) Mechanical work done per second = electrical power generated

$$F \times v = V \times I$$

$$F \times 0.10 = 3.0 \times 10^{-3} \times 60 \times 10^{-6}$$

$$F = 1.8 \times 10^{-6}\,\text{N}$$

Faraday's law

The equation $E = Blv$ can be re-written:

$$E = B\frac{dA}{dt}$$

where dA/dt is the 'area swept out' each second. The equation is a special case of Faraday's law of electromagnetic induction:

$$E = -\frac{d\phi}{dt} \qquad \text{where } \phi = BA$$

Faraday's law states that the induced e.m.f. in a circuit is proportional to the rate of change of flux linkage through the circuit.

Magnetic flux and flux linkage

Magnetic flux is defined by the equation:

$$\phi = BA$$

where A is the area perpendicular to the magnetic field B (see Figure 23.4). It is measured in webers (Wb)

Re-arranging this equation:

$$B = \frac{\phi}{A}$$

which explains why the magnetic field strength is also called the **magnetic flux density**, and can be measured in $Wb\,m^{-2}$. Note: $1\,Wb\,m^{-2} = 1\,T$

Magnetic flux can be thought of as the total number of magnetic field lines passing perpendicularly through an area. If the magnetic field lines are at an angle other than $90°$, the component of the magnetic field perpendicular to the area is used.

The magnetic flux in this case is $\phi = BA\cos\theta$ (see Figure 23.5).

The greater the density of magnetic field lines, the greater the magnetic field strength. The greater the number of field lines enclosed by an area, the greater the magnetic flux.

If there were two wires moving in Figure 23.1, the induced e.m.f. would be twice as much; if there were N wires, the induced e.m.f. would be N times as much. The general form of Faraday's law is:

$$E = -\frac{\Delta(N\phi)}{\Delta t}$$

where $N\phi$ is the **magnetic flux linkage**, measured in webers (Wb). The symbol Δ means 'change in'.

The e.m.f. E induced is equal to the **rate** at which the flux linkage changes. The minus sign in the equation indicates that the induced voltage acts in such a way as to oppose the change producing the voltage (see Lenz's law, below).

For an e.m.f. to be induced, the magnetic flux must be **changing**. In question 2 of the previous worked example, an e.m.f. only occurs when the total number of field lines enclosed by ABCD is increasing or decreasing – if all the loop is inside the magnetic field (or all outside) no e.m.f. is produced. The faster the magnetic flux is changing, the greater the e.m.f. induced. Faraday's law explains how generators, motors, and transformers work.

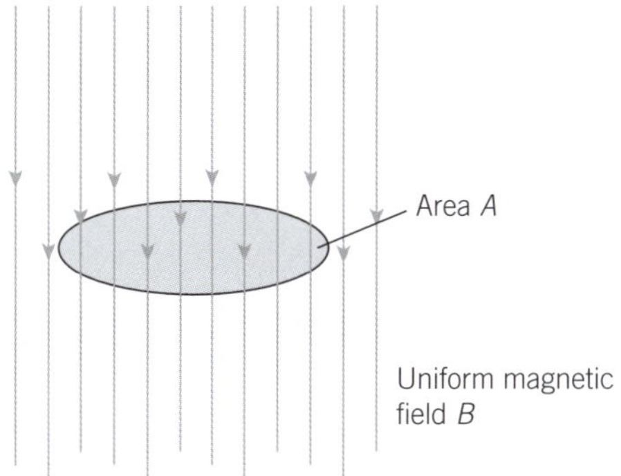

▲ **Figure 23.4** Magnetic flux $\phi = BA$

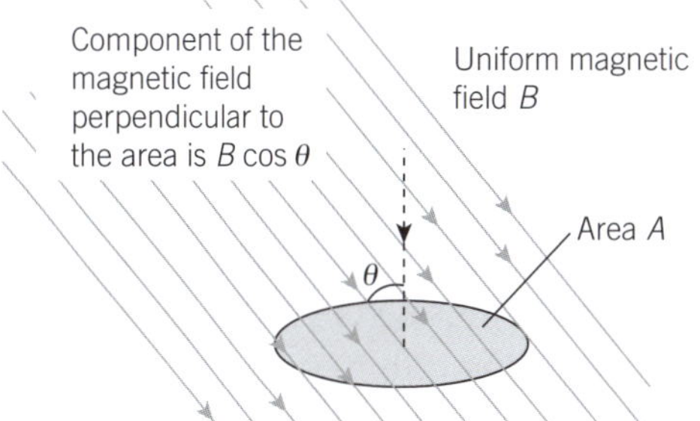

▲ **Figure 23.5** Magnetic flux $\phi = BA\cos\theta$

> **Remember**
>
> **magnetic flux** $\phi = BA$
>
> SI unit: Wb or $T\,m^2$
>
> **magnetic flux density**
>
> $B = \dfrac{\phi}{A}$
>
> SI unit: $Wb\,m^{-2}$ or T
>
> **magnetic flux linkage**
>
> $N\phi = NBA$
>
> SI unit: Wb

Worked example

A 20-turn circular coil of diameter 8.0 cm is held with its plane perpendicular to a uniform magnetic field of strength 1.4 T (see Figure 23.6). It is rotated 90° in 2.8 ms.

a) Calculate the flux linkage:

   i) before the coil is rotated

   ii) after the coil is rotated.

b) Determine the average e.m.f. between the ends of the coil while the coil is being rotated.

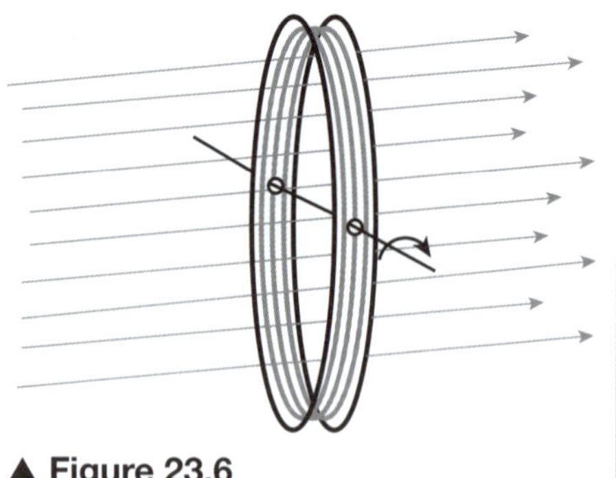

▲ Figure 23.6

Answer

a) **i)** flux linkage $= N\phi = 20 \times 1.4 \times \pi \times (4.0 \times 10^{-2})^2 = 0.14\,\text{Wb}$

   ii) flux linkage $= 0$

b) Average e.m.f. $= \dfrac{\Delta(N\phi)}{\Delta t} = \dfrac{0.14}{2.8 \times 10^{-3}} = 50\,\text{V}$

Lenz's law

Lenz's law states that the direction of the induced e.m.f. or current is such as to oppose the change that produces it.

It is a statement of the principle of conservation of energy applied to induced e.m.f.s.

In Figure 23.7a the north-seeking pole of a magnet is moving downwards and the flux inside the coil is increasing – more field lines are enclosed by ('link' with) the coil.

The induced voltage across the ends of the coil causes a current to flow anticlockwise when viewed from above. This makes the top of the coil a north pole, repelling the north pole of the magnet.

If the induced current flowed the other way, the top of the coil would become a south pole, attracting the magnet and causing it to move faster. The rate of change of flux linking the coil would increase, inducing a larger induced e.m.f. This in turn would create a larger current making the magnetic pole of the coil even stronger, causing the magnet to move even faster. The magnet would be gaining extra kinetic energy (more than it would gain by just falling without the coil of wire present); this would contradict the principle of conservation of energy.

In Figure 23.7b, as the north-seeking pole of the magnet moves up, the flux linking the coil decreases – fewer and fewer lines 'link' with the coil.

The e.m.f. induced causes a current to flow clockwise when viewed from above. This makes the top of the coil a south-seeking pole. This south pole tries to pull the magnet down; it tries to prevent the magnet moving up.

> **Remember**
>
> **Faraday's law:** the induced e.m.f. is proportional to the rate of change of flux linkage.
>
> **Lenz's law:** the direction of the induced e.m.f. or current is such as to oppose the change that produces it.

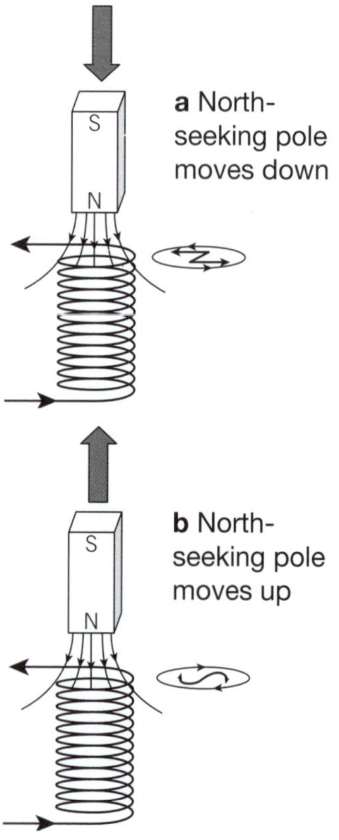

▲ **Figure 23.7** Lenz's law

'Flux cutting', 'flux linking' and induced e.m.f.

As the magnet falls into the coil, the field lines are being 'cut' by the coil (more and more field lines 'link' with the coil). The total magnetic flux inside the coil is **increasing** in order to produce an e.m.f. across the ends of the coil.

As the magnet is pulled up from the coil, the number of magnetic field lines linking with the coil is **decreasing**, and so an e.m.f. is again induced, but in the opposite direction.

If the magnet is moved more quickly into the coil, the rate of change of flux linking with the coil would be greater, and so the e.m.f. induced across the ends of the coil would be larger.

If a stronger magnet is used (more field lines per unit area), and the magnet moved into the coil at the same speed, the number of field lines linking with the coil would increase more rapidly, inducing a greater voltage.

Worked example

An aluminium disc of diameter 30.0 cm is connected to an electric motor and rotates at a constant speed of 120 r.p.m. in a vertical plane, as shown in Figure 23.8. A constant magnetic field of flux density 0.30 T acts horizontally and perpendicular to the plane of the disc. A voltmeter is connected to points O and P using sliding contacts.

a) Calculate the potential difference between points O and P.

b) Determine the p.d.:

   i) between O and Q

   ii) between P and Q.

   Justify your answers.

c) Describe and explain what happens to 'free' conduction electrons in the aluminium disc.

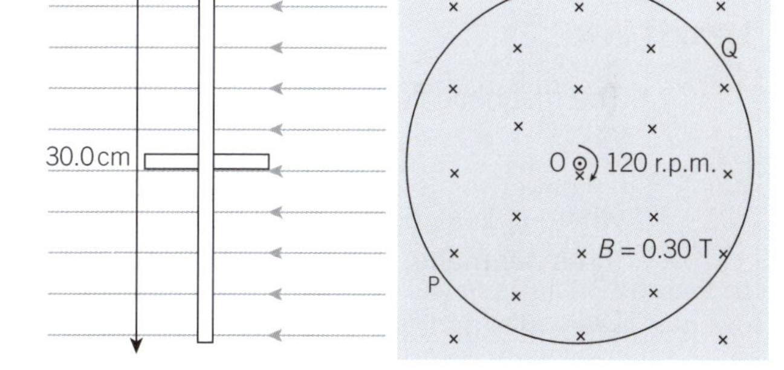

▲ **Figure 23.8** e.m.f. induced on a rotating aluminium disc

Answer

a) Consider OP as a thin metal strip. It completes two rotations each second (120 r.p.m.) so 'sweeps out' an area of 2 × area of the disc in 1 s.

e.m.f. between O and P = magnetic flux 'swept' per second

$$= 0.30 \times 2 \times (\pi \times 0.15^2) = 0.042 \text{ V}$$

b) i) OQ is 'cutting' or sweeping flux at the same rate as OP. The p.d. between O and Q will be the same as the p.d. between O and P.

   ii) P and Q will be at the same potential, so the p.d. between P and Q will be zero.

c) An electron moving downwards at one instant is equivalent to a conventional current moving upwards, as shown in Figure 23.9. Using Fleming's left-hand rule, the force on the electron is to the left (i.e., towards the centre). A similar argument applies to electrons in other positions on the disc – they all move towards the centre.

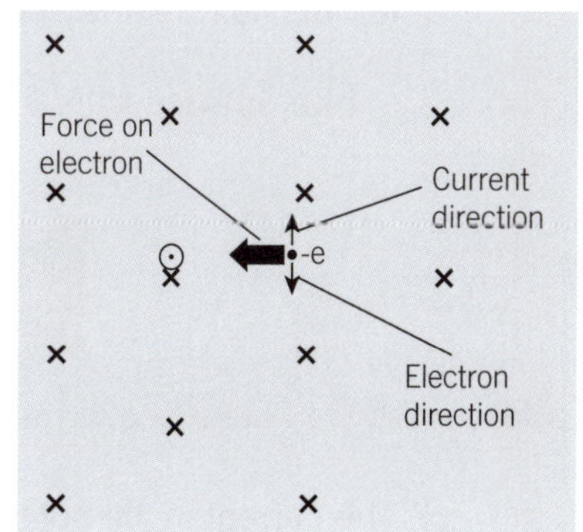

▲ **Figure 23.9**

(a) State Faraday's law of electromagnetic induction.

The induced emf is proportional to the flux linkage ✗ ✗

The correct answer is 'The induced e.m.f. is proportional to the **rate of change** of flux linkage'.

[2]

(b) A long solenoid of length 42.0 cm, has 200 equally-spaced turns. A search coil, of diameter 2.0 cm and consisting of 5000 turns of thin insulated copper wire, is placed at the centre of the solenoid, with its axis parallel to the axis of the solenoid, as shown.

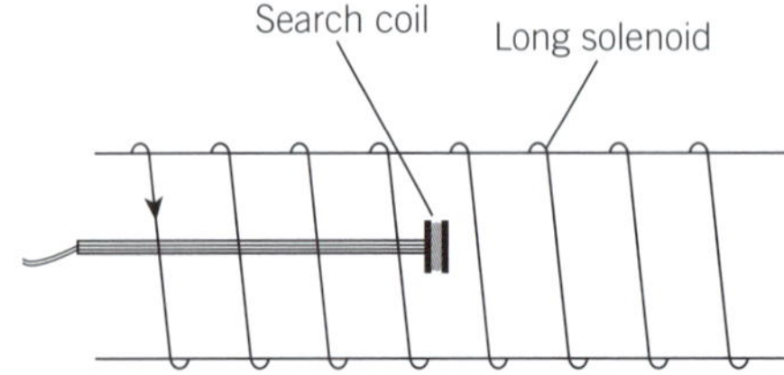

The magnetic flux density at the centre of the solenoid is given by the equation

$$B = \mu_0 n I$$

where μ_0 is a constant, called the permeability of free space ($4\pi \times 10^{-7}$ H m^{-1}), n is the number of turns per metre, and I is the current.

Calculate:

(i) the magnetic flux density at the centre of the coil when the current is 6.0 A

$$B = \mu_0 n I = 4\pi \times 10^{-7} \times \frac{200}{0.42} \times 6.0 = 3.6 \times 10^{-3} \text{ T} \quad ✔ ✔$$

Correct substitutions and calculation of n in turns / metre.

Correct calculation.

magnetic field strength $= 3.6 \times 10^{-3}$ T [2]

(ii) the flux linkage in the search coil.

$$\text{Flux linkage} = N\phi = NBA = 5000 \times 3.6 \times 10^{-3} \times \pi \times (1.0 \times 10^{-2})^2 = 5.7 \times 10^{-3} \text{ Wb} \quad ✔ ✔$$

Correct calculation of cross-sectional area of search coil.

Correct calculation of $N\phi$.

flux linkage $= 5.7 \times 10^{-3}$ Wb [2]

(c) The current in the solenoid now begins to decrease linearly with time, as shown in the graph.

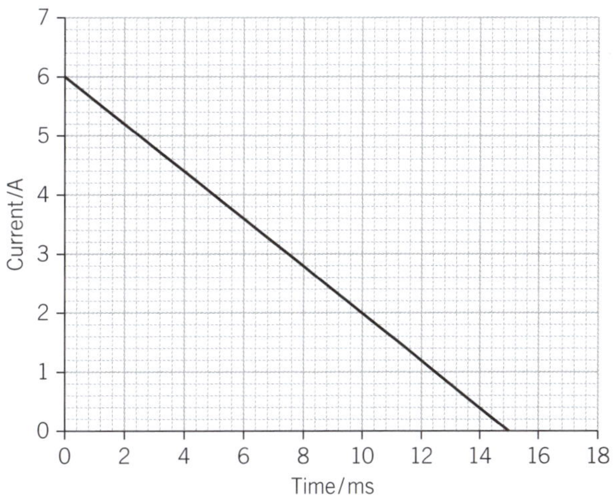

(i) Explain why an e.m.f. is induced across the ends of the search coil

As the current decreases, the magnetic flux linking with the search coil decreases

because the magnetic flux density (B) is decreasing. From Faraday's law, the induced

e.m.f. depends on the <u>rate</u> at which the magnetic flux decreases. ✔✔ .... [2]

> A good answer

(ii) Calculate the e.m.f. across the coil.

induced e.m.f. equals the rate of change of flux linkage

the flux linkage falls from 5.7×10^{-3} Wb to zero in 15×10^{-3} s. ✔

induced e.m.f. $= \dfrac{5.7 \times 10^{-3}}{15 \times 10^{-3}} = 0.38\,V$ ✔

> Correct method.
> Correct calculation.

$= \underline{0.38\,V}$ [2]

(d) The search coil is now moved to one end of the solenoid and the experiment repeated. Explain why the induced e.m.f. is now half the value calculated in (c)(ii).

The magnetic field strength at the end of the solenoid is half as big, so the

induced e.m.f. is half as big. ✔✗ .... [2]

> The statement is correct, but not a complete answer. The magnetic flux density at one end of the solenoid is half the magnetic flux density in the middle (imagine measuring the flux density in the middle of a long solenoid and then removing one half of the solenoid – the remaining half solenoid would produce half the flux density at the end).
>
> As the maximum magnetic field strength is half as much as before, and the current changes exactly as previously, the **rate of change** of the magnetic flux must be half as much, so the induced e.m.f. is half as much.

1 An aeroplane has a wingspan of 60 m and is travelling horizontally at a speed of 900 km h⁻¹. The aeroplane is flying in a region where the vertical component of the Earth's magnetic field is 6.4×10^{-5} T.

(a) Show that the e.m.f. induced between the two wing tips is 0.96 V. [2]

(b) Explain why a voltmeter connected to the two wing tips would read zero volts. [2]

2 A sliding rod AB, supported by two conducting rails, is pulled at a steady speed of 3.0 m s⁻¹ through a magnetic field of flux density 40 mT, as shown.

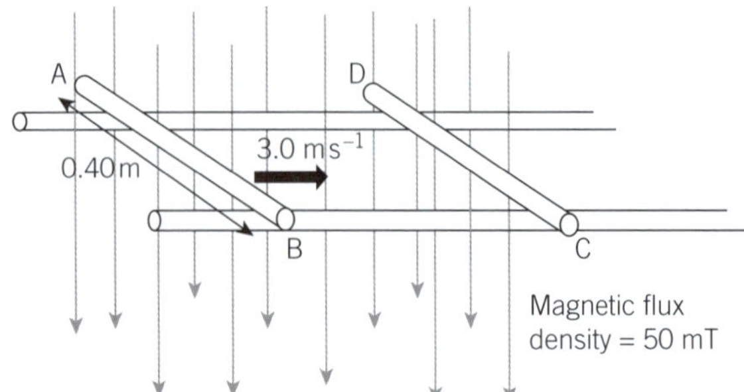

(a) Calculate:

(i) the area 'swept' by the rod in one second

(ii) the magnetic flux 'cut' by the rod each second

(iii) the induced e.m.f. across the rod. [4]

(b) A second rod CD is placed across the rails and an electric current is induced in the loop ABCD.

(i) State and explain the direction of the current in rod AB.

(ii) Suggest a reason why the current in AB increases as it approaches rod CD. [3]

3 A small bar magnet is dropped through a coil of wire which is connected to a voltage sensor and datalogger as shown.

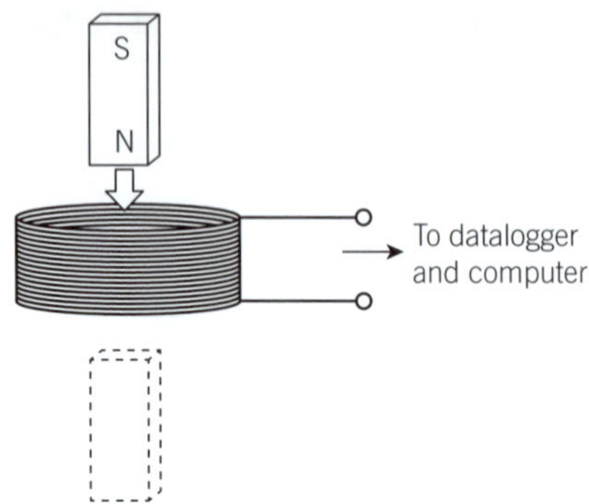

The voltage sensor records the induced e.m.f. across the coil, which is then displayed on a computer screen as shown.

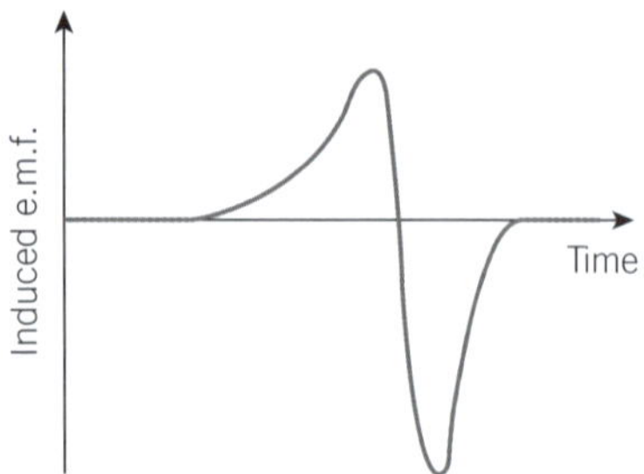

(a) Explain, with reference to Faraday's law:

(i) why the induced voltage is first positive and then negative

(ii) why the negative maximum voltage is larger than the positive maximum. [4]

(b) On a copy of the screen above, sketch the output that would be obtained if the magnet was dropped from a greater height. [3]

(c) Explain what the area under the graph represents. [2]

4 A horizontal flat circular coil of diameter 30.0 cm, consisting of 200 turns of wire, is perpendicular to a uniform magnetic field acting vertically downwards, as shown. The magnetic flux density increases linearly from 2.0 T to 5.0 T in 600 ms.

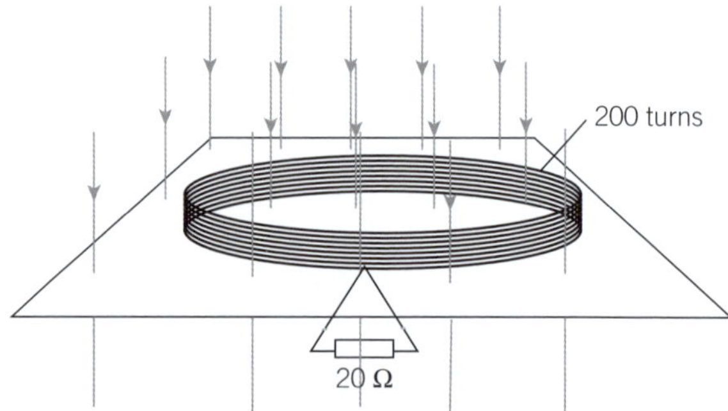

(a) Calculate the induced e.m.f. between the ends of the coil. [3]

(b) The ends of the coil are connected to a 20 Ω resistor.

(i) Calculate the current in the resistor.

(ii) Viewed from above, state the direction of the induced current in the coil. Explain your reasoning. [3]

(c) The current in the coil induces a magnetic field. State the direction of the induced magnetic field inside the coil. Explain your reasoning. [2]

Key points

- [ ] Understand and use the terms period, frequency, peak value, and root-mean-square value as applied to an alternating current or voltage.
- [ ] Deduce that the mean power in a resistive load is half the maximum power for a sinusoidal alternating current.
- [ ] Represent a sinusoidally alternating current or voltage by an equation of the form $x = x_0 \sin \omega t$.
- [ ] Distinguish between r.m.s. and peak values, and recall and solve problems using $I_{\text{r.m.s.}} = I_0/\sqrt{2}$.
- [ ] Understand the principle of operation of a simple laminated iron-cored transformer, and recall and solve problems using $\dfrac{N_S}{N_P} = \dfrac{V_S}{V_P} = \dfrac{I_P}{I_S}$ for an ideal transformer.
- [ ] Understand the sources of energy loss in a practical transformer.
- [ ] Appreciate the practical and economic advantages of alternating current and of high voltages for the transmission of electrical energy.
- [ ] Distinguish graphically between half-wave and full-wave rectification.
- [ ] Explain the use of a single diode for the half-wave rectification of an alternating current.
- [ ] Explain the use of four diodes (bridge rectifier) for the full-wave rectification of an alternating current.
- [ ] Analyse the effect of a single capacitor in smoothing, including the effect of the value of capacitance in relation to the load resistance.

Alternating currents and voltages (a.c.)

Alternating currents and voltages vary sinusoidally. They are used throughout the world when generating and transmitting electrical energy. Figure 24.1 shows an example of alternating current.

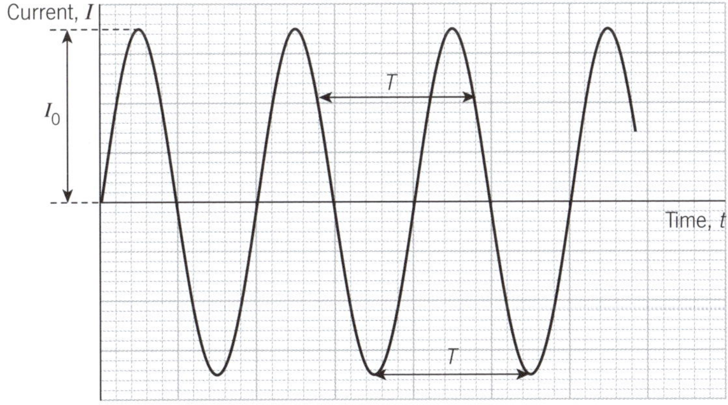

▲ **Figure 24.1** Alternating current (a.c.)

The **maximum (peak) value** of the current is I_0. The **frequency** f of the alternating current is $1/T$ where T is the **period**. The current I at time t is given by the equation:

$$I = I_0 \sin(2\pi ft) = I_0 \sin \omega t$$

where $\omega = 2\pi f$, and ω is the angular frequency in rad s^{-1}.

Similarly, for an alternating voltage:

$$V = V_0 \sin(2\pi ft)$$

> **Remember**
>
> Just as for SHM:
> $$\omega = 2\pi f = \frac{2\pi}{T}$$
> The general equation for a quantity x varying sinusoidally with time t is:
> $$x = x_0 \sin \omega t$$
> It was also met in Unit 13 *Oscillations*. This equation is provided in Exam Papers 1, 2, and 4.

> **Maths skills**
>
> Alternative expressions are:
> $$I = I_0 \cos(2\pi ft)$$
> $$V = V_0 \cos(2\pi ft)$$
> for the case when I and V are at their peak values when $t = 0$.

A Level

Worked example

Determine, from Figure 24.2:

a) the period **b)** the frequency **c)** the peak voltage **d)** the peak-to-peak voltage.

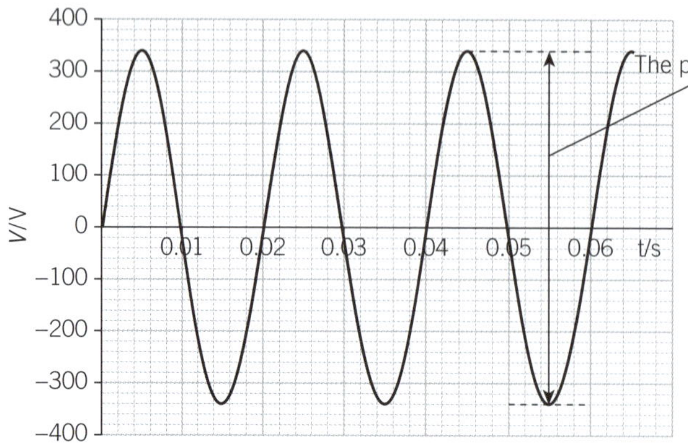

▲ **Figure 24.2**

Answer

a) period, $T = 0.02\,\text{s}$ **b)** frequency $f = \dfrac{1}{T} = 50\,\text{Hz}$ **c)** peak voltage $= 340\,\text{V}$

d) $\dfrac{\text{peak-to-peak}}{\text{voltage}} = \dfrac{\text{range of}}{\text{voltages}} = 680\,\text{V}$

If an alternating voltage $V = V_0 \sin(2\pi ft)$ is connected across a resistor of resistance R (see Figure 24.3 a and b), an alternating current passes through the resistor. At time t the current is $I = I_0 \sin(2\pi ft)$, so P, the power dissipated in the resistor, is:

$$P = VI = V_0 I_0 \sin^2(2\pi ft) = \frac{V_0 I_0}{2}\left[1 - \cos(4\pi ft)\right]$$

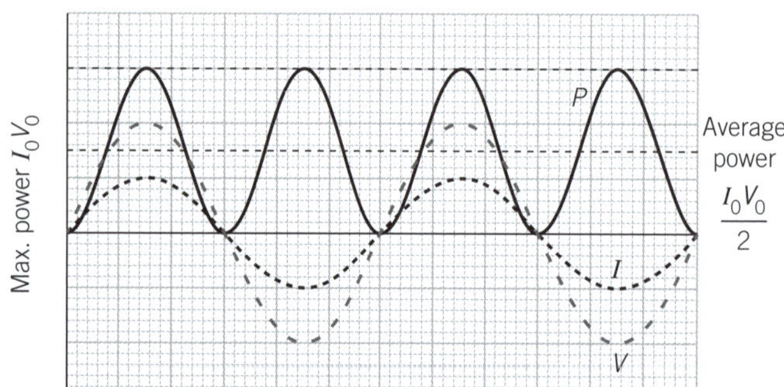

a Variation of current I, voltage V, and power P with time

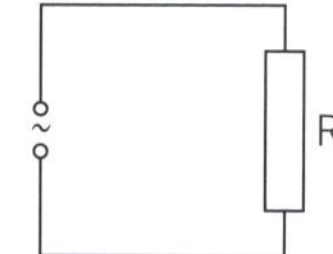

b a.c. current across a resistor

▲ **Figure 24.3** Power dissipated in a resistor R

The maximum power is $I_0 V_0$.

The average power over a complete cycle is half this: $\dfrac{I_0 V_0}{2}$.

The direct current and voltage that would give the same power output are called the **root-mean-square** (r.m.s.) values.

$$I_{\text{r.m.s.}} = \frac{I_0}{\sqrt{2}} \qquad V_{\text{r.m.s.}} = \frac{V_0}{\sqrt{2}}$$

a.c. ammeters and voltmeters are calibrated to measure the r.m.s. values of current and voltage. The r.m.s. values of current and voltage can be used in the same equations already used for analysing d.c. circuits, such as the equation for power.

Transformers

Transformers are used to increase (**step-up**) or decrease (**step-down**) a.c. voltages. If a voltage is increased, the alternating current will decrease, and vice versa. Figure 24.4 shows the main components of a transformer.

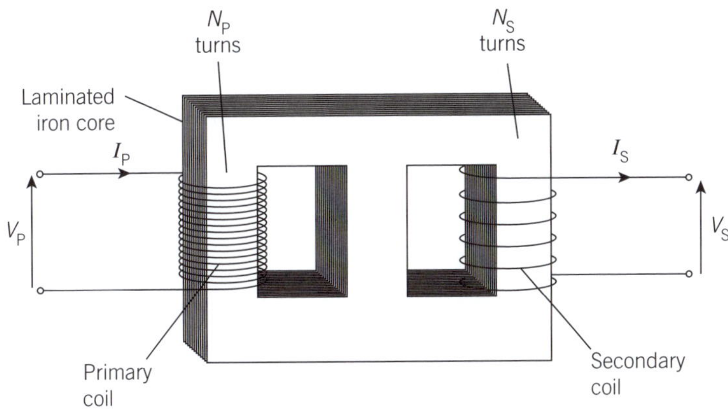

▲ **Figure 24.4** Transformer

An alternating voltage V_P across the primary coil causes an alternating current to pass through the primary coil which induces a changing magnetic flux inside the coil and the iron core. This changing flux passes around the iron core and 'links' with the secondary coil. The changing flux inside the secondary coil induces an alternating voltage V_S in the secondary coil.

From Faraday's law, it can be shown that:

$$\frac{V_S}{V_P} = \frac{N_S}{N_P}$$

Where N_P and N_S are the number of turns on the primary coil and the secondary coil respectively.

If the transformer is 100% efficient (i.e., no energy losses) the power out must equal the power in:

$$I_P V_P = I_S V_S$$

For an ideal transformer:
$$\frac{V_S}{V_P} = \frac{N_S}{N_P} = \frac{I_P}{I_S}$$

Energy losses in transformers

Practical transformers are almost 100% efficient, but there are some energy losses. Table 24.1 shows the main energy losses.

▼ **Table 24.1** Energy losses in transformers

Loss	Cause	Method of reducing
'Copper losses'	The coils of wire have resistance. Energy is lost through heating of the coils.	Using low resistance windings (e.g., large diameter copper wire).
'Iron losses'	Eddy currents ('loops' of current induced in the iron core by the changing magnetic field) cause heating of the core.	Laminating the iron core – the core is made of thin sheets of iron separated by layers of insulating material. The core then has a high resistance in the direction of the eddy currents.
Magnetic hysteresis	Energy lost through repeated magnetisation and demagnetisation of the iron core.	The core is made of 'soft iron' (iron with a low carbon content) which is easily magnetised and demagnetised without much energy loss.

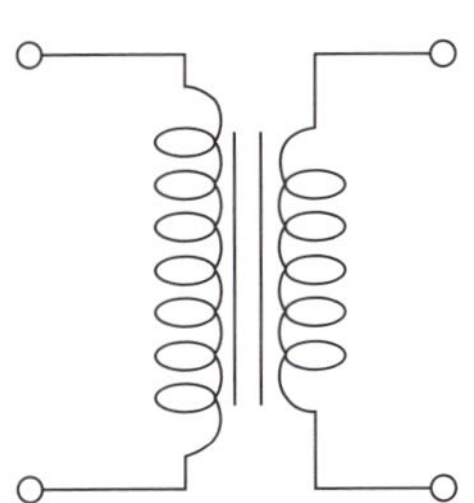

▲ **Figure 24.5** Symbol for a transformer

> **Remember**
>
> **Step-up** transformers increase (step-up) voltages (and decrease currents).
>
> **Step-down** transformers decrease (step-down) voltages (and increase currents).

> The core of a transformer is usually made of 'soft' iron (iron with a low carbon content) making it easy to magnetise and demagnetise, reducing **hysteresis** losses.

> **Remember**
>
> For an ideal transformer:
> $$\frac{V_S}{V_P} = \frac{N_S}{N_P} = \frac{I_P}{I_S}$$

Using transformers

Transformers are an essential part of an electricity supply system, enabling electrical energy to be transferred efficiently from power stations to consumers. Once electrical energy is generated at a power station, the potential difference of the electricity generated is then **increased** in a step-up transformer. This **decreases** the current in the distribution cables, making the energy lost in the cables much smaller. Near the consumer, the p.d. of the electrical supply is decreased (using a series of step-down transformers) to provide suitable voltages for consumers to use with electrical appliances.

Worked example

A power station generates 2000 MW at a p.d. of 25 kV (Figure 24.6). The cables connecting the power station to the consumer have a combined resistance of 0.2 Ω.

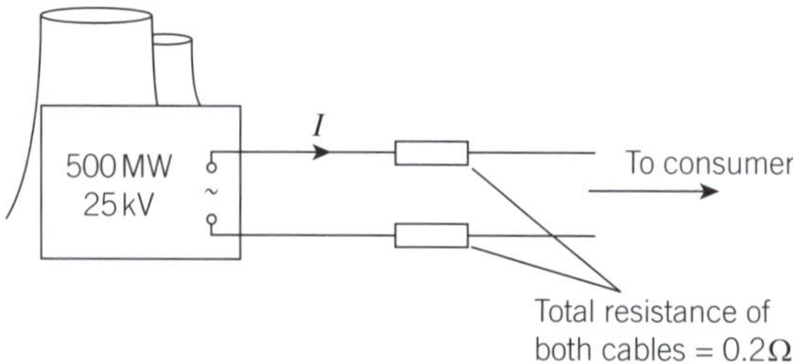

▲ Figure 24.6

a) Calculate:

   i) the current in the cables   **ii)** the power lost as heat in the cables.

b) The p.d. from the power station is now 'stepped-up' using a transformer with a turns ratio of 1:16.

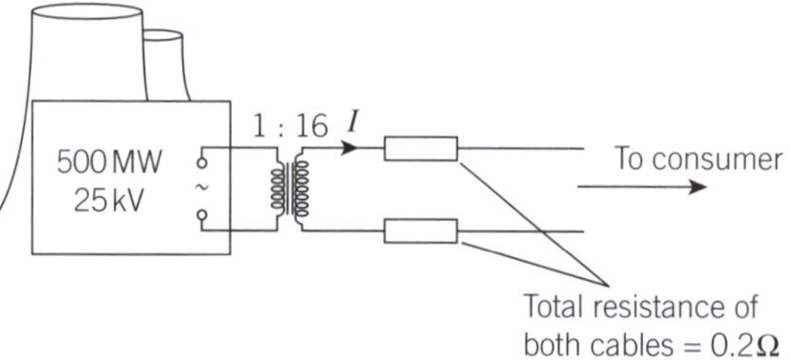

▲ Figure 24.7

Calculate:

   i) the current in the cables   **ii)** the power lost as heat in the cables.

Answer

a) **i)** $P = IV$ so $I = \dfrac{P}{V} = \dfrac{500 \times 10^6}{25 \times 10^3} = 2.0 \times 10^4\,\text{A}$

   ii) Power lost $= I^2 R = (2 \times 10^4)^2 \times 0.2 = 8.0 \times 10^7\,\text{W}$ (80 MW)

> 16% of the energy transmitted is lost as heat in the transmission cables.

b) **i)** $P = IV \Rightarrow I = \dfrac{P}{V} = \dfrac{500 \times 10^6}{16 \times 25 \times 10^3} = 1.25 \times 10^3\,\text{A}$

   ii) Power lost $= I^2 R = (1.25 \times 10^3)^2 \times 0.2 = 3.1 \times 10^5\,\text{W}$ (0.31 MW)

> Less than 0.1% of the energy from the power station is lost as heat in the transmission cables.

Rectification

Half-wave rectification

Some electrical appliances can operate using either a.c. or d.c. – for example electric heaters, toasters or irons – but many other appliances require a direct current supply.

A single diode can be used to convert a.c. into d.c (see Figure 24.8). The diode is connected between the a.c. input and the load resistor R. In the first half of the a.c. cycle the diode is forward-biased, a current flows in the circuit, and there is a p.d. across the resistor R. In the second half of the a.c. cycle the diode is reverse-biased and no current flows. The p.d. across R is zero (see Figure 24.9).

This is called **half-wave rectification**. Although the output p.d. varies, it is never negative, though for half the time the output p.d. and the current are zero.

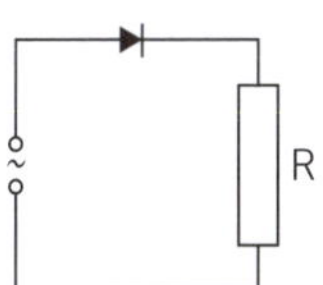

▲ **Figure 24.8** Half-wave rectification

▲ **Figure 24.9** Half-wave rectification output

Full-wave rectification (the bridge rectifier)

Full-wave rectification can be achieved by combining four diodes in the form of a square, as shown in Figure 24.10.

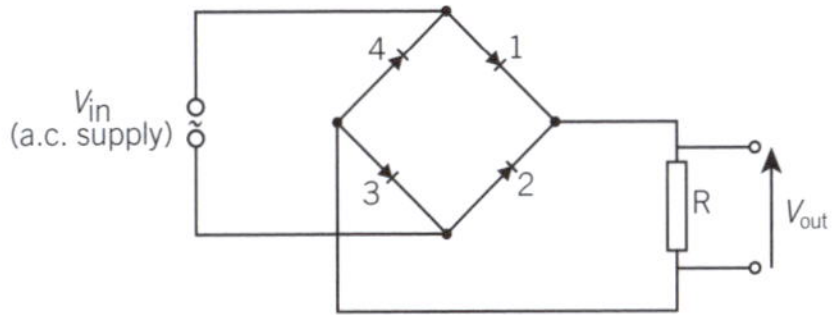

▲ **Figure 24.10** Full-wave rectification

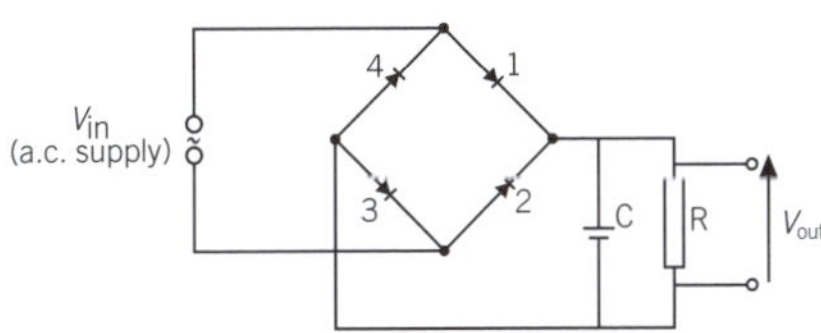

▲ **Figure 24.11** Full-wave rectification output

When V_{in} is positive, current flows through diode 1, downwards through resistor R and returns via diode 3; when V_{in} is negative, current flows through diode 2, downwards through resistor R and returns via diode 4. In both halves of the a.c. cycle the current is passing downwards through R. The output is shown in Figure 24.11.

Smoothing

Some electrical appliances, such as mobile phone chargers, require a steady d.c. supply to work properly. The 'bouncy' d.c. output from a half-wave or full-wave rectifier circuit can be 'smoothed' by the addition of a capacitor in parallel with the external load, as shown in Figure 24.12.

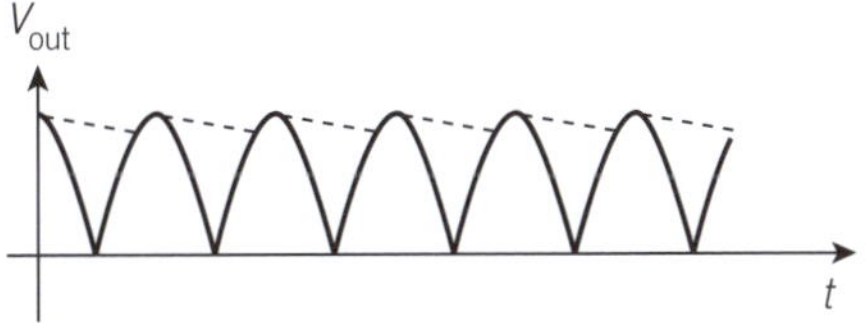

▲ **Figure 24.12** Smoothing

▲ **Figure 24.13** Ripple of smoothed output

When the p.d. across the load resistor is rising, the capacitor charges up. As the p.d. across the load resistor starts to fall, the capacitor maintains the output p.d. by only discharging slowly. When the p.d. from the rectifier rises again, the capacitor will charge up again and the process is repeated.

The output p.d. is not completely smooth – it has 'ripple', as shown in Figure 24.13. The amount of ripple depends on the value of CR (called the **time constant**) – the larger the value of CR, the smoother the output.

The value of the time constant CR should be much greater than the time period of the a.c. supply. If the load resistance is quite small the smoothing capacitor must be larger.

# ⬆ Raise your grade

A student investigating electromagnetic induction sets up the experiment shown.

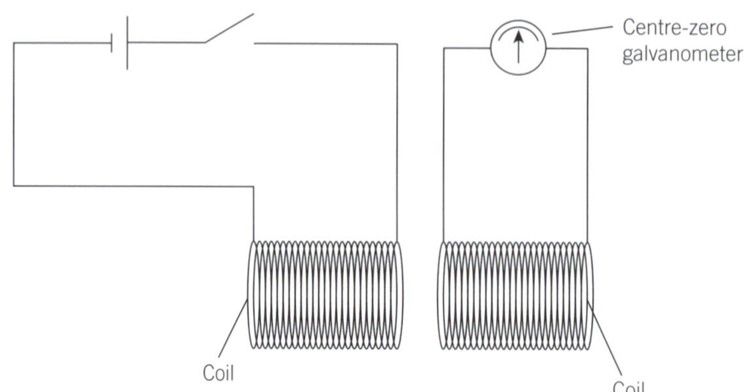

When the student closes the switch the galvanometer moves briefly to the right and then returns to zero.

(a) Explain, with reference to Faraday's law of electromagnetic induction, why the galvanometer briefly deflects one way and then returns to zero.

The first coil becomes magnetised. This magnetises the second coil, which creates a current.

✗ ✗

When the switch is pressed the current in the first coil **increases** (from zero) causing an **increasing** magnetic field/flux in the first coil. This flux 'links' with the second coil. From Faraday's law, a **changing flux** in the second coil will induce a p.d. across the coil, so the galvanometer deflects. When the current in the first circuit is constant, there is no changing flux, and so no induced e.m.f. across the second coil.

[2]

(b) Describe and explain what will happen when the switch is opened.

The first coil becomes demagnetised. This demagnetises the second coil – it is as if a magnet

had been pulled out of the second coil. The changing magnetic field induces a voltage ✔ ✗

The deflection of the galvanometer is in the opposite direction because the magnetic field inside the second coil is **decreasing** rather than increasing.

[2]

(c) The student now passes an iron bar between the two coils, as shown and repeats the experiment.

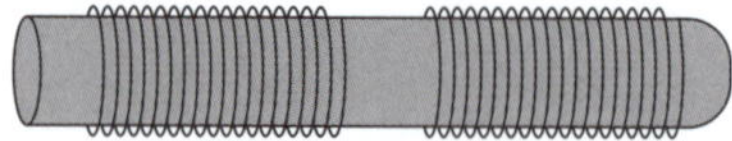

Explain why there is a much larger deflection on the galvanometer.

The iron bar makes the magnetic field stronger. This makes the induced e.m.f. larger. ✔ ✗

A better answer would be 'The iron increases the maximum magnetic field strength by several thousand times, so the **rate of change** of magnetic flux when the switch is opened or closed is much greater, inducing a much greater p.d. across the second coil'.

[2]

(d) The student now replaces the cell with a 'slow' a.c. supply from a signal generator, as shown. She decides to replace the centre-zero galvanometer with a double-beam oscilloscope.

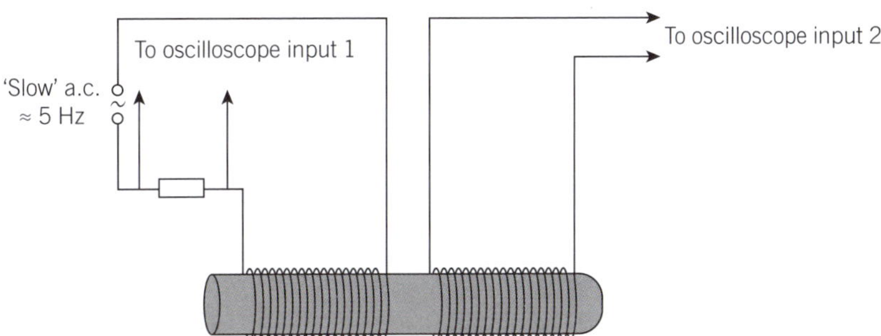

Suggest a reason why monitoring the p.d. across the second coil using an oscilloscope is preferable to using the centre-zero galvanometer.

The voltage is changing direction too fast – the galvanometer cannot keep up. ✔

The needle of the galvanometer has **inertia** – it does not reach its maximum value in one direction by the time the induced p.d. changes direction. [1]

(e) The output from the oscilloscope is shown.

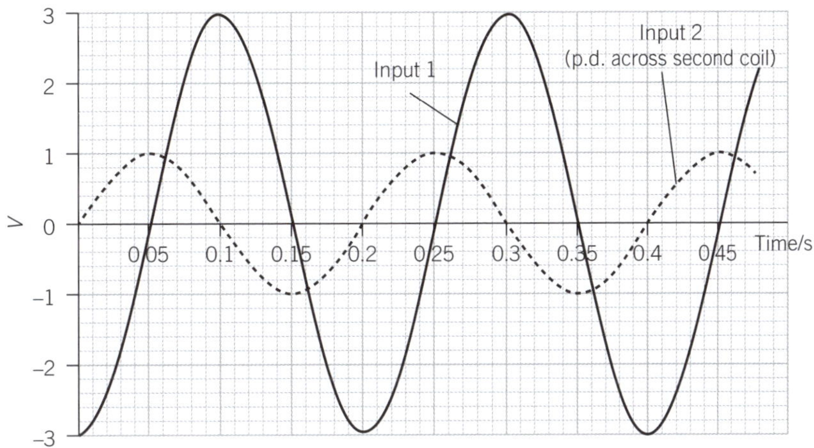

(i) Explain why input 1 is proportional to the current in the first coil.

Input 1 is displaying the p.d. across a resistor, which is proportional to the current ✔ [1]

(ii) Explain why the potential difference across the second coil reaches a maximum when the current in the first coil is zero.

When the current in the first coil is zero it is changing rapidly, so the magnetic field in the first coil is momentarily zero, but changing rapidly. This means that the magnetic field (flux) inside the second coil is changing rapidly so, from Faraday's law, there is a large e.m.f. induced across the second coil. ✔✔ [2]

A good answer.

1 An alternating voltage is displayed on an oscilloscope.

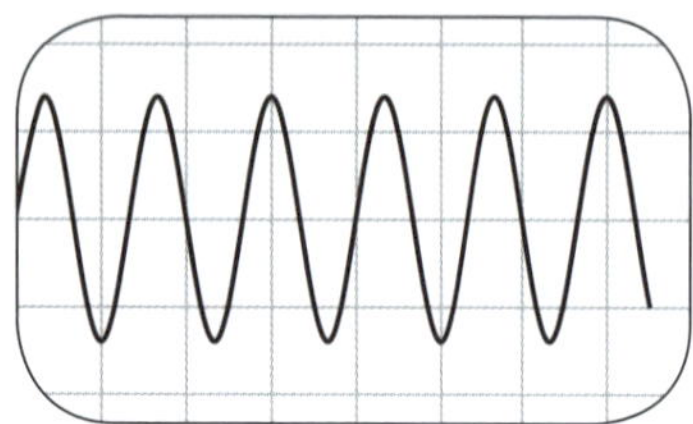

The y-gain of the oscilloscope is set to $0.1\,\text{V}/\text{div}$. The time-base setting is $100\,\mu\text{s}/\text{div}$.

(a) Determine:

  (i) the maximum voltage V_{max}

  (ii) the r.m.s. voltage $V_{r.m.s.}$

  (iii) the peak-to-peak voltage. [3]

(b) Calculate:

  (i) the period

  (ii) the frequency of the a.c. voltage. [2]

(c) Write an equation which describes the variation of voltage with time. [2]

2 An alternating potential difference has a peak value of $325\,\text{V}$ and a frequency of $50\,\text{Hz}$.

(a) Determine the r.m.s. voltage $V_{r.m.s.}$. [1]

(b) The alternating p.d. is connected to a $100\,\Omega$ resistor. Calculate:

  (i) the r.m.s. current in the resistor

  (ii) the average current

  (iii) the average power dissipated in the resistor. [3]

(c) Sketch a graph of the power dissipated in the resistor against time for two complete cycles of the alternating voltage. [3]

3 An ideal transformer has a primary coil of 1200 turns and a secondary coil of 300 turns. A resistor of resistance $48\,\Omega$ is connected to the secondary coil. An alternating p.d. of r.m.s. value $240\,\text{V}$ is connected to the primary coil.

Calculate:

(a) the r.m.s. p.d. across the resistor [1]

(b) the secondary current I_s [1]

(c) the primary current I_p [1]

(d) the power dissipated in the resistor. [1]

4 The diagram shows the electrical circuit of a voltage adaptor used for charging a laptop computer. The diagram has been divided into three sections.

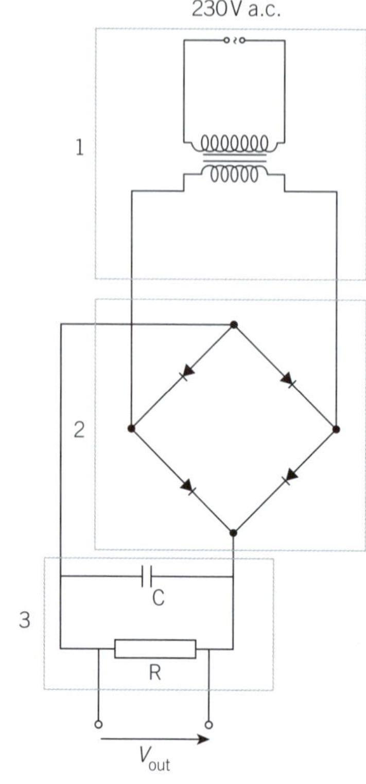

(a) Describe the purpose of section 1 of the circuit. [2]

(b) Section 2 of the circuit is a *bridge rectifier*.

  (i) State one advantage of using a bridge rectifier rather than a single diode

  (ii) Explain how the bridge rectifier works. [3]

(c) Section 3 of the circuit is responsible for smoothing the voltage output.

  (i) Describe what is meant by *smoothing*.

  (ii) Explain how the capacitor achieves smoothing. [3]

(d) The output p.d. will have some ripple.

  (i) Describe what is meant by *ripple*.

  (ii) Explain how the amount of ripple can be reduced. [3]

Key points

- [ ] Understand the particulate nature of electromagnetic radiation.
- [ ] Recall and use $E = hf$ as the energy of a photon.
- [ ] Understand that the photoelectric effect provides evidence for a particulate nature of electromagnetic radiation, while phenomena such as interference and diffraction provide evidence for a wave nature.
- [ ] Recall the significance of threshold frequency.
- [ ] Explain photoelectric phenomena in terms of photon energy and work function energy.
- [ ] Explain why the maximum photoelectric energy is independent of intensity, whereas the photoelectric current is proportional to intensity.
- [ ] Recall, use, and explain the significance of $hf = \phi + \frac{1}{2}mv^2_{max}$.
- [ ] Describe the evidence provided by electron diffraction for the wave nature of particles.
- [ ] Recall and use the relation for the de Broglie wavelength $\lambda = h/p$.
- [ ] Show an understanding of the existence of discrete electron energy levels in isolated atoms (e.g., atomic hydrogen), and deduce how this leads to spectral lines.
- [ ] Distinguish between emission and absorption line spectra.
- [ ] Recall and solve problems using the relation $hf = E_1 - E_2$.
- [ ] Understand that, in a simple model of band theory, there are energy bands in solids.
- [ ] Understand the terms valence band, conduction band, and forbidden band (band gap).
- [ ] Use simple band theory to explain the temperature dependence of the resistance of metals and of intrinsic semiconductors.
- [ ] Use simple band theory to explain the dependence on light intensity of the resistance of an LDR.
- [ ] Explain the principles of the production of X-rays by electron bombardment of a metal target.
- [ ] Describe the features of an X-ray tube, including control of the intensity and hardness of the X-ray beam.
- [ ] Understand the use of X-rays in imaging internal body structures, including a simple analysis of the causes of sharpness and contrast in X-ray imaging.
- [ ] Recall and solve problems by using the equation $I = I_0 e^{-\mu x}$ for the attenuation of X-rays in matter.
- [ ] Understand the purpose of computed tomography or CT scanning.
- [ ] Understand the principles of CT scanning.
- [ ] Understand how the image of an 8-voxel cube can be developed using CT scanning.

Photoelectric effect

Electromagnetic radiation incident on a metal can cause electrons (called **photoelectrons**) to be emitted from the surface. This is known as the **photoelectric effect**. For electrons to be emitted from the metal, the frequency of the electromagnetic radiation must be above a certain frequency, known as the **threshold frequency** f_0. The effect can be demonstrated using a gold-leaf electroscope as shown in Figure 25.1.

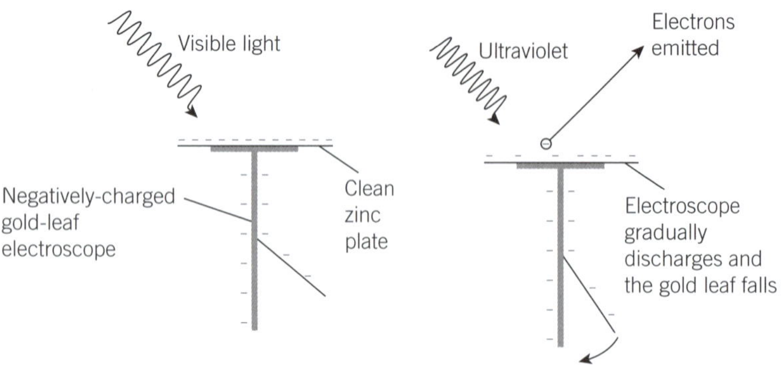

a Visible light incident on a zinc plate produces no photoelectrons. The negatively-charged gold-leaf electroscope remains charged and the gold leaf does not fall.

b Ultraviolet light causes electrons to be emitted (called photoelectrons). The negatively-charged electroscope discharges and the gold leaf falls.

▲ **Figure 25.1** Photoelectric effect demonstrated using a gold-leaf electroscope

Key points about the photoelectric effect:

- Below the threshold frequency f_0, no electrons are emitted; this frequency depends on the metal being used.

- The greater the frequency of the radiation, the greater the **maximum** kinetic energy of the photoelectrons.

- A more intense (brighter) source of radiation produces **more** photoelectrons but does not change the maximum energy of the photoelectrons.

- Photoelectric emission occurs **instantaneously**, regardless of the intensity of the source of radiation provided that the frequency is above the threshold frequency for the metal.

These observations cannot be explained using the wave model of light. With the wave model, eventually enough energy would arrive to be able to free an electron from the surface, regardless of the frequency of the radiation or the type of metal being used.

To explain the photoelectric effect, Einstein proposed that light could be thought of as a stream of particles called **photons**, each with a packet or **quantum** of energy hf, where f is the frequency of the radiation and h is the Planck constant ($6.63 \times 10^{-34}\,\mathrm{J\,s}$).

An incident photon delivers an amount of energy hf to an electron on the surface of the metal. The electron needs a minimum amount of energy ϕ (called the **work function**, which is different for different metals) to escape from the surface of the metal; the remaining energy appears as kinetic energy of the photoelectron.

The **maximum** kinetic energy of the photoelectrons (see Figure 25.2) is found from the equation:

$$E_{k(max)} = hf - \phi$$

Some of the photoelectrons will have less than the maximum kinetic energy because they need more than the minimum energy ϕ to escape from the surface of the metal.

The threshold frequency f_0 which allows electrons to just escape from the surface without any additional kinetic energy is given by:

$$hf_0 - \phi = 0$$
$$f_0 = \frac{\phi}{h}$$

The wavelength corresponding to the threshold frequency f_0 is called the **threshold wavelength** $\lambda_0 = c/\lambda_0$ where c is the speed of light. **Above** the maximum wavelength λ_0 no photoelectrons will be emitted.

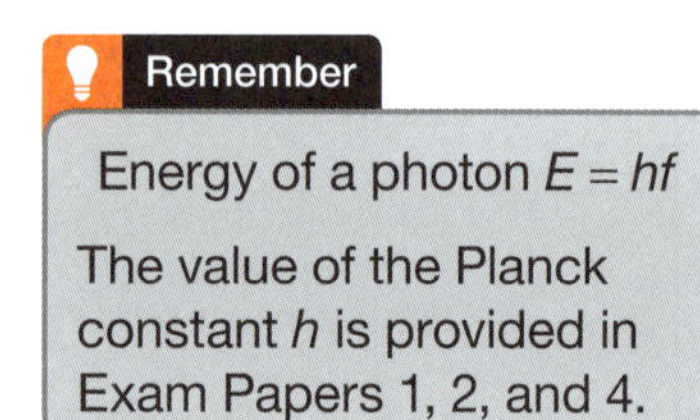
Remember

Energy of a photon $E = hf$

The value of the Planck constant h is provided in Exam Papers 1, 2, and 4.

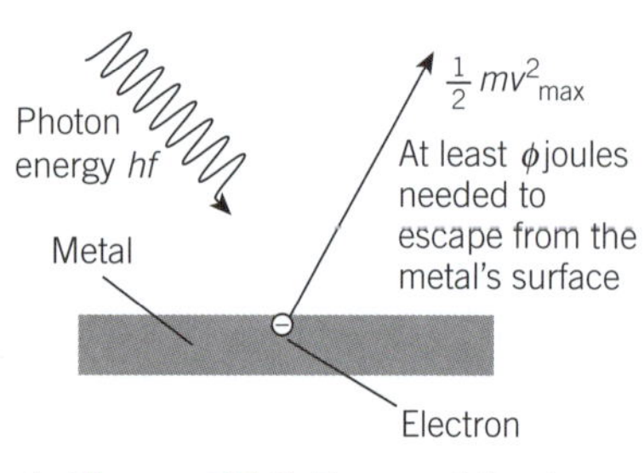

▲ **Figure 25.2** $E_{k(max)} = hf - \phi$

Worked example

The work function ϕ for aluminium is 6.85×10^{-19} J. The speed of light is 3.00×10^8 m s^{-1} and Plank's constant is 6.63×10^{-34} J s.

a) Calculate: **i)** the threshold frequency **ii)** the threshold wavelength.

b) If radiation of wavelength 200 nm is incident on a sheet of aluminium, calculate the maximum kinetic energy of the electrons.

Answer

a) i) $f_0 = \dfrac{\phi}{h} = \dfrac{6.85 \times 10^{-19}}{6.63 \times 10^{-34}} = 1.03 \times 10^{15}$ Hz

ii) $\lambda_0 = \dfrac{c}{f_0} = \dfrac{3.0 \times 10^8}{1.03 \times 10^{15}} = 2.91 \times 10^{-7}$ m

b) $E_{k(max)} = hf - \phi = 6.63 \times 10^{-34} \times \left(\dfrac{3.0 \times 10^8}{200 \times 10^{-9}}\right) - 6.85 \times 10^{-19}$

$= 3.1 \times 10^{-19}$ J

Light – particle or wave?

The photoelectric effect can only be satisfactorily explained by treating light as a stream of photons, delivering energy in 'lumps'; the diffraction and interference of light can only be explained by treating light as waves. Light exhibits either wave-like or particle-like behaviour, according to circumstances – it has a dual nature. This is described as **wave–particle duality**.

Matter waves

Light has particle-like properties, and matter has wave-like properties. Evidence for this includes a beam of electrons passing through a thin crystal of graphite – the electrons appear to 'diffract', producing a pattern of rings (see Figure 25.3).

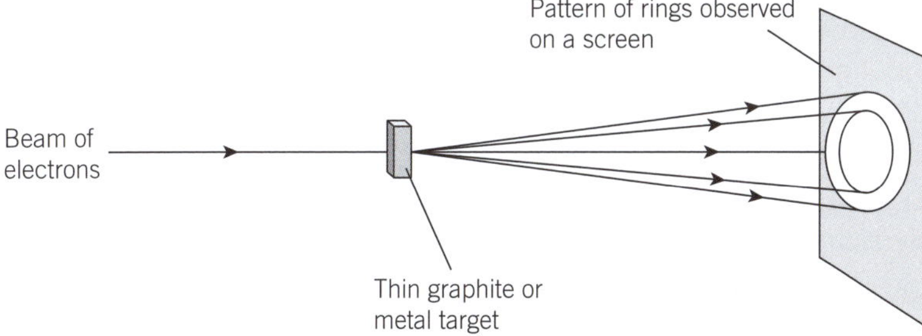

▲ **Figure 25.3** Diffraction of electrons

The wavelength associated with the electrons is found from **de Broglie's equation**:

$$mv = \frac{h}{\lambda}$$

where mv is the momentum of the electron and h is Planck's constant.

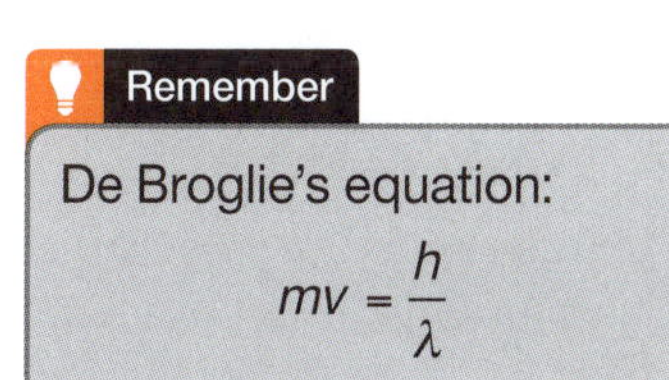

When electrons are detected, they are detected as particles, with mass and velocity. The de Broglie wavelength associated with the electron enables the location of the electrons to be calculated. When the p.d. accelerating the electrons in the diffraction experiment is increased, the momentum of the electrons increases, and so the wavelength associated with the electrons decreases. The diameters of the rings observed when electrons are diffracted are seen to decrease.

Worked example

Calculate the de Broglie wavelength for:

a) an electron ($m = 9.11 \times 10^{-31}\,\text{kg}$)

b) an alpha particle ($m = 6.64 \times 10^{-27}\,\text{kg}$)

c) a tennis ball ($m \approx 60\,\text{g}$)

moving at a speed of $1.5 \times 10^6\,\text{m}$.

Answer

From de Broglie's equation $\quad \lambda = \dfrac{h}{mv} = \dfrac{6.6 \times 10^{-34}}{m \times 1.5 \times 10^6} = \dfrac{4.4 \times 10^{-40}}{m}$

a) $\lambda = \dfrac{4.4 \times 10^{-40}}{m} = \dfrac{4.4 \times 10^{-40}}{9.11 \times 10^{-31}} = 4.8 \times 10^{-10}\,\text{m}$

b) $\lambda = \dfrac{4.4 \times 10^{-40}}{m} = \dfrac{4.4 \times 10^{-40}}{6.64 \times 10^{-27}} = 6.6 \times 10^{-14}\,\text{m}$

c) $\lambda = \dfrac{4.4 \times 10^{-40}}{m} = \dfrac{4.4 \times 10^{-40}}{6.0 \times 10^{-2}} = 7.3 \times 10^{-39}\,\text{m}$

Electron energy levels

An electron in an isolated atom (e.g. an atom of hydrogen or helium gas) can only have certain specific amounts of energy; the energy of the electron is **quantised**. This can be represented on an energy level diagram – Figure 25.4 shows the different energy levels for the electron in a hydrogen atom. By convention, an electron which is completely free of the atom has zero energy. An electron inside the atom requires energy to 'escape' so it has less than zero energy (it has negative energy).

An electron with the lowest possible energy is in its **ground state** ($n = 1$). If an electron absorbs energy it can jump to a higher energy level – it is then in an **excited state**, as shown in Figure 25.5. After a short time, an electron in a higher energy state (E_1) will fall back down to a lower energy level (E_2). In doing so it emits an amount of energy $E_1 - E_2$ as a photon of frequency f, where:

$$hf = E_1 - E_2.$$

The movement of an electron from one energy level to another is called a **transition**.

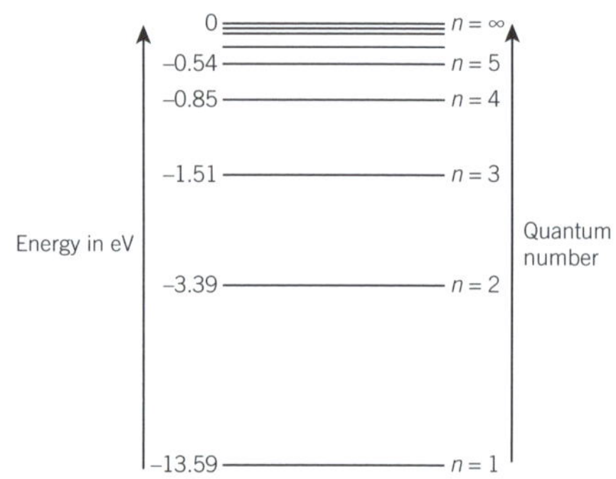

▲ **Figure 25.4** Electron energy levels in a hydrogen atom

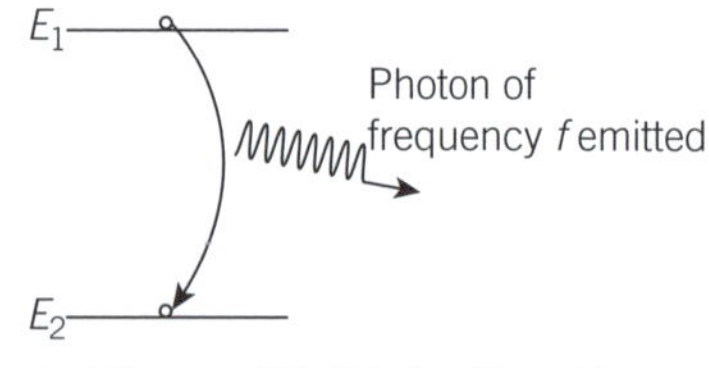

▲ **Figure 25.5** $hf = E_1 - E_2$

Energy levels of electrons in atoms are usually measured in electronvolts (eV).

1 electronvolt is the energy gained by an electron in moving across a p.d. of 1 V.

$$1\,\text{eV} = 1.6 \times 10^{-19}\,\text{J}$$

Worked example

Determine the wavelength of light emitted when the electron in a hydrogen atom falls from the $n = 4$ state to the $n = 2$ state.

Answer

Using Figure 25.4: $\quad hf = E_1 - E_2 = [(-0.85) - (-3.39)] \times 1.6 \times 10^{-19}$

$$f = \frac{4.06 \times 10^{-19}}{6.63 \times 10^{-34}} = 6.12 \times 10^{14}\,\text{Hz}$$

$$\lambda = \frac{c}{f} = \frac{3.0 \times 10^8}{6.12 \times 10^{14}} = 4.90 \times 10^{-7}\,\text{m}\,(490\,\text{nm})$$

If an electron absorbs enough energy to reach the zero energy level ($n = \infty$) it will have escaped the atom (the atom has been **ionised**).

Emission spectra and absorption spectra

The electron in an atom of hydrogen can make a number of different transitions from a higher energy level to a lower energy level. Each transition corresponds to a specific wavelength of electromagnetic radiation being emitted, as illustrated in Figure 25.6.

Each chemical element has its own specific set of electron energy levels and consequently a unique set of wavelengths of electromagnetic radiation.

Emission spectra

If the light emitted from excited hydrogen gas atoms (e.g., from a gas discharge tube, as shown in Figure 25.7) passes through a diffraction grating or a prism, a series of bright lines is observed, called an **emission line spectrum** (Figure 25.8).

▲ **Figure 25.8** Emission line spectrum of hydrogen

Absorption spectra

If white light passes through hydrogen gas, for example, some frequencies of the white light will be absorbed by the gas. The frequencies are the same as those observed in an emission spectrum, but in this case electrons are absorbing energy in moving from a lower energy level to a higher energy level. The spectrum observed is called an **absorption spectrum** and the frequencies corresponding to the electron transitions appear as dark lines in an otherwise continuous spectrum (Figure 25.9).

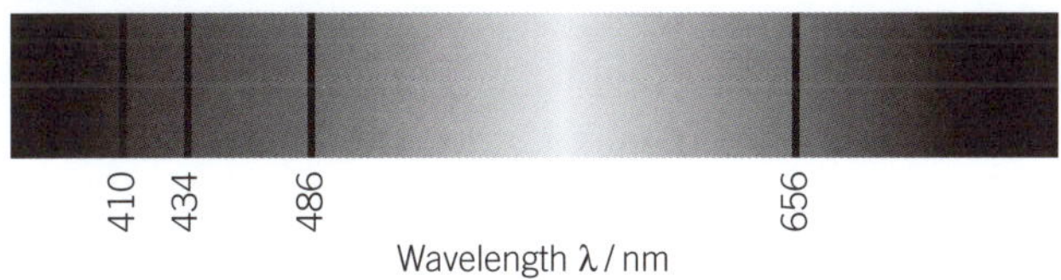

▲ **Figure 25.9** Absorption line spectrum of hydrogen

Line spectra are only produced by isolated atoms such as the atoms of a gas. When atoms combine to form molecules, or are closer together (e.g, in a solid or liquid, or a gas at high pressure) the interactions produce a greater number of possible electron energy levels leading to an energy level diagram with many energy levels close together, and gaps where there are no permitted energy levels, as shown in Figure 25.10a. The energy level diagram is usually drawn as a number of bands, as shown in Figure 25.10b.

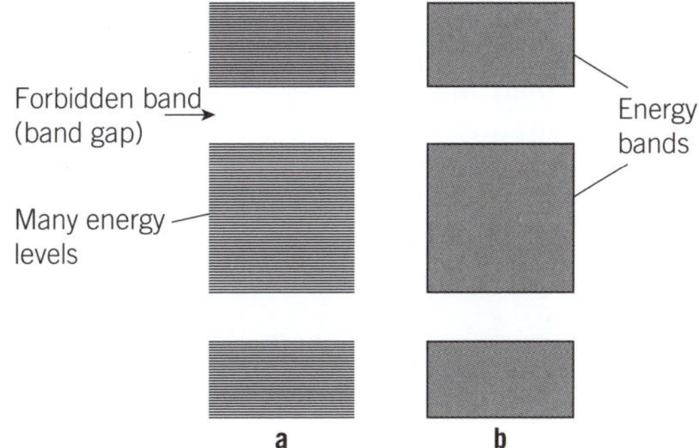

▲ **Figure 25.10** Energy bands

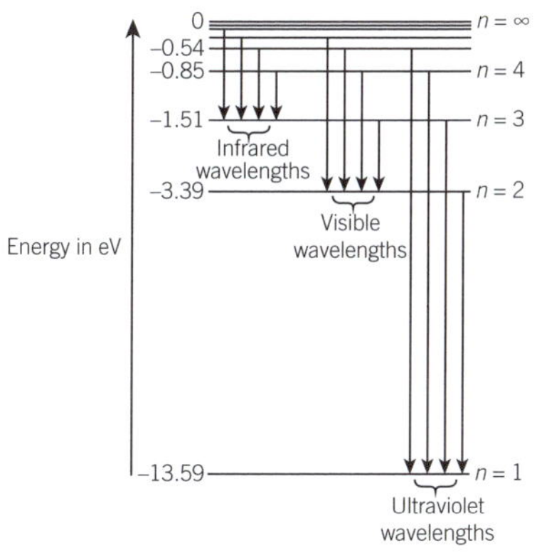

▲ **Figure 25.6** Electron transitions in hydrogen

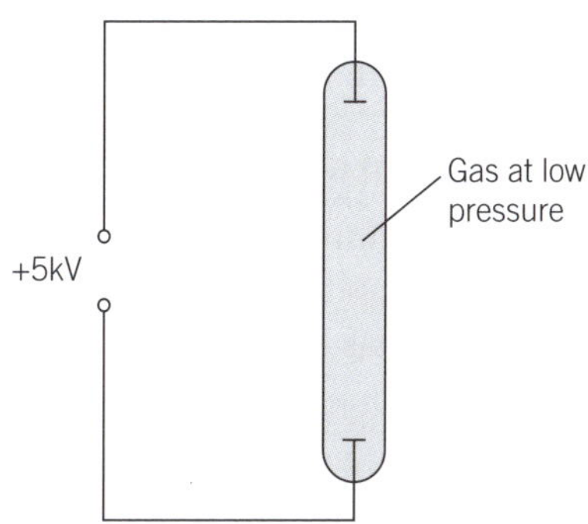

▲ **Figure 25.7** Gas discharge tube

Band theory

The highest energy band is called the **conduction band**; the band below the conduction band is referred to as the **valence band**. The gap between these two bands is called the **energy gap** or the **forbidden band**. Electrons cannot have energies in the forbidden band, in the same way as an electron in an isolated atom cannot have energies between two adjacent energy levels.

Metals

In **metals**, the conduction band is only partially filled (see Figure 25.11). The electrons in this band are the 'free' electrons responsible for the conduction of electricity. When a metal is connected to an electrical supply these electrons gain energy (they move into higher energy levels within the conduction band) – they now have enough energy to break away from atoms and can move through the metal. The electrons in the valence band are firmly attached to individual atoms.

When a metal is heated, its resistance increases. No more electrons are released to become conduction electrons (electrons in the valence band are still tightly held by atoms). The **charge density** does not change, but the increased vibrations of the metal atoms cause the conduction electrons to collide more frequently with them.

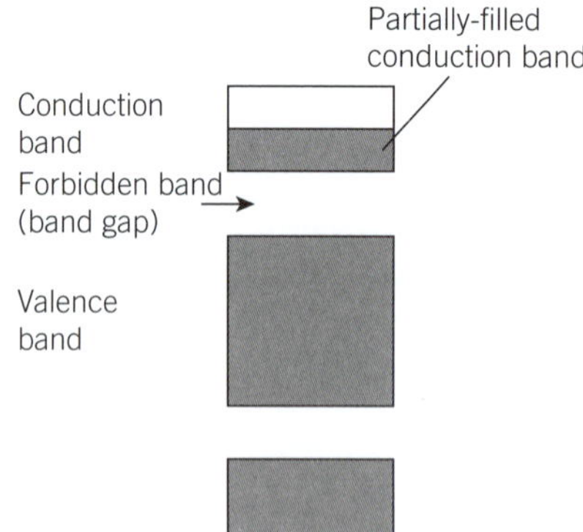

▲ **Figure 25.11** Energy bands in metals

Insulators

In **insulators** there are no electrons in the conduction band and there is a large gap between the conduction band and the valence band (the forbidden band), as shown in Figure 25.12.

When an electrical supply is connected to an insulator there is insufficient energy to 'lift' any electrons from the valence band into the conduction band (the energy gap is too large), so there are no electrons free to move through the material – it is does not conduct electricity.

Semiconductors

Semiconductor materials such as silicon and germanium only conduct a very small amount of electricity. As with insulators, its conduction band is empty and its valence band is full, but the gap between the valence band and the conduction band is much smaller (Figure 25.13).

When a potential difference is applied across a semiconductor, a few of the most energetic electrons have enough energy to move from the valence band to the conduction band. These electrons are then free to form an electric current.

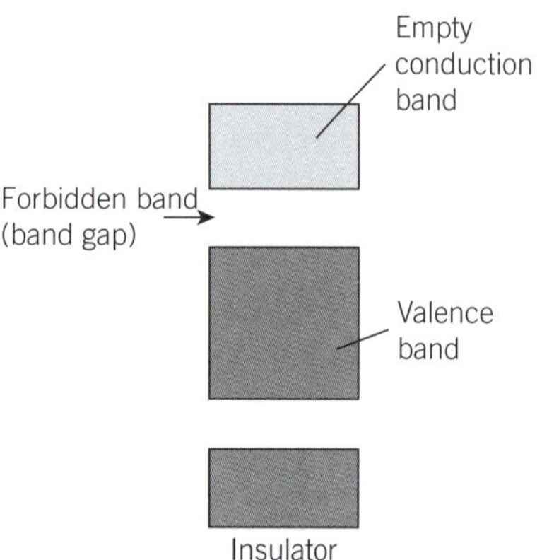

▲ **Figure 25.12** Energy bands in insulators

If a semiconductor is heated, more of the valence band electrons can acquire enough extra energy to move into the conduction band so the current increases. This is the reason why the resistance of many thermistors decreases as their temperature increases.

A light-dependent resistor (LDR) is made from a semiconductor material such as cadmium sulphide. In the dark it can have resistances of several megaohms. In bright light, photons of light energy give electrons in the valence band of the semiconductor enough energy to move into the conduction band, and so the resistance of the LDR decreases.

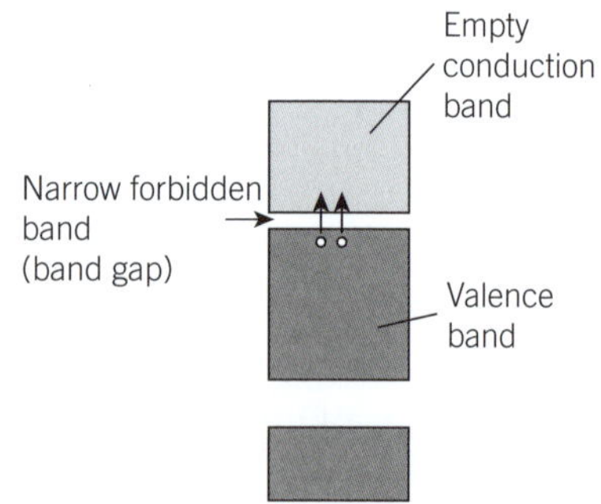

▲ **Figure 25.13** Energy bands in semiconductors

X-rays

X-rays are high-frequency electromagnetic waves, with wavelengths in the range 10^{-8} m to 10^{-11} m. They are produced by bombarding a metal target with high-energy electrons – the rapid deceleration of the electrons causes the emission of X-rays. Figure 25.14 illustrates how an X-ray tube works.

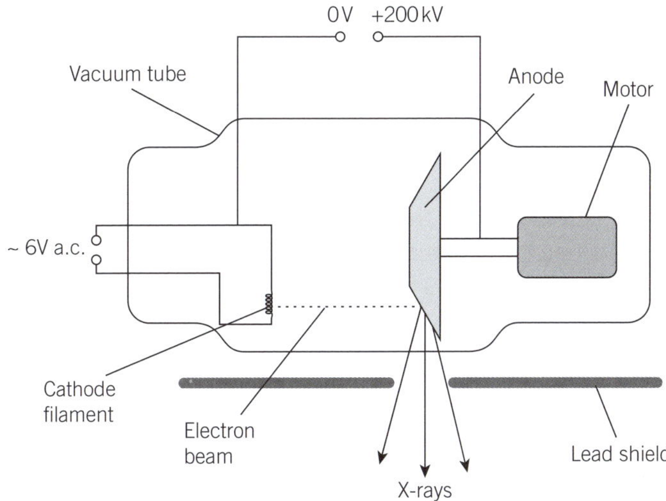

▲ **Figure 25.14** X-ray tube

Electrons are emitted from the hot filament (the **cathode**) by thermionic emission. They are then accelerated towards the **anode** by a p.d. of up to 200 kV. The anode is a small target made of tungsten (or other metal with a high melting point).

Approximately 1% of the kinetic energy of the electrons is converted into X-rays; the rest is converted to heat energy in the metal target. The tungsten target can be rotated rapidly by a motor so that a much greater area of tungsten is heated.

The X-rays are emitted through a thin 'window' surrounded by lead shielding. Metal tubes beyond the window **collimate** the beam (to make it parallel and not spread out like a fan).

The X-rays emitted have a continuous range of frequencies up to a maximum frequency f_{max}, determined by the magnitude of the accelerating potential. If an electron is accelerated across a p.d. V, the energy of the electron is eV, where e is the electronic charge. If all this energy is converted into a single X-ray photon:

$$hf_{max} = eV$$

$$f_{max} = \frac{eV}{h}$$

Worked example

An X-ray machine has an accelerating potential of 50 kV. Calculate the shortest possible X-ray wavelength that can be emitted. [The speed of light is 3.00×10^8 m s^{-1}, e is 1.6×10^{-19} C, and Plank's constant is 6.63×10^{-34} J s.]

Answer

$$\lambda_{min} = \frac{c}{f_{max}} = \frac{hc}{eV} = \frac{6.63 \times 10^{-34} \times 3.0 \times 10^8}{1.6 \times 10^{-19} \times 50 \times 10^3} = 2.5 \times 10^{-11} \text{ m}$$

Intensity and 'hardness' of X-rays

The **intensity** of an X-ray beam is a measure of the amount of energy emitted per second per unit area. The intensity can be increased by:

* increasing the accelerating potential

* increasing the number of electrons hitting the metal target. This can be achieved by increasing the current in the filament. The filament gets hotter, releasing more electrons each second.

The **hardness** of X-rays is a measure of their penetrating power. 'Hard' X-rays have higher energies (shorter wavelengths) than 'soft' X-rays, and so are more penetrating. When used to produce an X-ray image of an internal body structure, the soft X-rays are more easily absorbed, increasing the exposure of the patient to hazardous radiation – it is often better to use hard X-rays, using a metal filter to absorb the longer wavelength X-rays.

The hardness of X-rays can be increased by:

* using a filter to absorb the low energy, 'softer' X-rays

* increasing the accelerating potential. This increases the relative amounts of higher frequency (shorter wavelength) X-rays produced, as shown in Figure 25.15.

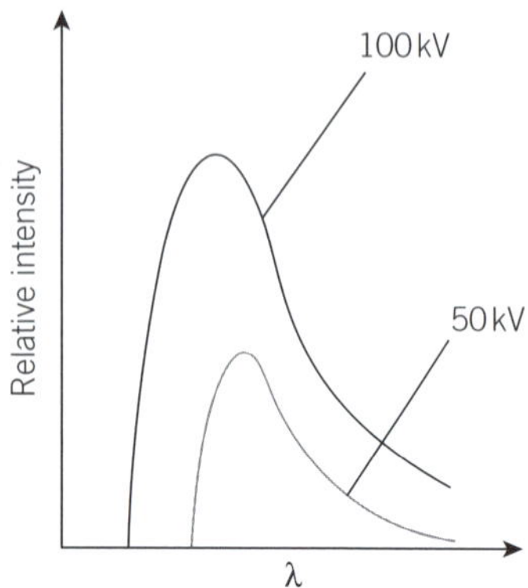

▲ **Figure 25.15** X-ray spectra

X-ray imaging

Sharpness

The **sharpness** of an X-ray image is a measure of how clearly the edges of an object are defined. This depends on the area of the tungsten anode, as shown in Figure 25.16.

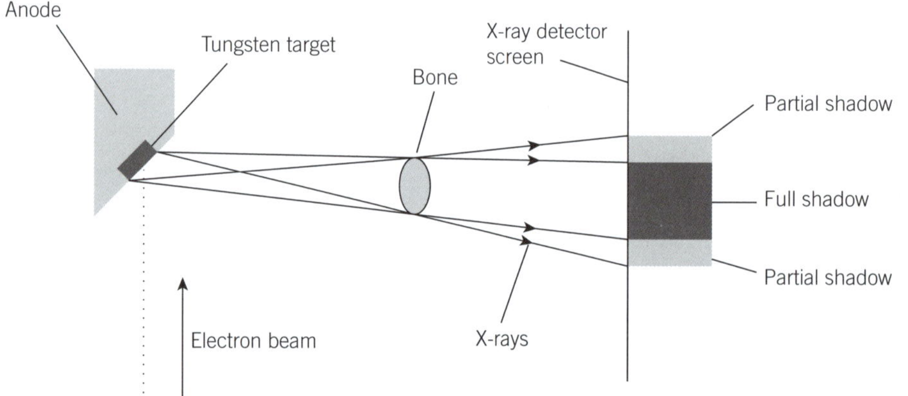

▲ **Figure 25.16** Sharpness

If the target area is too large, some of the X-ray detector screen (or film) will only be in partial shadow from the X-rays, and so the image appears blurred. If the target is too small, the target can overheat. To prevent this, the tungsten target can be mounted on a copper block with pipes containing a cooling liquid such as oil or water.

> **Link**
>
> Compare the advantages and disadvantages of X-ray imaging with NMR imaging in Unit 22 *Magnetic fields*.

Some of the X-rays are scattered in different directions by atoms of body tissue. These X-rays may fall on the shadow areas, blurring the image. To prevent this, a lead grid is placed between the patient and the detector screen, as shown in Figure 25.17. Lead is a good absorber of X-rays and any scattered X-rays will be absorbed by the grid – only the unscattered X-rays can reach the detector screen.

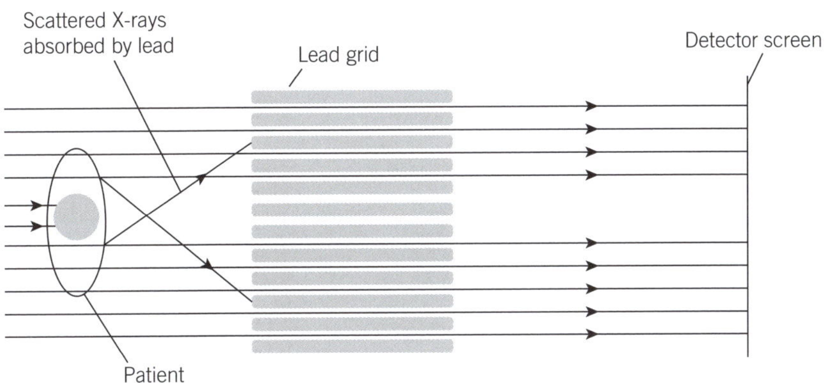

▲ **Figure 25.17** Using a lead grid to absorb scattered X-rays

Contrast

X-ray images of bones and teeth have good contrast – bones and teeth are good absorbers of X-rays so there is a clear difference between the exposed parts of the detector screen and the unexposed parts.

In order to obtain a clear image of softer body tissues such as the stomach, a **contrast medium** is used. If a patient has a drink containing barium sulphate before a stomach X-ray, the image of the stomach is much clearer as barium sulphate is a good absorber of X-rays. A contrast medium can also be injected, enabling blood flow to be observed more clearly.

The contrast of an X-ray image is reduced if:

* the exposure time is too long – the light and dark areas both become darker

* the X-rays are too penetrating – more X-rays would pass through the denser tissues, reaching the shadow areas

* too much scattering of the X-rays occurs.

Absorption of X-rays

As an X-ray beam passes through matter it is **attenuated** – the intensity of the beam decreases. For a parallel beam with initial intensity I_0, the transmitted intensity I after passing through a thickness x of a material is given by:

$$I = I_0 e^{-\mu x}$$

where μ is the attenuation coefficient of the material. (The SI unit of μ is m^{-1} though it is often given in cm^{-1}.)

The attenuation coefficient is proportional to the density of the material.

> **Remember**
>
> The intensity of an X-ray beam decreases exponentially with thickness of absorber.
>
> $$I = I_0 e^{-\mu x}$$
>
> The **half-thickness** of a material is the thickness of the material needed to halve the intensity.

Worked example

A metal plate of thickness 5.0 mm reduces the intensity of an X-ray beam by 40%. Determine:

a) the attenuation coefficient   **b)** the half-thickness of the metal.

Answer

a) As the intensity has been reduced by 40%, the intensity must be $0.6I_o$, where I_o is the initial intensity:

$$0.6I_o = I_o e^{-0.005\mu}$$

Re-arranging this equation and 'taking natural logs' of both sides:

$$-0.005\mu = \log_e 0.6$$

$$\mu = 100\,\text{m}^{-1}\ (1.0\,\text{cm}^{-1})$$

b) When the thickness of the metal is the 'half thickness' $x_{1/2}$, $I = \dfrac{I_o}{2}$

so:

$$\frac{I_o}{2} = I_o e^{-\mu x_{1/2}}$$

Re-arranging this equation:

$$e^{\mu x_{1/2}} = 2,\ \text{so}\ldots\ldots x_{1/2} = \frac{\log_e 2}{\mu} = 0.69\,\text{cm}$$

See Appendix: *Maths skills* for more about exponential functions.

In any calculations involving half thickness and attenuation coefficients, always ensure that the units you are using are *consistent*.

Notice the similarities between this equation and the equation for the half-life of radioactive isotopes:

$$T_{1/2} = \frac{\log_e 2}{\lambda}$$

Computed tomography (CT) scanning

Ordinary X-ray images are two-dimensional 'shadow' pictures of a patient. The medical imaging technique called **CT scanning** (also known as CAT scanning – computed axial tomography) is a way of obtaining much more detailed images (see Figure 25.18), including 3D images.

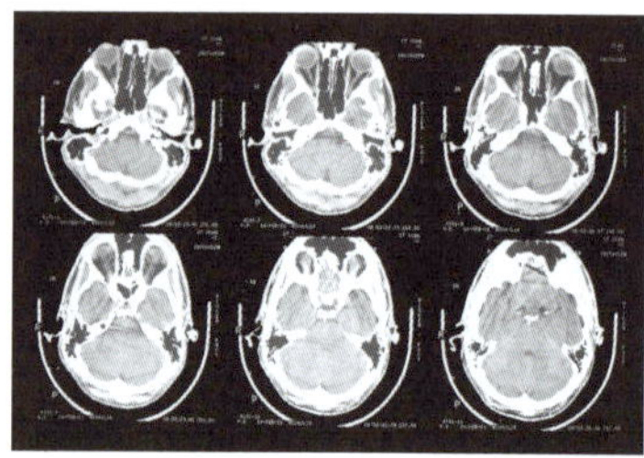

▲ **Figure 25.18** CT images of a brain

Figure 25.19 illustrates the main principles of CT scanning. An X-ray tube mounted on a gantry is able to rotate 360° around a patient. Several hundred X-ray sensors are mounted on a ring. The X-ray tube produces a narrow beam of X-rays which are detected by the sensors opposite the tube and the data sent to a computer. As the X-ray tube rotates, a detailed image is gradually built up and a cross-section of the patient can be displayed on a computer screen. If the patient is gradually moved along the axis of the ring, several 'slices' of the body can be obtained and combined to give a 3D image.

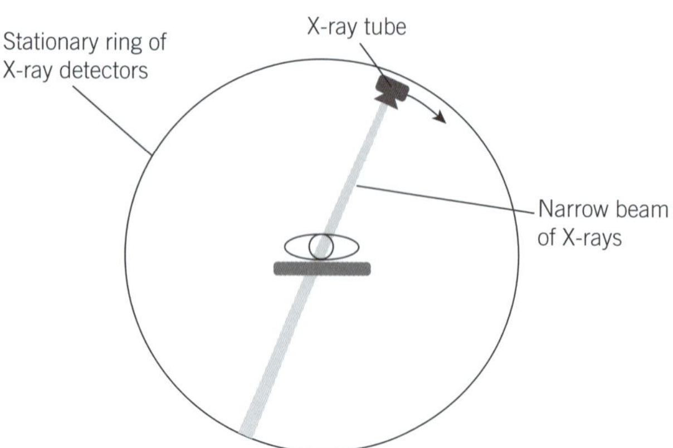

▲ **Figure 25.19** CT scanning

For a CT scanner to produce a clear image:

- the X-ray beam must be collimated (not spread out like a fan)

- the detecting elements must be as small as possible.

Constructing an image

The section (or slice) through a body is divided up into a series of small cubes called **voxels**. The intensity detected by a detector depends on the absorption of X-rays as they pass through different voxels.

Imagine an object consisting of four voxels with absorption 'densities' 2, 3, 5, and 1, as shown in Figure 25.20a. The beam of X-rays is directed horizontally at the object and the detector readings recorded (see Figure 25.20b).

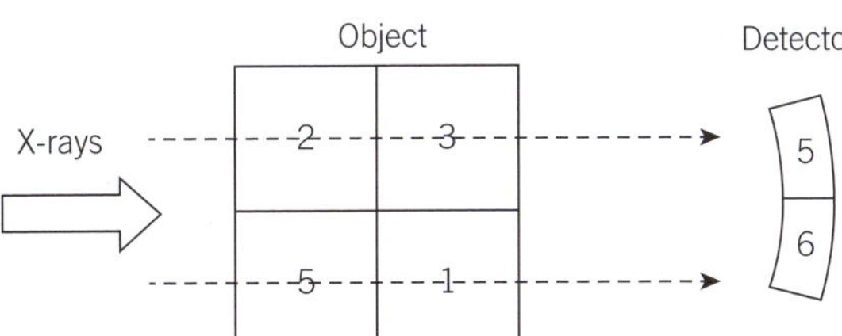

a X-rays are directed at the body from one direction and the detectors record the intensity

b The intensities are matched to the voxels (the amount each individual voxel contributes is unknown at this stage)

▲ **Figure 25.20** Voxels: first scan

This process is repeated from a new direction and the results added on to the earlier values in the image voxels, as shown in Figure 25.21.

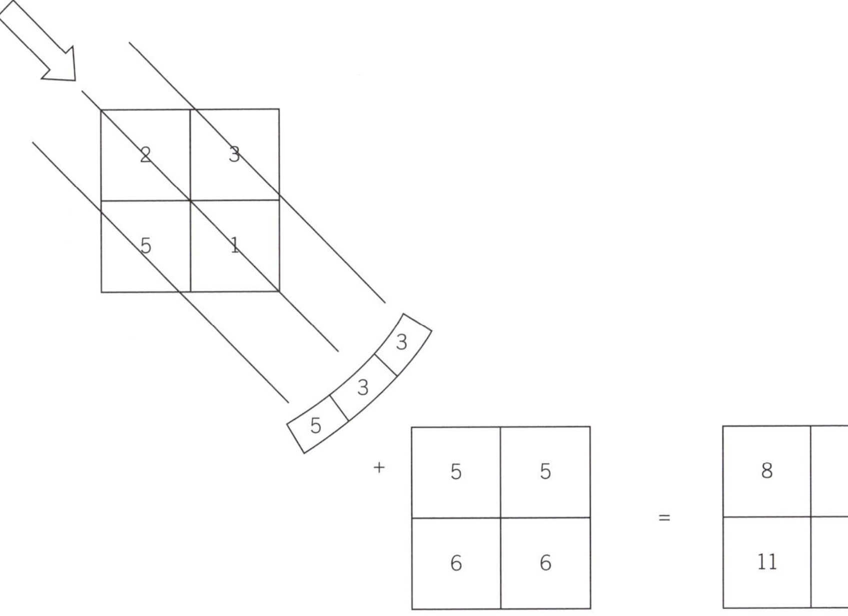

▲ **Figure 25.21** Voxels: second scan

The process is repeated in two more directions, as shown in Figure 25.22, each time adding the detector values to the image voxels.

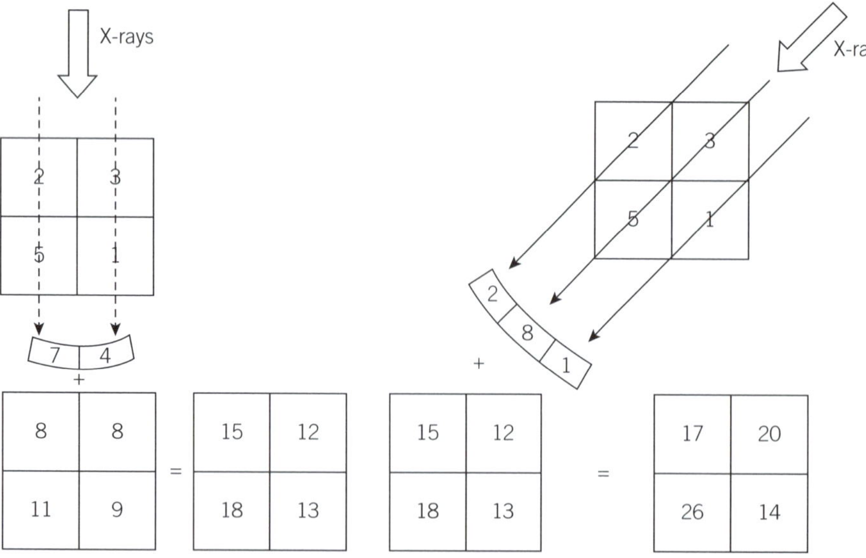

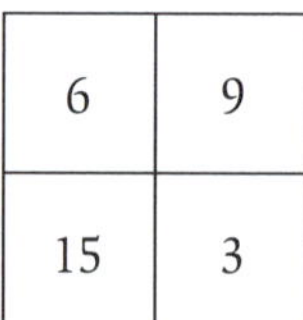

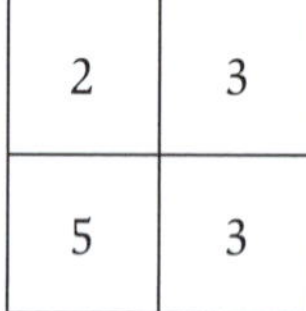

▲ **Figure 25.22** Voxels: third and fourth scans

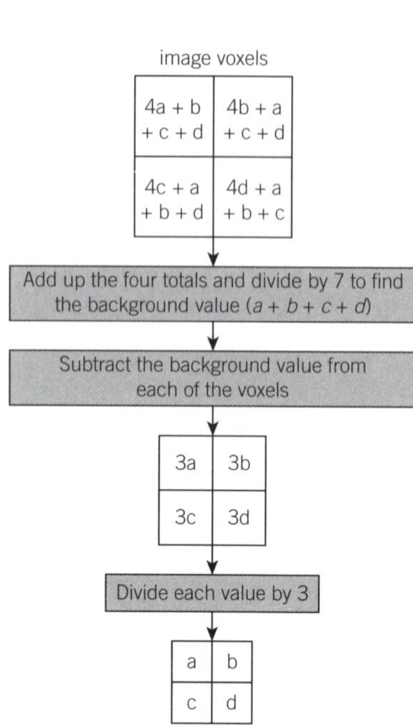

▲ **Figure 25.23** Finding the values of the object voxels

The final image voxels contain the information to be able to 'work back' to find the values of the object voxels. Each image voxel has the value of its corresponding object voxel added to it four times (once for each direction) and the value of each of the other object voxels added once.

The values of the image voxels added up is $7(a + b + c + d)$. In this example,

$$7(a + b + c + d) = 17 + 20 + 26 + 14 = 77 \quad \text{so} \quad a + b + c + d = 11$$

This value is the **background** value for each of the four image voxels.

Subtracting the background from the values of the image voxels leaves:

6	9
15	3

▲ **Figure 25.24**

These values are three times the pixel values, so the values of the object voxels must be:

2	3
5	3

▲ **Figure 25.25**

This 2×2 voxel array is in two dimensions; for a real object we would need to consider a three-dimensional $2 \times 2 \times 2$ array, called an 8-voxel cube. The object is divided into a very large number of small **8-voxel cubes**, which are exposed to X-rays from multiple directions. The large amount of numerical information obtained is then analysed by powerful computers to obtain the detailed images we see in CT scans.

 **Raise your grade**

A student is investigating the photoelectric effect using a photocell.

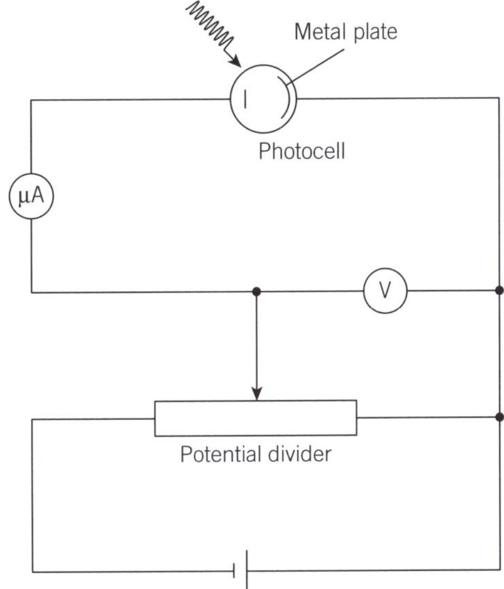

Electromagnetic radiation is incident on a metal plate inside an evacuated tube. Photoelectrons are emitted and a small current flows. A potential divider circuit provides a potential difference opposing this current. As the p.d. is increased, the current decreases and eventually falls to zero. The minimum potential difference needed to reduce the current to zero is called the stopping potential.

The student measures the stopping potential for different wavelengths of electromagnetic radiation:

Wavelength $\lambda/10^{-7}$ m	Stopping potential V_s/V	$1/\lambda/\times 10^6$ m^{-1}
5.00	0.6 ± 0.05	2.00
4.29	1.0 ± 0.05	2.33
3.75	1.4 ± 0.05	2.67
3.33	1.8 ± 0.05	3.00
3.00	2.2 ± 0.05	3.33

✔ Column heading correct with quantity and unit separated by /.

✔ Calculations correct and recorded to same number of sig. figs. as the raw data.

(a) Complete the third column in the table by calculating values of $1/\lambda$. [2]

(b) **(i)** Plot a graph of V_s on the y-axis against $1/\lambda$ on the x-axis.

Include error bars for V_s. [2]

(ii) Draw the straight line of best fit and a worst acceptable straight line on your graph.

Both lines should be clearly labelled. [2]

(iii) Determine the gradient and the y-intercept of the line of best fit.

$$\text{gradient} = \frac{2.100 - 0.725}{(3.25 - 2.1) \times 10^6} = 1.20 \times 10^{-6} \text{ Vm} ✔✔$$

Read-offs correct and substituted into gradient calculation correctly. Hypotenuse of triangle for calculating gradient larger than half the length of the line drawn. Units correct.

Using point $(3.00 \times 10^6, 1.8)$ in $y = mx + c$

gradient $\underset{\cdots\cdots\cdots\cdots\cdots\cdots}{1.20 \times 10^{-6} \text{ Vm}}$ [2]

$y\text{-intercept} = c = y - mx$

Read-off correct and substituted into $y = mx + c$ correctly.

$$= 1.8 - (1.20 \times 10^{-6}) \times (3.00 \times 10^6)$$
$$= -1.8 \text{ V} ✔✔$$

y-intercept $\underset{\cdots\cdots\cdots\cdots}{-1.8 \text{ V}}$ [2]

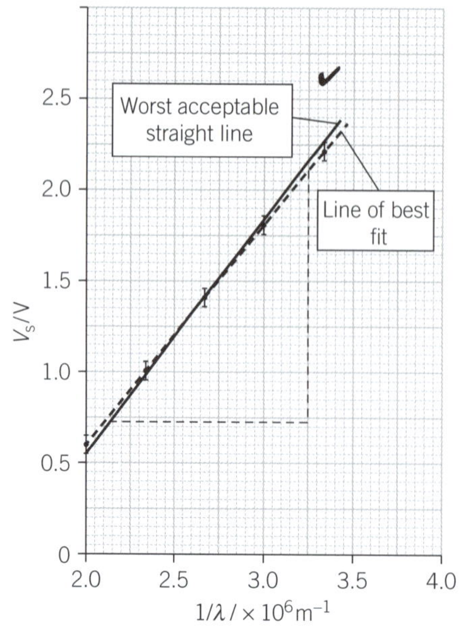

✔ Points plotted correctly.

✔

✔ Error bars drawn correctly.

The 'worst acceptable' straight line is the steepest (or the least steep) straight line which still passes through all the error bars.

(c) Theory predicts that:

$$V_s = \frac{hc}{e\lambda} - \frac{\phi}{e}$$

where e is the charge on the electron (1.6×10^{-19} C) and c is the speed of light (3.0×10^8 m s^{-1}).

Use your answers to (b)(iii) to determine h and ϕ.

$$\text{gradient} = \frac{hc}{e} = 1.20 \times 10^{-6} \; ✔$$

$$h = 1.20 \times 10^{-6} \times \frac{e}{c}$$

$$= 1.20 \times 10^{-6} \times \frac{1.6 \times 10^{-19}}{3 \times 10^8}$$

$$h = 6.4 \times 10^{-34} \; ✔$$

$$\text{y-intercept} = -\frac{\phi}{e} = -1.8 \text{ V} \; ✔$$

$$\phi = 1.8 \times 1.6 \times 10^{-19} = 2.88 \times 10^{-19} \; ✗$$

Gradient $= \dfrac{hc}{e}$.
Correct calculation of h.

$h = $ 6.4 × 10⁻³⁴ Js [2]

y-intercept $= -\dfrac{\phi}{e}$

$\phi = $ 2.88 × 10⁻¹⁹ [2]

Units for ϕ omitted (J or V C).

$[c = 3.0 \times 10^8 \, \mathrm{m\,s^{-1}}; \ e = 1.6 \times 10^{-19} \, \mathrm{C};$
$h = 6.63 \times 10^{-34} \, \mathrm{J\,s}]$

1 Electromagnetic radiation of wavelength 4.5×10^{-7} m is incident upon a metal surface which then emits electrons with a maximum kinetic energy of 4.2×10^{-20} J. Radiation of wavelength 3.0×10^{-7} m incident on the same metal produces electrons with a maximum kinetic energy of 2.63×10^{-19} J.

Calculate:

(a) the frequencies of the two radiations [3]

(b) the value of Planck's constant. [2]

2 Electromagnetic radiation of frequency 2.5×10^{15} Hz is incident on a clean magnesium surface. The work function for magnesium is 3.68 eV.

(a) Calculate:

(i) the maximum kinetic energy of the photoelectrons emitted

(ii) the threshold frequency of magnesium. [3]

(b) Determine the *stopping potential* (the p.d. needed to prevent the photoelectrons from escaping). [2]

3 Electrons, initially at rest, are accelerated across a potential difference of 40 kV.

(a) Calculate:

(i) the speed of the electrons

(ii) their momentum. [3]

(b) State what is meant by the *de Broglie wavelength*. [2]

(c) Calculate the de Broglie wavelength associated with these electrons. [2]

4 The energy levels for electrons in a helium atom are given by the equation:

$$E_n = \frac{-54.4}{n^2} \, \mathrm{eV}$$

(a) Explain why electron energy levels are negative. [1]

(b) Calculate the wavelength of the radiation emitted by an electron falling from level $n = 5$ to $n = 2$. [3]

(c) State which region of the electromagnetic spectrum this wavelength belongs to. [1]

5 The diagram shows the energy levels for the electron in a hydrogen atom.

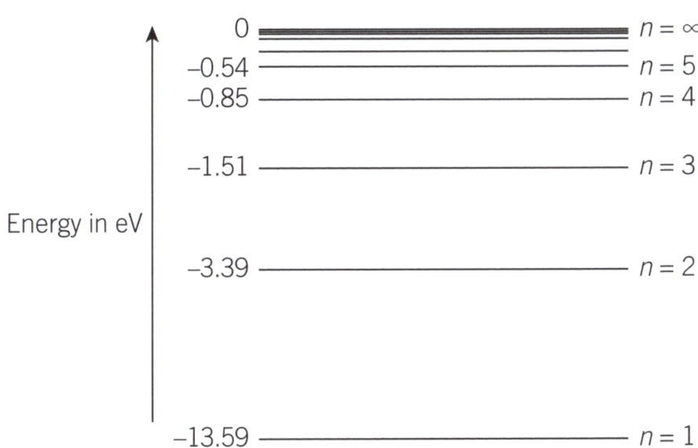

(a) Calculate the longest and shortest possible wavelengths that could be produced by a transition from an excited state to the ground state ($n = 1$). [3]

(b) Determine the number of spectral lines produced by transitions between the lowest four states. [1]

6 (a) State the difference between a gamma-ray of wavelength 10^{-11} m and an X-ray of wavelength 10^{-11} m. [1]

(b) Describe how X-rays can be produced. Include a labelled diagram in your answer. [4]

7 (a) Describe what happens to:

(i) the hardness   **(ii)** the intensity

of X-rays emitted by an X-ray tube if the accelerating potential is increased. [2]

(b) Explain why longer wavelength X-rays are filtered out when using X-rays for medical imaging. [2]

8 (a) Show that the 'half-thickness' $x_{1/2}$ of an absorber of X-rays is related to the attenuation coefficient μ by the equation:

$$x_{1/2} = \frac{\ln 2}{\mu}$$
[2]

(b) The attenuation coefficient of bone for one frequency of X-rays is $0.60 \, \mathrm{cm^{-1}}$.

(i) Determine the half thickness of bone for these X-rays.

(ii) Calculate the thickness of bone needed to absorb 90% of these X-rays. [3]

Key points

- ☐ Infer from the results of the α-particle scattering experiment the existence and small size of the nucleus.

- ☐ Describe a simple model for the nuclear atom to include protons, neutrons, and orbital electrons.

- ☐ Distinguish between nucleon number and proton number.

- ☐ Understand that different isotopes of an element have different numbers of neutrons.

- ☐ Use the usual notation for the representation of nuclides.

- ☐ Know that nucleon number, proton number, and mass-energy are all conserved in nuclear processes.

- ☐ Show an understanding of the nature and properties of α-, β-, and γ-radiations (including β^- and β^+).

- ☐ State that (electron) antineutrinos and (electron) neutrinos are produced during β^-- and β^+-decay.

- ☐ Know that protons and neutrons are not fundamental particles since they consist of quarks.

- ☐ Describe a simple quark model of hadrons in terms of up, down, and strange quarks, and their antiquarks.

- ☐ Describe protons and neutrons in terms of a simple quark model.

- ☐ Know that there is a weak interaction between quarks, giving rise to β-decay.

- ☐ Describe β^-- and β^+-decay in terms of a simple quark model.

- ☐ Know that electrons and neutrinos are leptons.

- ☐ Know the link between energy and mass, $E = mc^2$. Recall and use this relationship.

- ☐ Understand the significance of the terms mass defect and mass excess in nuclear reactions.

- ☐ Represent simple nuclear reactions by nuclear equations of the form $^{14}_{7}\text{N} + ^{4}_{2}\text{He} \rightarrow ^{17}_{8}\text{O} + ^{1}_{1}\text{H}$.

- ☐ Define and understand the terms mass defect and binding energy.

- ☐ Sketch the variation of binding energy per nucleon with nucleon number.

- ☐ Explain what is meant by nuclear fusion and nuclear fission.

- ☐ Explain the relevance of binding energy per nucleon to nuclear fusion and to nuclear fission.

- ☐ Infer the random nature of radioactive decay from the fluctuations in count rate and be aware of the spontaneous and random nature of nuclear decay.

- ☐ Define the terms activity and decay constant, and recall and solve problems using $A = \lambda N$.

- ☐ Infer and sketch the exponential nature of radioactive decay and solve problems using the relationship $x = x_0 e^{-\lambda t}$, where x could represent activity, number of undecayed nuclei, or measured count rate.

- ☐ Define half-life.

- ☐ Solve problems using the relation $\lambda = \dfrac{0.693}{t_{\frac{1}{2}}}$.

Atoms, nuclei, and radiation

Alpha-particle scattering – developing the nuclear model of the atom

At the beginning of the 20th century it was known that atoms contained both positively and negatively-charged parts, but not how these were organised inside the atom.

An experiment carried out by Geiger and Marsden was key to beginning to reveal the structure of the atom.

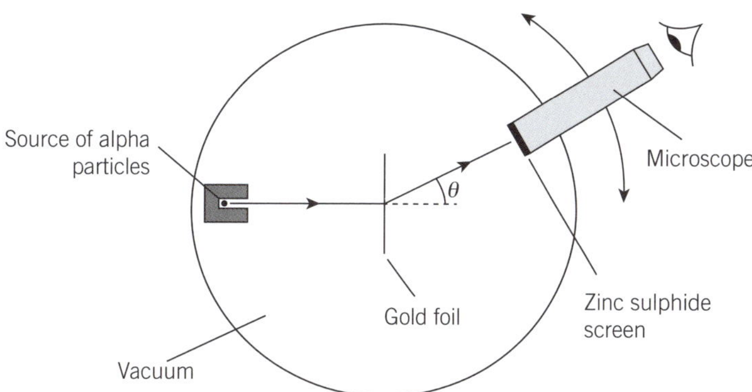

▲ **Figure 26.1** Geiger and Marsden's experiment

A thin piece of gold foil was bombarded by a **collimated** (narrow and parallel) beam of alpha particles in an evacuated chamber (see Figure 26.1). Geiger and Marsden counted the number of alpha particles deflected at different angles by the gold foil by observing small flashes of light as the alpha particles hit a zinc sulphide screen. They found that:

- most of the alpha particles passed straight through the gold foil undeflected, or only deflected by a small angle

- a few were deflected by a large angle, sometimes greater than 90°.

Rutherford described the large angle deflection of a small number of alpha particles as 'quite the most incredible event that has ever happened to me in my life. It was almost as incredible as if you fired a 15-inch shell at a piece of tissue paper and it came back and hit you.' He showed that these observations could be explained by the atom being mostly empty space, with most of the mass of the atom concentrated in a (positively) charged nucleus (see Figure 26.2).

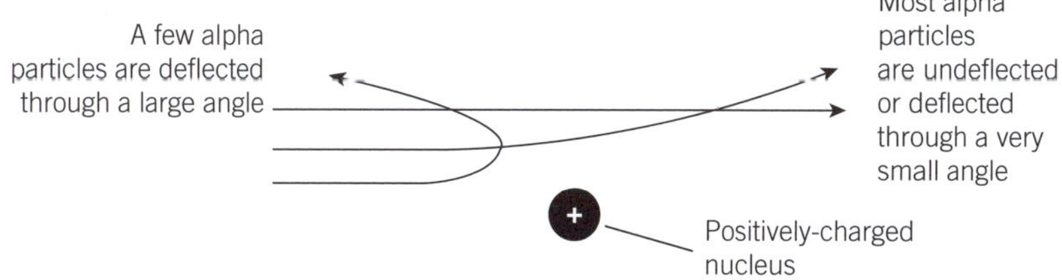

▲ **Figure 26.2** Alpha particle scattering

Later experiments showed that there were two types of particle in the nucleus: positively-charged protons and uncharged neutrons with electrons 'in orbit' around the nucleus. There were an equal number of protons and electrons, with the charge on each proton being $+e$.

Nucleon number, proton number and isotopes

The planetary model of the atom consists of a nucleus formed of **protons** and **neutrons**, surrounded by **electrons**, as shown in Figure 26.3.

> **Remember**
>
> **nucleon number** A: the total number of particles in the nucleus (neutrons + protons).
>
> **proton number** Z: the number of protons in the nucleus (equal to the number of electrons orbiting the nucleus).

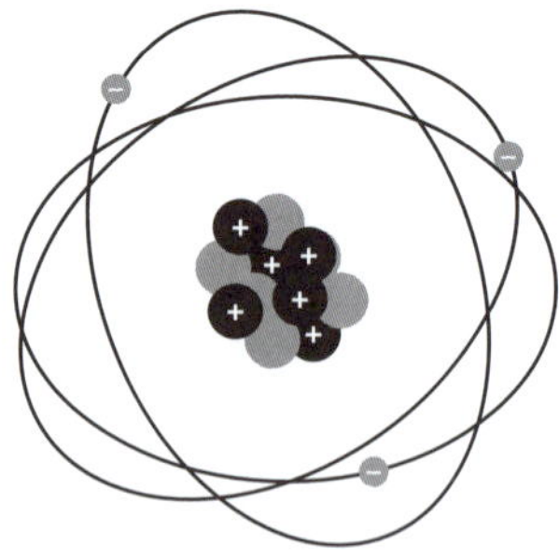

▲ **Figure 26.3** Planetary model of the atom

The number of neutrons in the nucleus is $A - Z$. The proton number Z is different for each element (Figure 26.4). For example, if Z is 3, the element is lithium.

Isotopes are different versions of the same element. Their nuclei have the same number of protons but different numbers of neutrons. For example, $^{12}_{6}C$ and $^{13}_{6}C$ are different isotopes of carbon with exactly the same chemical properties – they both have six protons in their nuclei, but $^{12}_{6}C$ has six neutrons and $^{13}_{6}C$ has seven neutrons.

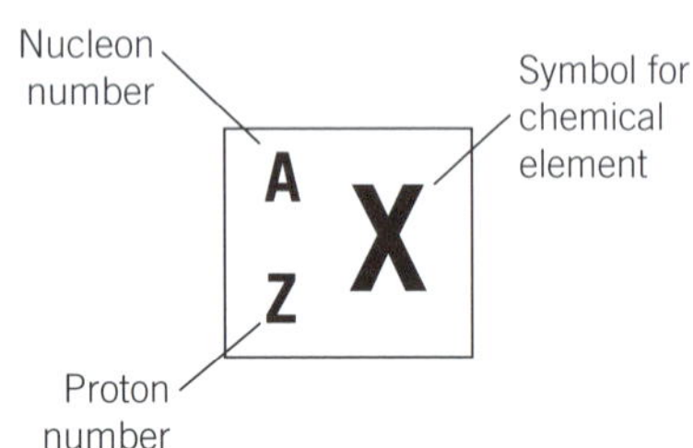

▲ **Figure 26.4** Proton number and nucleon number

Worked example

$^{60}_{27}Co$ is an isotope of cobalt. State how many:

a) protons   **b)** neutrons   **c)** nucleons   **d)** electrons there are in an atom of this isotope.

Answer

a) There are 27 protons (the proton number).   **b)** There are $60 - 27 = 33$ neutrons.

c) The total number of particles in the nucleus is 60.   **d)** The number of electrons is the same as the proton number, 27.

In any nuclear process both the nucleon number and the proton number are conserved; for example:

$$^{9}_{4}Be + ^{4}_{2}He \rightarrow ^{12}_{6}C + ^{1}_{0}n$$

The total number of nucleons remains constant (13) as does the total number of protons (6). The nuclei of some isotopes are unstable – at some point they will 'decay' by emitting an alpha particle, beta particle or gamma ray. Some of the mass of the nucleus is converted to energy of the emitted radiation (e.g. the kinetic energy of the alpha particle). This process is called **radioactive decay**.

α-, β-, and γ-decay

Alpha (α) decay

An **alpha particle** consists of two protons and two neutrons (identical to a $^{4}_{2}H$ nucleus). The proton number of a nuclide emitting an alpha particle will decrease by two and the nucleon number will decrease by four.

For example, radium-226 decays into radon-222 by emitting an alpha particle:

$$^{226}_{88}Ra \rightarrow ^{222}_{86}Rn + ^{4}_{2}He + energy$$

The mass of the radon-222 nucleus added to the mass of the helium-4 nucleus is slightly less than the mass of the radium-226 nucleus. The mass lost (called the **mass defect**) has been 'converted' into energy (the α-particle has kinetic energy, for example), but the total **mass–energy** is the same.

> **Remember**
>
> Both nucleon number and proton number are conserved in nuclear reactions, as is mass–energy.

Beta (β) decay

There are two types of **beta decay**:

- **β⁻-decay** occurs when a neutron in the nucleus changes into a proton and an electron. The electron is emitted as a fast-moving β⁻-particle. A very light, electrically-neutral **antiparticle**, called an electron antineutrino, is also emitted.

  For example, carbon-14 decays or 'transmutes' into nitrogen-14 by β⁻-decay:

$$^{14}_{6}\text{C} \rightarrow {}^{14}_{7}\text{N} + {}^{0}_{-1}\text{e} + \bar{\nu} + \text{energy}$$

> ν is the symbol for an electron neutrino, a neutral particle with almost no mass. $\bar{\nu}$ is the symbol for an electron antineutrino.

- **β⁺-decay** occurs when a proton decays into a neutron and a positron $(^{0}_{+1}\text{e})$ and an electron neutrino (ν) are emitted. The positron is emitted as a fast-moving β⁺-particle. The electron neutrino is a very light, electrically neutral particle.

  A positron is the antimatter equivalent of an electron – if a positron and an electron were to meet they would annihilate each other, becoming electromagnetic energy.

  For example, carbon-10 decays into boron-10 by β⁺-decay:

$$^{10}_{6}\text{C} \rightarrow {}^{10}_{5}\text{B} + {}^{0}_{+1}\text{e} + \nu + \text{energy}$$

Gamma (γ) decay

Gamma decay occurs when a γ-ray is emitted (a high-frequency electromagnetic wave). There is no change in either the proton number or the nucleon number. Gamma decay can occur alongside alpha or beta decay, when an unstable nucleus adjusts to a more stable energy level.

Properties of alpha, beta, and gamma radiations

The Table 26.1 lists some important properties of α-, β-, and γ-radiations.

▼ **Table 26.1** Properties of α-, β-, and γ-radiations

	Alpha (α)	Beta (β)	Gamma (γ)
Nature	2 protons + 2 neutrons ($^{4}_{2}$He nucleus)	fast-moving electron (or positron for β⁺-decay)	high-frequency electromagnetic wave
Charge	+2e	−e (or +e for β⁺-decay)	no charge
Range in air	a few cm	~ 1 m	unlimited
Stopped by …	a few sheets of paper	several millimetres of aluminium	several centimetres of lead

> **Remember**
>
> Matter and antimatter:
>
> - For every known type of particle, such as an electron or proton, there is a corresponding **antiparticle** with the same mass and opposite charge (e.g., an antielectron called a positron) and an antiproton.
>
> - If a particle collides with its antiparticle, they annihilate each other, producing photons. Equally, a photon of sufficient energy can create a particle and its corresponding antiparticle.

Alpha particles and beta particles are deflected in opposite directions in a uniform electric field, but gamma rays are undeflected as they have no charge (see Figure 26.5). The force on an alpha particle is twice as large as the force on an electron moving at the same speed, but the deflection of the alpha particle is much smaller due to its much greater mass ($m_\alpha \approx 7000\,m_e$.)

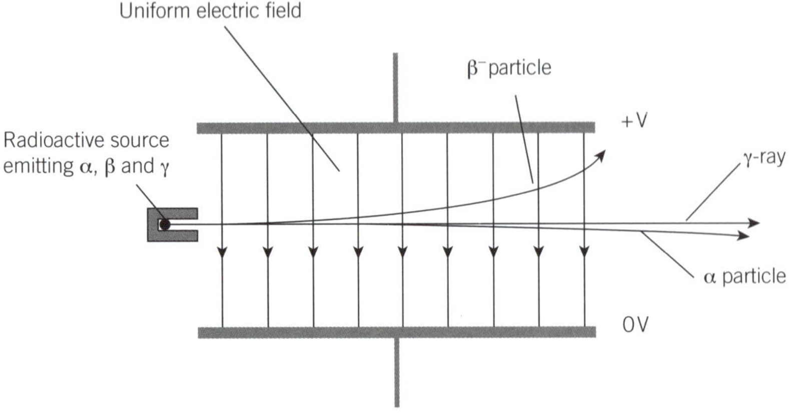

▲ **Figure 26.5** Deflections of α, β-, and γ radiation in an electric field

When alpha particles and beta particles enter a magnetic field they are deflected in opposite directions; gamma rays are undeflected, as shown in Figure 26.6.

Fundamental particles

Electrons, neutrons, and protons were once thought to be **fundamental particles** (i.e., they did not consist of combinations of other particles). It was later discovered that, although electrons are still believed to be fundamental particles, protons and neutrons consist of combinations of smaller particles. These particles were given the name **quarks**.

The **standard model of particle physics** asserts that there are 12 fundamental particles, which can be divided into two groups, according to their properties, as shown in Table 26.2.

Quarks: there are six types of quark. Protons and neutrons are made up of different combinations of quarks.

Leptons: there are six types of lepton. An electron is one example of a lepton. All leptons have very small masses (lepton means light in Greek).

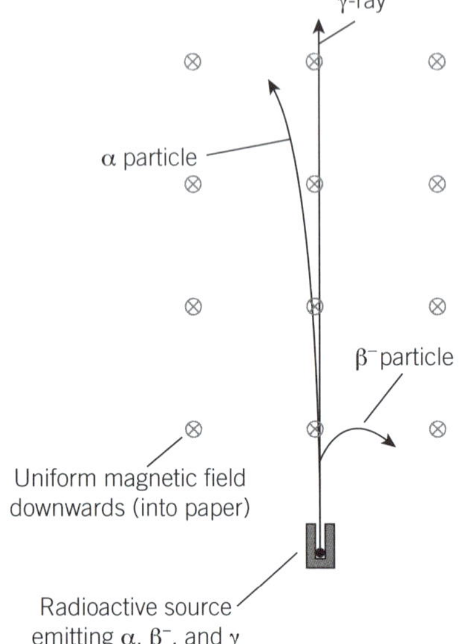

▲ **Figure 26.6** Deflections of α, β-, and γ radiation in a magnetic field

Link

See Unit 22 *Magnetic fields* for more about the deflections of moving charged particles in magnetic fields.

▼ **Table 26.2** Quarks and leptons

				Charge / e
Quarks	up, u	charm, c	top, t	$+\dfrac{2}{3}$
	down, d	strange, s	bottom, b	$-\dfrac{1}{3}$
Leptons	electron, e	muon, μ	tau, τ	-1
	electron-neutrino, v_e	muon-neutrino, v_μ	tau-neutrino, v_τ	0

Quarks occur in groups of two or three, never separately. The top quark is the heaviest with a mass approximately 200 times the mass of a proton. As well as the 12 fundamental particles, there are 12 equivalent antiparticles.

There are four fundamental forces that control the interactions between fundamental particles, as shown in Table 26.3.

▼ **Table 26.3** Fundamental forces

Force	Range	Acts on
Gravity	no limit	all objects
Electromagnetic	no limit	charged objects
Strong nuclear force	10^{-15} m	quarks and antiquarks
Weak nuclear force	10^{-18} m	fundamental particles

A proton consists of two up quarks and one down quark (uud), held together by the **strong nuclear force**. A neutron consists of one up quark and two down quarks (udd). Particles that consist of combinations of quarks and antiquarks are called **hadrons** (hadrons are defined as particles held together by the strong nuclear force). Baryons are particles consisting of three quarks (Figure 26.7). They include protons and neutrons. Mesons (Figure 26.8) are particles consisting of one quark and one antiquark. Antibaryons consist of three antiquarks (Figure 26.9).

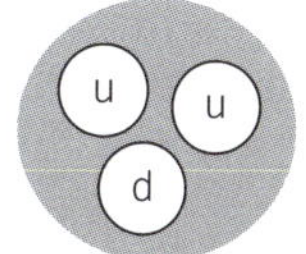
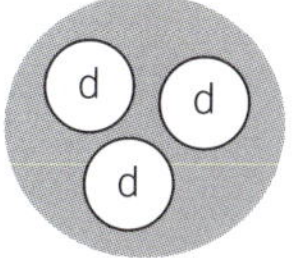
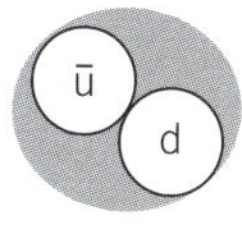

▲ **Figure 26.7** Some baryons

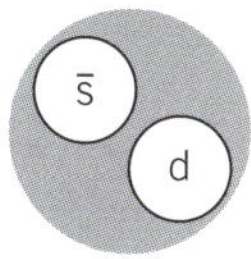

▲ **Figure 26.8** Some mesons

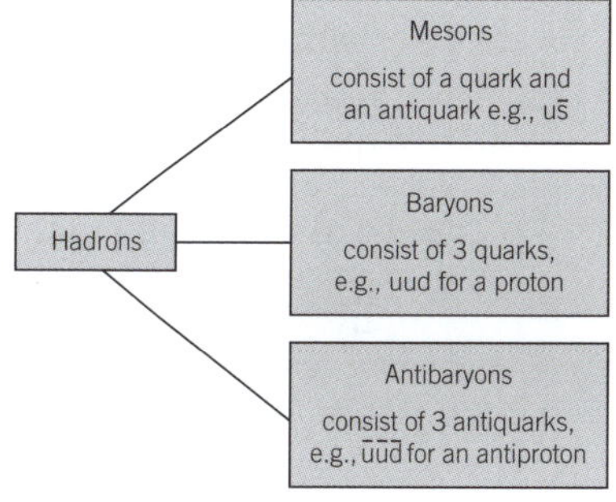

▲ **Figure 26.9** Hadrons

> **Worked example**
>
> One type of hadron consists of two down quarks and one strange quark. State the charge on this hadron.
>
> **Answer**
>
> Both the down quark and the strange quark have a charge $-\frac{1}{3}e$, so the total charge must be $-e$.

Quarks and beta decay

In β^--decay, a down quark changes into an up quark in one of the neutrons in a nucleus, making it a proton, and in doing so emits an electron (the β^--particle) and an electron antineutrino. In β^+-decay, one of the protons in a nucleus changes into a neutron by one of the up quarks changing into a down quark, emitting a positron (the β^+-particle) and an electron neutrino in the process.

The force (or interaction) responsible for beta decay, causing a neutron to change into a proton (or a proton into a neutron), is the **weak nuclear force** (or **weak interaction**).

Mass defect, mass excess, and binding energy

The mass of an atomic nucleus is slightly less than the total mass of the separate nucleons. For example, a helium-4 nucleus consists of two protons and two neutrons. Table 26.4 gives the masses of a proton, a neutron, and a helium-4 nucleus in u (atomic mass units).

▼ **Table 26.4** Masses of proton, neutron, and a helium-4 nucleus

	Mass/u
Proton	1.00728
Neutron	1.00867
Helium-4 nucleus	4.00150

> 💡 **Remember**
>
> The **unified atomic mass unit** (u) is defined as $\frac{1}{12}$ of the mass of a ^{12}C atom (including its electrons).
>
> $1\,\text{u} = 1.661 \times 10^{-27}\,\text{kg}$

The mass of a helium-4 nucleus (Figure 26.10) is less than the combined mass of 2 protons and 2 neutrons. The difference is called the **mass defect** Δm.

$$\Delta m = 2 \times 1.00728 + 2 \times 1.00867 - 4.00150 = 0.0304\,\text{u}$$
$$= 5.046 \times 10^{-29}\,\text{kg}$$

Einstein showed that mass m and energy E are related by the equation:

$$E = mc^2$$

where c is the speed of light ($3.0 \times 10^8\,\text{m s}^{-1}$). For a mass change Δm, the corresponding change in energy is:

$$\Delta E = c^2(\Delta m)$$

For a helium nucleus, $\Delta E = (3.0 \times 10^8)^2 \times 5.046 \times 10^{-29} = 4.54 \times 10^{-12}\,\text{J}$

This is the energy that is needed to completely separate a helium nucleus into individual protons and neutrons, and is called the **binding energy**. The value of the binding energy expressed in joule is very small, and so binding energies are usually expressed in electronvolts (eV).

Binding energy of a helium-4 nucleus in eV $= \dfrac{4.54 \times 10^{-12}}{1.6 \times 10^{-19}} = 2.84 \times 10^7\,\text{eV}$

The binding energy per nucleon $= \dfrac{2.84 \times 10^7}{4} = 7.1\,\text{MeV}$

Figure 26.11 shows how the binding energy/nucleon varies with nucleon number.

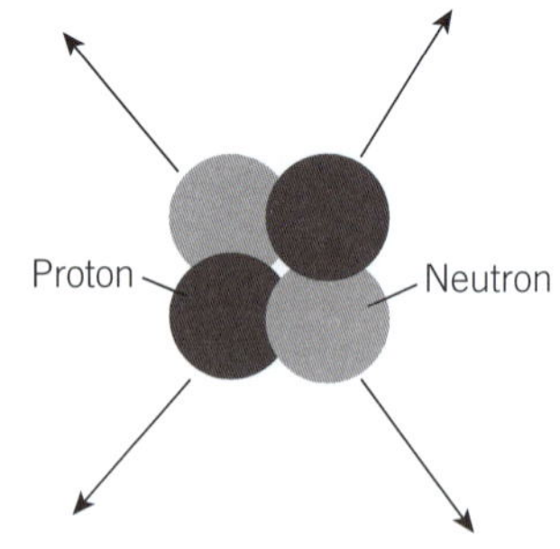

▲ **Figure 26.10** Binding energy of helium-4

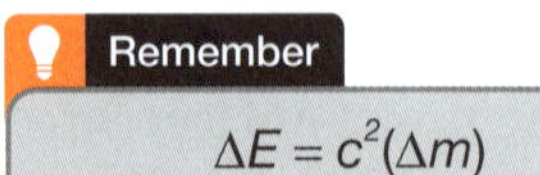

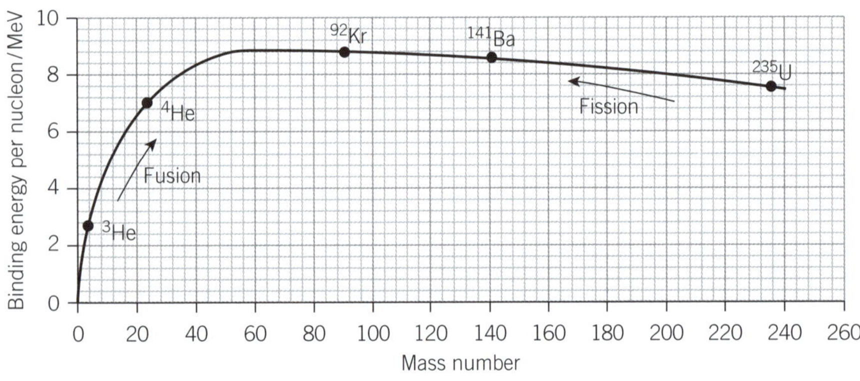

▲ **Figure 26.11** Binding energy per nucleon

The greater the binding energy per nucleon, the more stable the nucleus because more energy is needed to remove a nucleon from the nucleus.

Nuclear fission

In **nuclear fission** a heavy nucleus splits into two lighter nuclei, together with a few individual neutrons. A large amount of energy is released, principally as kinetic energy of the fission fragments.

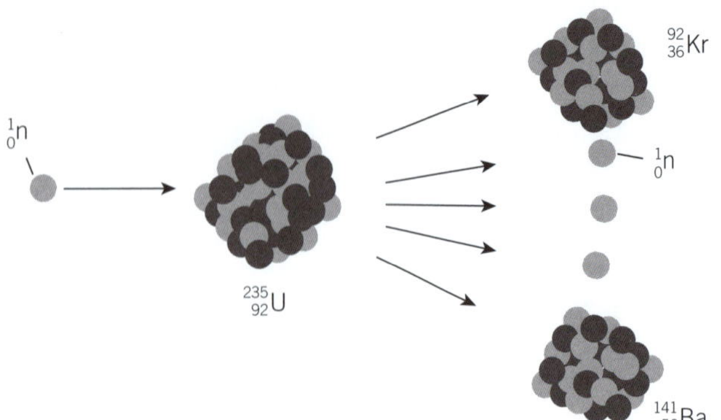

▲ **Figure 26.12** Nuclear fission

Figure 26.12 illustrates one example of nuclear fission. A ^{235}U nucleus absorbs a slow-moving ('thermal') neutron, momentarily becoming ^{236}U which is unstable, and splits into a nucleus of ^{92}Kr and ^{141}Ba, together with three neutrons. The binding energy/nucleon for ^{92}Kr and ^{141}Ba is greater than the binding energy per nucleon for ^{235}U. Each nucleon is now more tightly bound to the nucleus, requiring more energy to 'escape' from the nucleus than before the fission reaction, and so these nucleons must have lost energy (only the three 'free' neutrons have gained energy). This energy appears mainly as kinetic energy of the fission fragments (they are 'hot'!).

Energy released in nuclear fission

Worked example

Calculate the energy released by one fission reaction in the example given in Figure 26.12, using the BE per nucleon values given in Table 26.5.

Answer

Energy released per fission reaction:

$$\Delta E = 141 \times (8.3 - 7.6) + 92 \times (8.5 - 7.6) - 2 \times 7.6 = 166\,\text{MeV}$$

▼ Table 26.5

	BE per nucleon/ MeV
^{235}U	7.6
^{141}Ba	8.3
^{92}K	8.5

Mass excess

The mass of a nuclide in atomic mass units is very close to the nucleon number A. For example, the mass of U-238 is slightly more than $238\,\text{u}$ ($238.050788\,\text{u}$). The difference between the two is known as the **mass excess**.

$$\text{mass excess} = \text{mass (in u)} - \text{nucleon number}$$

Nuclear fusion

Nuclear fusion occurs when two light nuclei join together to form a larger, more stable nucleus. In one of the fusion reactions taking place in the Sun, two ^{3}He nuclei combine to form ^{4}He as shown in Figure 26.13.

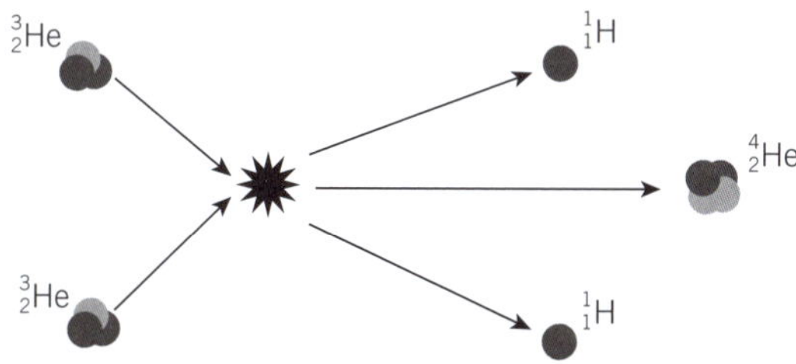

▲ **Figure 26.13** Nuclear fusion

Energy released in nuclear fusion

Four of the nucleons finish in a more stable nucleus so have lost energy; the other two have become single, 'free' protons so have gained energy.

Worked example

Calculate the energy released per fission reaction in the example given in Figure 26.13, using the binding energy per nucleon values given in Table 26.6.

Answer

Energy released per fission reaction:

$$\Delta E = 4 \times (7.1 - 2.6) - 2 \times 2.6 = 12.8\,\text{MeV}$$

▼ Table 26.6

	BE per nucleon/ MeV
^{3}He	2.6
^{4}He	7.1

Radioactive decay

The exact time an unstable nucleus will decay cannot be predicted – radioactive decay is a **random** process – but the rate of decay of a very large number of atoms of any particular radioactive isotope can be calculated accurately. The activity A (the number of decays per second) is given by:

$$A = \lambda N$$

where N is the number of undecayed nuclei and λ is a constant called the **decay constant**. λ is the probability of a nucleus decaying in unit time (the fraction of atoms that decay in unit time) and has a different value for different radioactive isotopes.

The **activity** A is the number of nuclei decaying per second (the count rate) $\dfrac{dN}{dt}$, so:

$$\frac{dN}{dt} = -\lambda N$$

The minus sign occurs because the number of undecayed nuclei is decreasing with time. The solution to this equation is:

$$N = N_0 e^{-\lambda t}$$

where N_0 is the number of undecayed atoms at time $t = 0$ (see Figure 26.14).

As the rate of decay is proportional to N:

$$A = A_0 e^{-\lambda t}$$

> **Remember**
>
> The activity is measured in **bequerels** (Bq).
>
> 1 Bq = 1 disintegration per second

> **Remember**
>
> The **decay constant** is the probability of a nucleus decaying per unit time.

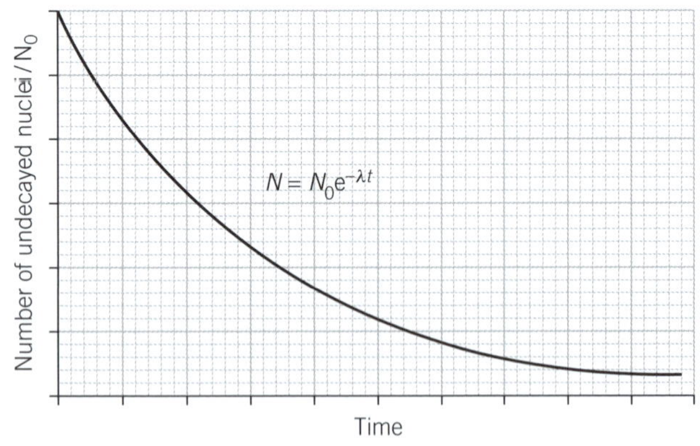

▲ **Figure 26.14** Exponential decay of the number of undecayed nuclei

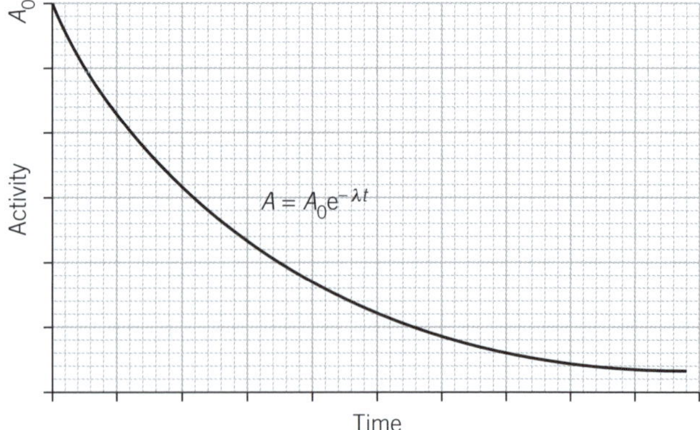

▲ **Figure 26.15** Exponential decay of activity

The activity A (the number of disintegrations per second) also decreases **exponentially** (Figure 26.15).

A useful measure of the rate of decay is the **half-life** $t_{1/2}$ – the time it takes for the number of undecayed nuclei to fall by half (and for the activity to halve).

From $N = N_0 e^{-\lambda t}$:

$$\frac{N_0}{2} = N_0\, e^{-\lambda t_{1/2}} \text{ so,} \qquad e^{\lambda t_{1/2}} = 2$$

$$\lambda t_{1/2} = \ln 2 = 0.693$$

$$t_{1/2} = \frac{0.693}{\lambda}$$

> **Remember**
>
> The **half-life** $t_{1/2}$ is the time it takes for half the radioactive nuclei to decay (and the time it takes for the activity to halve).

In Figure 26.16 the activity falls from 12 kBq to 6 kBq in 10 s. It then falls from 6 kBq to 3 kBq in the next 10 s, from 3 kBq to 1500 Bq in the next 10 s, and so on.

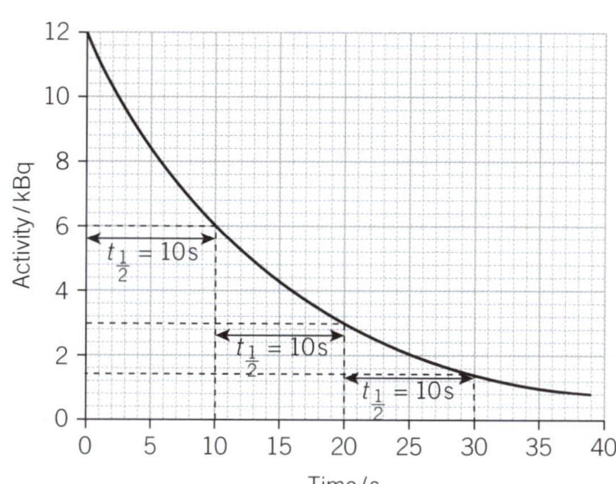

▲ **Figure 26.16** Half-life

Half-lives of radioactive isotopes vary from fractions of a second to many millions of years.

Worked examples

1 An isotope of radon has a half-life of 3.83 days. If a sample of the gas has a mass of 12.0 g at time $t = 0$, determine how much of the sample remains after:

   a) 1 day   **b)** 1 week   **c)** 1 year.

Answer

The decay constant $\lambda = \dfrac{0.693}{t_{1/2}} = \dfrac{0.693}{3.83} = 0.181 \text{ days}^{-1}$

a) Fraction remaining after 1 day $= \dfrac{N}{N_0} = e^{-\lambda t} = e^{-0.181 \times 1} = 0.834$

   Mass of radon gas remaining $= 12.0 \times 0.834 = 10.0 \text{ g}$

b) Fraction remaining after 1 week $= e^{-\lambda t} = e^{-0.181 \times 7} = 0.282$

   Mass of radon gas remaining $= 12.0 \times 0.282 = 3.4 \text{ g}$

c) Fraction remaining after 1 year $= e^{-\lambda t} = e^{-0.181 \times 365} = 2.03 \times 10^{-29}$

   Mass of radon gas remaining $= 12.0 \times 2.03 \times 10^{-29} = 2.44 \times 10^{-28} \text{ g}$.

2 Uranium-235 has a half-life of 7.1×10^8 years. Calculate the activity of a sample of 1.0 g of U-235.

Answer

1.0 g of U-235 contains $1.0 \times \dfrac{6.02 \times 10^{23}}{235} = 2.56 \times 10^{21}$ atoms of uranium.

The decay constant $= \lambda = \dfrac{0.693}{7.1 \times 10^8} = 9.76 \times 10^{-10} \text{ years}^{-1}$

The activity $A = 2.56 \times 10^{21} \times 9.76 \times 10^{-10} = 2.50 \times 10^{12}$ disintegrations/year $= 8.1 \times 10^4$ Bq

1 In Geiger and Marsden's α-particle scattering experiment a collimated beam of alpha particles is fired at a thin gold foil. Most of the α-particles pass straight through the foil, or are deflected by a small angle. About one in 8000 α-particles are deflected by an angle greater than 90°.

(a) (i) State what is meant by *collimated*.

A parallel beam of α-particles ✔

> i.e., not 'fanning out' in all directions. [1]

(ii) Suggest why a *thin* foil is used.

If the foil was too thick the α-particles would not go through ✗

> Though the statement is true, more detail is needed. Ideally the foil is a few atoms thick – otherwise the alpha particles would be deflected several times. [1]

(b) State two deductions about the structure of the atom that can be drawn from the results of Geiger and Marsden's experiment.

1. The atom is mostly empty space ✔

> A good answer.

2. Most of the mass of the atom is concentrated in the nucleus ✔ ✗

> The candidate should have added '… which is (positively) charged …' for the second mark. [3]

(c) Alpha particles of fixed energy are used in the experiment. Suggest what would happen if alpha particles with greater energy are used.

More of the α-particles would be deflected by more than 90° ✗ [1]

> Just the opposite – a smaller proportion of the α-particles would be deflected by a large angle. As they have more kinetic energy, they are moving faster, and so the repulsive force from a gold nucleus will act for a shorter time.

2 A neutron may decay into a proton together with two other particles.

(a) (i) The equation for this decay is shown below. Complete the equation.

$$^1_0 n \rightarrow\ ^1_1 p + ^{\ \ 0}_{-1} e + ^{\ 0}_{\ 0} \bar{\nu}$$
       ✔   ✔ ✔ [3]

(ii) State the name of the particle $\bar{\nu}$.

An electron neutrino ✗

> $\bar{\nu}$ is the symbol for an electron antineutrino – the bar above the symbol indicates an antiparticle. [1]

(iii) State two quantities that are conserved during the decay.

1. The proton number ✔

2. The nucleon number ✔

> Other valid answers: mass–energy, charge, and momentum. [2]

(iv) State the force that gives rise to this decay.

The strong nuclear force ✗

> The **weak nuclear force** (or interaction) is linked to β-decay. [1]

(b) State the quark composition of

(i) a neutron udd ✔   **(ii)** a proton uud ✔ [2]

? Exam-style questions

1 Which statement is supported by the scattering of alpha particles by gold foil? [1]

   A The atom contains electrons in orbit around the nucleus.

   B The atom is mostly empty space.

   C The nucleus is held together by strong forces.

   D The nucleus of an atom contains positively charged particles and electrically neutral particles.

2 Which statement about the nuclei of two different isotopes of a substance is correct? [1]

   A The two nuclei contain equal numbers of neutrons.

   B The two nuclei contain equal numbers of nucleons.

   C The two nuclei contain equal numbers of protons and neutrons.

   D The two nuclei contain equal numbers of protons.

3 ^{218}Po, an isotope of polonium, decays into an isotope of lead by emitting an alpha particle. The lead decays into an isotope of bismuth by emitting a β^- particle:

$$^{218}_{84}\text{Po} \rightarrow {}^{X}_{82}\text{Pb} \rightarrow {}^{214}_{Y}\text{Bi}$$

What are the values of X and Y? [1]

	X	Y
A	214	81
B	214	83
C	218	81
D	218	83

4 How many nucleons are there in a nucleus of an atom of $^{23}_{11}$Na? [1]

   A 11   **B** 12   **C** 23   **D** 34

5 The isotope niobium-86 ($^{86}_{41}$Nb) decays by β^+-decay into an isotope of zirconium (Zr). Which equation describes the decay? [1]

   A $^{86}_{41}\text{Nb} \rightarrow {}^{86}_{40}\text{Zr} + {}^{0}_{-1}e + v$

   B $^{86}_{41}\text{Nb} \rightarrow {}^{86}_{40}\text{Zr} + {}^{0}_{+1}e + v$

   C $^{86}_{41}\text{Nb} \rightarrow {}^{86}_{40}\text{Zr} + {}^{0}_{-1}e + \bar{v}$

   D $^{86}_{41}\text{Nb} \rightarrow {}^{86}_{40}\text{Zr} + {}^{0}_{+1}e + \bar{v}$

6 The Δ^{++} hadron has a charge $+2e$ and consists of three quarks. If two of the quarks are up quarks what is the third quark? [1]

   A d   **B** $\bar{\text{d}}$ **C** u   **D** $\bar{\text{u}}$

7 Which one of the particles listed is not a fundamental particle? [1]

   A an electron   **B** a neutrino

   C a quark   **D** a proton

8 The half-life of an isotope of radium is 1620 years.

   (a) Explain what is meant by *isotope*. [2]

   (b) Define radioactive *half-life*. [2]

   (c) Show that the decay constant is 4.28×10^{-4} per year. [2]

   (d) A sample of this isotope has an activity of 6.2×10^{9} Bq. Calculate the number of radium atoms in the sample. [2]

9 A thermal neutron collides with a uranium nucleus $^{235}_{92}$U, which undergoes nuclear fission. The fission products are $^{93}_{38}$Sr and $^{140}_{54}$Xe, together with two neutrons:

$$^{235}_{92}\text{U} + {}^{1}_{0}\text{n} \rightarrow {}^{93}_{38}\text{Sr} + {}^{140}_{54}\text{Xe} + 3\,{}^{1}_{0}\text{n} + \text{energy}$$

   (a) Explain what is meant by a *thermal neutron*. [1]

   (b) State what is meant by *nuclear fission*. [2]

   (c)

	Mass/u
$^{235}_{92}$U	235.04393
$^{93}_{38}$Sr	92.91399
$^{140}_{54}$Xe	139.92162

   Using the masses given in the table, calculate:

   (i) the mass defect, in atomic mass units

   (ii) the energy released in the reaction, in MeV. [5]

10 An isotope of rubidium, $^{87}_{37}$Ru, decays by β^--decay into $^{87}_{38}$Sr, which is a stable isotope. The half-life of $^{87}_{37}$Ru is 4.9×10^{10} years.

The ratio of ^{87}Sr to ^{87}Ru in a sample of rock is found to be 0.0060.

Assuming that there was no ^{87}Sr when the rock was formed, calculate the age of the rock. [4]

27 Practical assessment

Paper 3: Advanced practical skills

There are two questions on Paper 3 *Advanced practical skills*, each question lasting 1 hour and each worth 20 marks. The two questions are set in different areas of physics. They are designed to test your practical skills – no prior knowledge of the theory is needed.

Question 1

The first question usually requires you to set up some simple apparatus, such as an electrical circuit, or a beam supported by springs or an oscillating system. You will then make a number of simple measurements, such as measuring length using a rule with a millimetre scale, angle using a protractor, or time using a stopwatch.

Once you've taken your measurements (the **raw data**) you will be asked to calculate other quantities from them, and then to plot a graph. You will then find values from the graph, such as the gradient and the *y*-intercept, and use these values to find constants in an equation.

Question 2

The second question involves carrying out an experiment using apparatus that you may need to assemble, and taking a number of measurements. You will be asked to estimate the percentage uncertainty in one these measurements and to consider the appropriate number of significant figures for any calculated values.

You will record a set of measurements for two values of the **independent variable**, and be asked whether the results you've obtained support a hypothesis. Some of the readings are designed to be difficult to measure accurately, and you will be asked to identify any limitations or sources of uncertainty in the experiment, and to suggest improvements.

Marking and assessment

Tables 27.1 and 27.2 give details of how marks are allocated for each question.

▼ **Table 27.1** Mark allocation for Question 1

Question 1 (20 marks)			
Skill	**Mark**	**Skills needed**	**Mark allocation***
Manipulation, measurement and observation	7 marks	Collecting data successfully	5
		Selecting a suitable range of values	1
		Ensuring good quality data	1
Presentation of data and observations	6 marks	Compiling a table of results	1
		Recording data, observations, and calculations	2
		Drawing a graph	3
Analysis, conclusions and evaluation	4 marks	Interpreting the graph	2
		Drawing conclusions	2

*The remaining 3 marks are allocated across the skills in the table.

You could also be asked to measure force using a newton-meter, a volume of liquid using a measuring cylinder, or temperature with a thermometer. Other quantities, such as the mass on a mass hanger, or the number of paperclips used to balance a wooden beam, may also need to be recorded.

> **Remember**
>
> The **independent variable** is the one **you change** and control; for example, the length of a resistance wire or the number of masses hung on a spring.
>
> The **dependent variable** is the one that is **altered** by the change you make; for example, the current in a circuit or the extension of a spring.

▼ **Table 27.2** Mark allocation for Question 2

Question 2 (20 marks)			
Skill	**Mark**	**Skills needed**	**Mark allocation***
Manipulation, measurement and observation	5 marks	Collecting data successfully	4
		Ensuring good quality data	1
Presentation of data and observations	2 marks	Recording data, observations and calculations	2
Analysis, conclusions and evaluation	10 marks	Drawing conclusions	1
		Estimating uncertainties	1
		Identifying limitations	4
		Suggesting improvements	4

*The remaining 3 marks are allocated across the skills in the table.

Key skills

Measuring length

Using a metre rule

When using a metre, half-metre, or 30 cm rule with a millimetre scale, **all** the raw measurements should be recorded to the nearest millimetre. The mean value should normally be recorded to the same number of significant figures as the raw measurements (or one significant figure more).

Diameters can be measured using a rule together with two set squares, as shown in Figure 27.1.

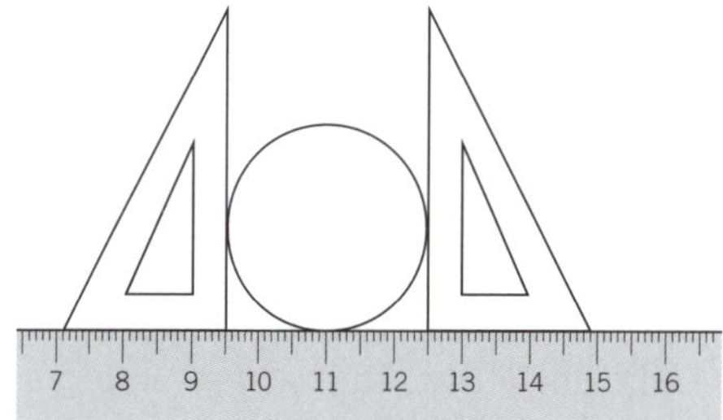

▲ **Figure 27.1** Measuring diameters

Using vernier calipers

If the measurement to be made is less than 15–20 cm in length, then a pair of vernier calipers can be used to obtain a more accurate value. The value of a length measured using vernier calipers can usually be recorded to the nearest 0.1 mm (0.01 cm). Very high quality vernier calipers can give readings accurate to the nearest 0.02 mm.

To read a vernier scale (see Figure 27.2):

- first read the value on the main scale that is just before the zero line on the vernier scale

- then read the value on the vernier scale in line with a marking on the main scale and add this to the first reading.

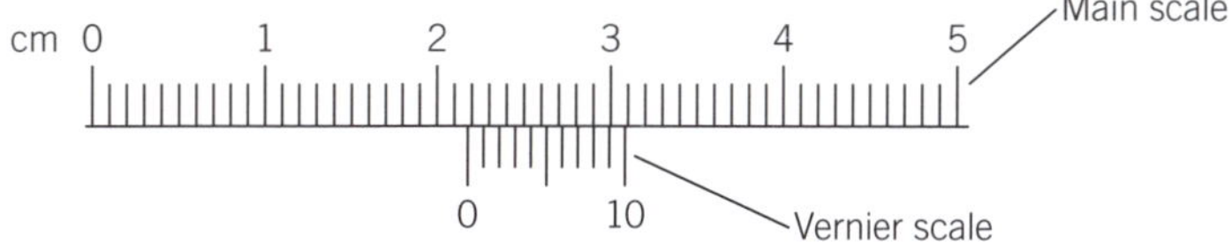

▲ **Figure 27.2** Using a vernier scale

In Figure 27.2:

- main scale reading: 2.1 cm

- reading on vernier scale coinciding with reading on main scale: 0.8 mm (0.08 cm)

- final reading = 2.1 + 0.08 = 2.18 cm.

Using a micrometer

If available, a screw-gauge micrometer (see Figure 27.3) can be used to measure lengths up to 2–3 cm, to an accuracy of 0.01 mm. The resistance wires used in electrical circuits often have diameters less than 1.00 mm, and so particular care needs to be taken to read the micrometer correctly:

- the reading on the barrel gives the length to the nearest 0.5 mm

- the reading where the centre line meets the thimble gives the value to be added on between 0.00 mm and 0.50 mm (one complete rotation of the thimble moves one jaw of the micrometer by 0.50 mm).

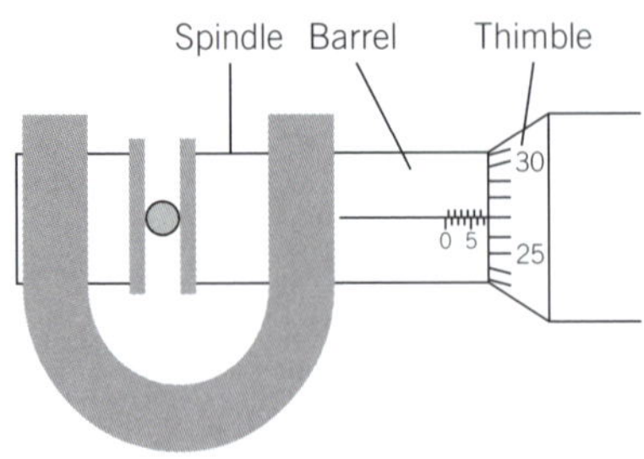

▲ **Figure 27.3** Reading a micrometer scale

In Figure 27.3:

- reading on the barrel is 7.00 mm

- reading on the thimble is 0.27 mm

- final reading = 7.00 + 0.27 = 7.27 mm.

Measuring angle

Readings of angle using a protractor should normally be recorded to the nearest degree (°). For example, the angle a beam makes with the bench can be measured by sliding the protractor along the bench so that the ⊥ of the protractor is level with the lower edge of the beam, as shown in Figure 27.4. The angle the beam makes with the horizontal is 47°.

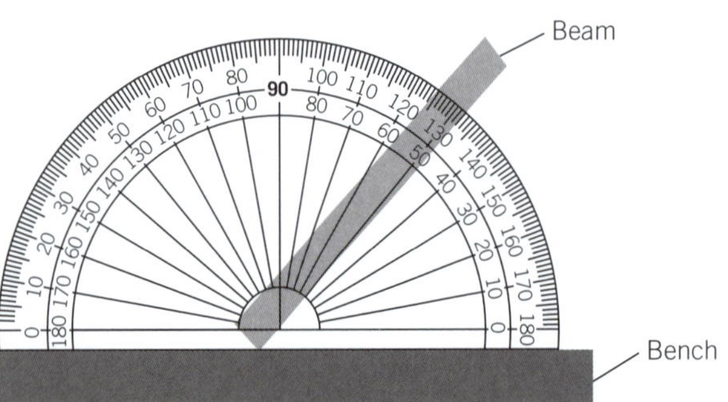

▲ **Figure 27.4** Measuring angle using a protractor

In Figure 27.4, the angle the beam makes with the horizontal is 47°.

Measuring time

Measurements of time usually need a judgement about when a particular event or cycle has been completed. For this reason, time measurements should always be repeated and a mean value calculated. The precision of many stopwatches is ±0.01 s, but this does not mean the accuracy of the measurement is ±0.01 s. The accuracy depends on the reaction time of the experimenter (typically ±0.2 s) and the ability to judge, for example, the start and finish of a complete oscillation of a mass oscillating on a spring.

The raw readings of time can be recorded to the precision of the stopwatch and then a mean value calculated.

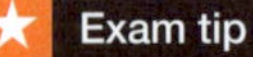

★ **Exam tip**

A stop clock showing:

00:05:28

is reading 5.28 s, not 0.0528 s

Precision and accuracy

The **accuracy** of a measurement is an indication of how close the measurement is to the 'true' value. Accurate values are obtained by using properly calibrated instruments correctly and adjusting the readings for any systematic errors such as zero error. The accuracy can be improved by repeating readings and finding a mean value. Calculating half the difference between the smallest and largest values obtained (half the range) is a reasonable estimate of the random uncertainty in the measurement.

The **precision** of a measurement is indicated by the 'exactness' of the measurement – the smallest division that can be read on the instrument. A value of 0.53 mm for the diameter of a wire measured using a micrometer is a very precise value (it is to the nearest 0.01 mm) but it might not be very accurate, particularly if the micrometer has a zero error.

Uncertainty

The **uncertainty** in a reading is an indication of the confidence in the reading. If a reading from an ammeter is recorded as 23.7 ± 0.2 mA, this indicates that the reading lies somewhere between 23.9 mA and 23.5 mA.

> **★ Exam tip**
>
> Never add extra zeros to your data to make the readings look more precise.

> **💡 Remember**
>
> The percentage uncertainty can be found from the equation:
>
> $$\text{percentage uncertainty} = \frac{\text{absolute uncertainty}}{\text{mean value}} \times 100\%$$

Significant figures

Any calculated values from your raw data should be recorded to the same number of significant figures as, or one more than, the significant figures of the quantity used in the calculation with the least number of significant figures.

Significant figures: a special case

In Table 27.3, although h_1 and h_2 are both recorded to three significant figures, their **difference** is used in the calculation, and $(h_1 - h_2)$ is only to one significant figure. This means the values of $1/(h_1 - h_2)$ can only be expressed to one or two significant figures.

▼ Table 27.3

h_1 / cm	h_2 / cm	$\dfrac{1}{(h_1 - h_2)}$ /cm^{-1}
23.5	22.9	1.7

Calculations

Any calculated values should be rounded correctly, and to an appropriate number of digits. For example, 3.745 can be recorded as 3.75, 3.7 or 4.

When carrying out a series of calculations it is important to carry forward an appropriate number of digits at each stage; a calculator normally has more than enough digits for this. Do not round intermediate steps in your calculation, otherwise you may lose precision in your final answer.

> **Worked example**
>
> Calculate the density of steel from measurements of the diameter d and mass m of a ball-bearing, using the equation:
>
> $$\rho = \frac{6m}{\pi d^3}$$
>
> **Answer**
>
> ▼ Table 27.4
>
> | m / g | d / cm | ρ / g cm^{-3} |
> |---|---|---|
> | 12.8 | 1.4 | 8.91 |
>
> d is recorded to the smallest number of significant figures (2) so ρ (the density) can be expressed to two or three significant figures.

Good practice

Repeating measurements

There are many occasions when measurements should be repeated and mean values calculated. For example, if you are measuring the period of oscillation of a pendulum, it is difficult to judge exactly when a complete oscillation begins and ends. It makes sense to measure the time for, say, 10 oscillations at least three times and then calculate a mean value (don't forget to divide the mean value by 10 to find the mean time for **one** oscillation!).

In general, measurements that are only available to measure momentarily (and may need good reactions) such as the time it takes for a ball-bearing to fall a fixed distance through a liquid, or the height an object bounces, should be repeated. Likewise, any measurement which could vary significantly, such as the diameter of a metal wire, should be repeated two or three times and a mean value found.

Recording results

Results should always be collated and displayed in a clear table, with columns of readings (not rows). Key points to remember:

- **Range of values:** The range of values of the independent variable (e.g., the length of a resistance wire or the mass suspended on a spring) should be as large as possible, with the measurements at approximately equal intervals.

- **Column headings:** Each column heading should have the quantity that has been measured with the appropriate unit.

- **Consistency: Raw** readings (the measurements you actually make) should all be recorded to the same degree of precision. This sometimes means that a set of data will include values with different numbers of significant figures (e.g., 11.7 cm and 5.3 cm). The precision is determined by the instrument you are using.

You may be penalised one or, at most, two marks for any help you receive (out of a total of 40 for the paper), but it will help you to complete the experiment and its follow-up successfully.

Measurements from a metre rule should be recorded to the nearest millimetre, but measurements using vernier calipers should all recorded to the nearest 0.1 mm.

The quantity and its unit should be separated by a '/' (called a 'solidus') or by placing the units in brackets (e.g., length / cm, time / s or θ / °, or length (cm), time (s), or angle θ (°)).

Worked example

This experiment investigates how the angle of tilt of a runway affects the time taken for a cylinder to roll down it (Figure 27.5).

Adjust the position of the clamp on the stand so that θ, the angle the runway makes with the horizontal, is 30°.

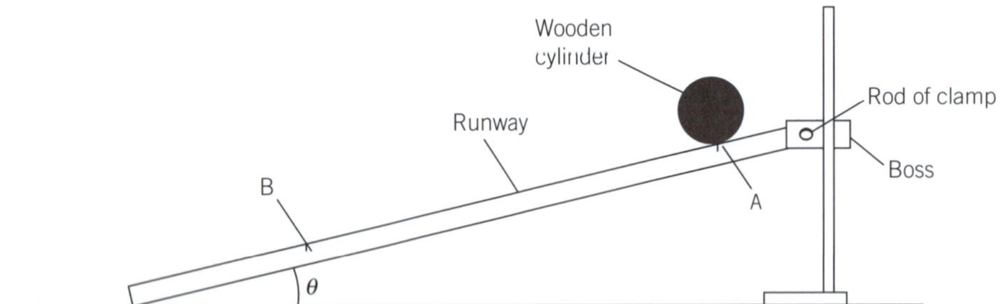

▲ Figure 27.5

Place the wooden cylinder at point A, release the cylinder, and measure the time T taken to reach point B.

Change θ and repeat the experiment until you have six values of θ and T. Record your results in a table and include values of $1/\sqrt{\sin\theta}$ in your table.

Values of θ should be no greater than 60°.

Answer

▼ **Table 27.5**

θ / °	T_1 / s	T_2 / s	T_3 / s	$\bar{T}$ / s	$\dfrac{1}{\sqrt{\sin\theta}}$
7	1.15	1.10	1.11	1.12	2.9
15	0.78	0.77	0.74	0.76	1.97
29	0.58	0.53	0.52	0.54	1.44
43	0.46	0.43	0.49	0.46	1.21
51	0.40	0.47	0.44	0.44	1.13
60	0.41	0.40	0.43	0.41	1.07

★ **Exam tip**

To achieve high marks:

- **Column headings:** All the quantities have the correct unit with the quantity and the unit separated by /. $1/\sqrt{\sin\theta}$ has no units; note also that the angle θ has a unit (°).
- **Collecting data:** Six sets of values of θ and T are recorded, showing the correct trend (T decreases as θ increases).
- **Range:** A good range of values of θ chosen, including a very small value and a very large value.
- **Consistency:** All the values of θ are recorded to the nearest degree; all the raw values of T are recorded to the same precision (the nearest 0.01 s).
- **Significant figures:** The mean values of T are recorded to the same number of significant figures as the raw values of T. The values of $1/\sqrt{\sin\theta}$ are recorded to one significant figure more than the significant figures of θ.

Drawing graphs

Choosing scales

When choosing scales for the x- and y-axes:

- choose scales so that the plotted points occupy more than half the grid of the graph

- scales do not have to start at zero – in many cases it is important that they do **not** start at zero, otherwise the plotted points will be compressed into a small area of the graph paper

- the scales must be linear

- the scales must be 'sensible' (e.g., increasing in twos, fives or tens). Scales that increase by a factor of three or seven, for example, make plotting the points accurately much more difficult and can lead to errors in taking read-offs for calculating a gradient or intercept value

- scale markings should occur no more than three large squares apart. A good rule is to put a scale value every two large squares

- the scales should be labelled with the quantities being plotted

- avoid 'springs' on your graphs (see Figure 27.6).

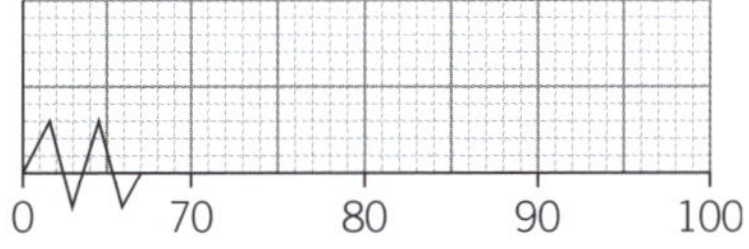

a Avoid a 'break' in the scale to include zero. This is an example of a **false origin**

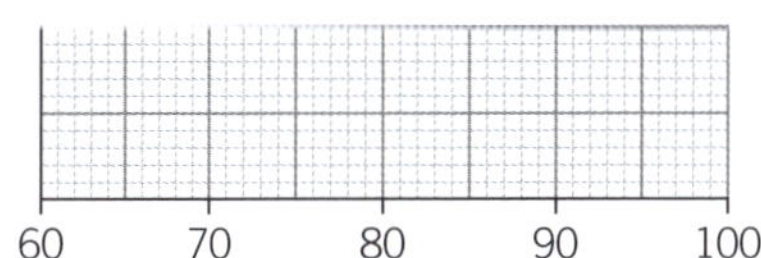

b The scale is linear and continuous. A 'zero' value at the origin is not necessary

▲ **Figure 27.6** Avoid 'springs' on your axes

Plotting points

Points should be small, sharp pencil crosses (×), or dots with a circle around them (⊙). If one of the points appears to be **anomalous** – not following the trend of the other points – it is a good idea to repeat the measurements you made for this value to check, for example, that you didn't mis-read a meter.

Drawing lines of best fit

Drawing a good line of best fit (almost always a straight line) can be challenging, but can be achieved with practice. Look at the examples in Figure 27.7.

Always use a 30 cm rule in good condition and a sharpened pencil. Try to ensure that there is a good balance of points above and below the line. The first and last points are of no greater importance than the points in between – sometimes the best line does not go through any of the points that have been plotted.

If one point appears to be anomalous, draw a ring around it and then draw the line of best fit based on the remaining points.

Calculating a gradient

To calculate the gradient, mark two points **on the line** you have drawn. The distance between the two points should be greater than half the length of the line drawn – the bigger, the better.

If possible, it is a good idea to choose points, like point **A** in Figure 27.8, which lie on the line, but also lie on both a horizontal and a vertical grid line. This makes reading the x and y values correctly a lot easier.

Calculating an intercept

The y-intercept can be found in one of two ways:

- **Directly from the graph**: but **only** if the scale on the x-axis does not have a false origin (i.e., the scale on the x-axis starts at zero).

- **Using $y = mx + c$**: Select a point on the line you have drawn (again it is easier to judge the read-off if you select a point that lies on both a horizontal and a vertical grid line). Read off the x and y values and substitute them into the equation for a straight line, together with your value for the gradient (m). Then rearrange the equation to find the intercept c.

See the Raise your grade example on pp. 226–7 for more on calculating gradients and intercepts.

Answering Question 2

Identifying uncertainties

A physical quantity can never be measured exactly. **Uncertainties** arise due to the limitations of the instrument you are using (how accurate it is), your skill at using it, and changes in the environment (e.g., the effects of draughts or a change in temperature). Random fluctuations in what you are trying to measure (e.g. the activity of a radioactive source) will also give rise to uncertainty and the act of measuring can affect the measurement itself – when a thermometer at room temperature is placed in a warm liquid it will cool the liquid slightly. Physicists try to estimate these uncertainties to calculate the degree of confidence they can have in a calculated value.

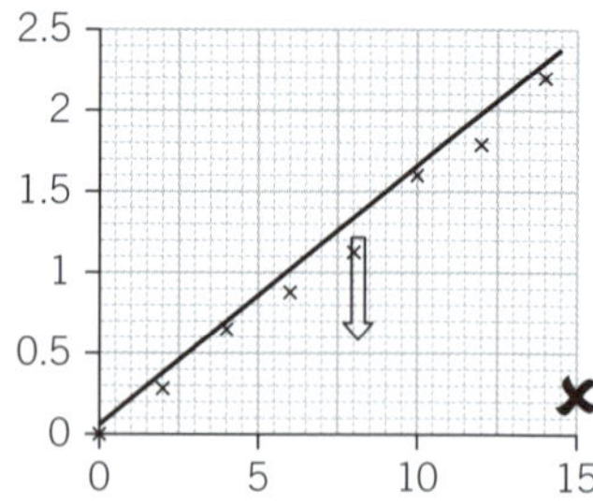

a A better line can be drawn by moving the candidate's line downwards

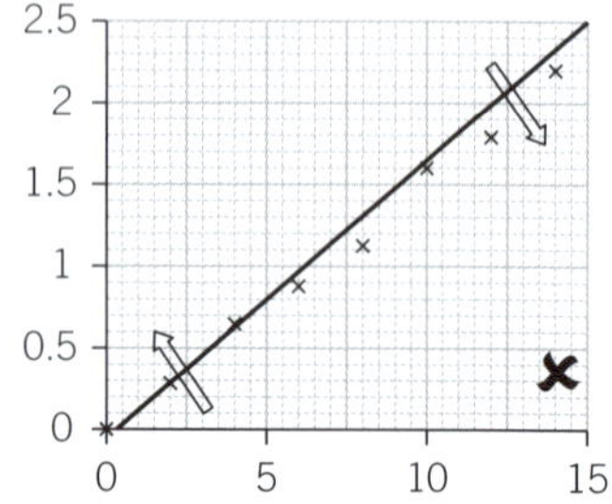

b A better line can be drawn by rotating the line clockwise. The candidate has joined the first and last points without taking into account the other points

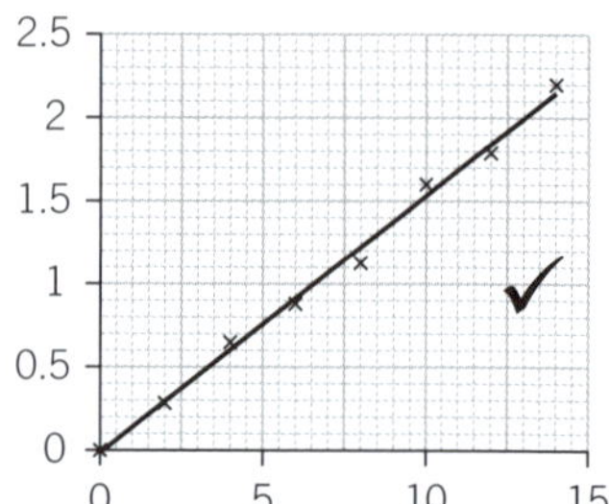

c A good attempt at drawing the straight line of best fit, with points above and below the line

▲ **Figure 27.7** Plotting the line of best fit

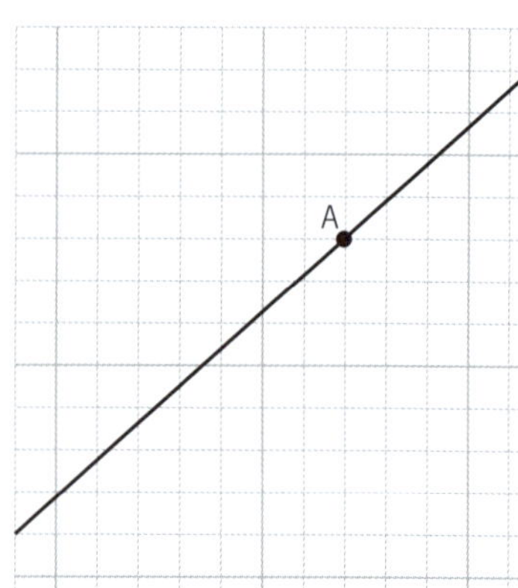

▲ **Figure 27.8** Choosing a point

Worked example

a) **i)** Measure the diameter d of a small ball made of modelling clay, using a 30 cm rule.

ii) Estimate the percentage uncertainty in your value.

b) Measure the mass m of the ball (an electronic balance is available).

c) **i)** Calculate the density ρ of the modelling clay, using the equation:

$$\rho = \frac{6m}{\pi d^3}$$

ii) Justify the number of significant figures you have given for your value of ρ.

Answer

a) **i)** d / mm: 26   27   30

Mean value for $d = 28$ mm

ii) $\Delta d = 4$ mm, percentage uncertainty $= \dfrac{4}{28} \times 100 = 14\%$

> **Exam tip**
>
> The diameter of the ball is likely to vary in different directions, and so it is sensible to measure the diameter in three different directions and find a mean.

> **Remember**
>
> The precision of the rule is ± 1 mm, but the accuracy of the measurement is much less than this. The accuracy is affected by:
>
> - parallax error due to the distance between the rule and the edges of the ball
>
> - variation in the diameter of the ball as it is a rolled piece of modelling clay which is not perfectly spherical.
>
> For these reasons, a better estimate of the absolute uncertainty is 2–5 mm. The percentage uncertainty is found from the equation:
>
> $$\text{percentage uncertainty} = \frac{\text{absolute uncertainty}}{\text{mean diameter}} \times 100$$

b) $m = 21$ g

c) **i)** $\rho = \dfrac{6m}{\pi d^3} = \dfrac{6 \times 21}{\pi \times 2.8^3} = 1.8\,\text{g cm}^{-3}$

ii) The value of ρ can be quoted to the same number of significant figures as the least number of significant figures in the raw data (or one more). In this example, ρ is calculated from the raw data of m (2 sig. figs.) and d (2 sig. figs.), and so ρ can be quoted to two or three significant figures.

> **Remember**
>
> Use consistent units in calculations.

> **Exam tip**
>
> It's not enough to say that the answer is consistent with the 'raw data'.

Testing hypotheses, sources of uncertainty and limitations of the procedure

The experiment in Question 2 will normally require you to obtain two sets of experimental results which are used to test the validity of a theoretical equation. The last section of the next worked example looks at sources of uncertainty and possible improvements.

Worked example

A student investigates the terminal speed of a ball-bearing of diameter d falling through a liquid (see Figure 27.9). He measures the time t taken for each ball to fall from A to B, repeating each measurement and finding a mean value for t (see Table 27.6).

▼ Table 27.6

Diameter d of ball / mm	Time to fall from A to B / s
30	2.2
38	1.4

It is suggested that the relationship between t and d is:

$$t = \frac{k}{d^2} \quad \text{where } k \text{ is a constant}$$

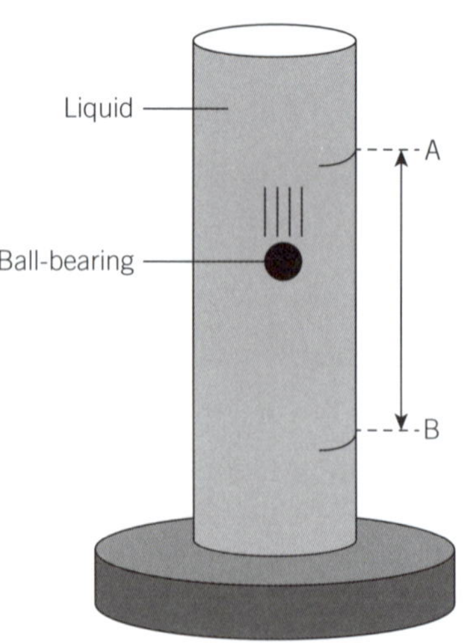

▲ **Figure 27.9** Ball-bearing falling in liquid

a) Using your data, calculate two values of k.

b) Explain whether your results support the suggested relationship.

Answer

a) $k = t\,d^2$

so

$$k_1 = 2.2 \times (30 \times 10^{-3})^2 = 1.98 \times 10^{-3}\,\text{m}^2\,\text{s}$$

$$k_2 = 1.4 \times (38 \times 10^{-3})^2 = 2.02 \times 10^{-3}\,\text{m}^2\,\text{s}$$

b) percentage difference between k values $= \dfrac{2.02 - 1.98}{2.00} \times 100 = 2\%$

This is less than my estimate of the overall uncertainty in the experiment of 10%, and so the two results are consistent with the suggested relationship.

> ★ **Exam tip**
>
> Answers such as 'k values are quite close' or 'values of k are far apart' are not good enough.

Worked example

Continues from the previous worked example

c) Describe four sources of uncertainty or limitations of the procedure.

Answer

c) 1. While two readings may be consistent with a particular relationship, they are not conclusive proof. More readings are needed using ball-bearings with other diameters.

2. The diameters were measured with a 30 cm rule, leading to parallax error.

3. It is not certain that the ball-bearings had reached their terminal velocity by the time they reached point A.

4. The time taken to fall from A to B is very short, and so there is a large percentage uncertainty in the value of t.

Identifying improvements

The last section of Question 2 looks at improvements. These can address the problems identified earlier, but this is not essential.

Some of the limitations, sources of error and possible improvements are specific to a particular experiment. Table 27.7 lists some potential limitations and sources of error that you may be able to use, together with possible improvements.

▼ **Table 27.7** Sources of error and suggestions for improvements

Limitations and sources of error	Improvements
Only two sets of data recorded – not enough to draw a valid conclusion.	Take more readings (with different values of the independent variable), and either calculate further values of k and compare them, or plot a graph.
Large percentage uncertainty in using a 30 cm or 0.5 m rule to measure small distances such as the diameter of a wire or the thickness of a coin.	Use vernier calipers or a micrometer.
Difficult to measure a small change in length (e.g. when stretching a metal wire).	Use a travelling microscope.
Difficult to check whether a rule is vertical.	Hold a set-square on the bench against the rule.
Difficult to check whether a wooden strip is horizontal.	Use a spirit level, or measure the height of the strip above the bench at both ends and check they are the same.
Difficult to hold a rule or a protractor steady when making a measurement.	Clamp the rule or protractor using a clamp and stand.
Difficult to judge the start or end of an oscillation.	Place marker (e.g. a pencil mounted on a stand) at the centre of the oscillation (the equilibrium position). Start and stop a stopwatch when the oscillator passes the equilibrium position.
Difficult to time an event accurately because the time is short or there is a large percentage uncertainty in the value of time.	Video the experiment, then play back frame-by-frame and use the camera's clock (or a stopwatch filmed next to the experiment) to measure the time. It should be clear how the measurement is to be made (e.g. include a metre rule in the picture when filming so that changes in height can be measured).
Difficult to release an object (e.g. a ball) without exerting some external force.	Hold the object against a stop (e.g. a piece of card) and release.
Difficult to calculate the volume of a ball as its diameter varies.	Partially fill a measuring cylinder with water, place the sphere in the water and measure the change in volume.
Difficult to measure volume of liquid in beaker/ measuring cylinder as liquid is clear/ transparent/ difficult to see meniscus.	Add coloured dye to the liquid.

★ **Exam tip**

- Avoid describing improvements that you should do as part of a good scientific technique, such as repeating measurements or reading instruments by viewing in a direction perpendicular to the scale (to reduce parallax errors).

- Answers such as using 'better equipment', 'computers' or 'requesting an assistant' do not receive any marks.

★ **Exam tip**

- Avoid 'inventing' problems such as a damaged rule, or 'it was difficult to drop the ball-bearing and measure the time at the same time'.

- Vague answers, such as 'systematic error', 'random error', or 'parallax error' do not gain credit without further explanation.

- 'Measurements not repeated' is not a valid answer as, where appropriate, measurements should have been repeated. Similarly, 'zero error in a micrometer' is not valid, as a correction should have been made for this.

Raise your grade

1 (a) Set up the circuit shown.

Make sure the positive terminal of the multimeter is connected to the positive side of the d.c. supply.

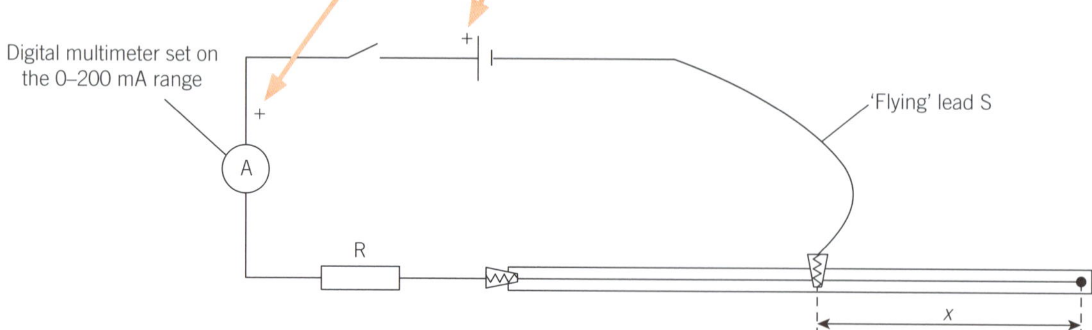

'Approximately' means within 0.5 cm.

Attach the flying lead S so that the distance x is approximately 50 cm.

(b) Measure and record x.

Close enough to 50 cm and has the correct unit.

$x =$ 49.7 cm ✔ [1]

(c) (i) Close the switch.

(ii) Measure and record the reading I on the ammeter.

$I =$ 136 A ✗ [1]

(iii) Open the switch.

I recorded in A not mA.

(d) Repeat (b) and (c) for different values of x until you have six values of x and I.

Include the values of $\dfrac{1}{I}$ in your table. [10]

✔ ✔ ✔ ✔ ✔ 6 sets of readings of x and I showing the correct trend (I increasing as x increases)

✔ A good **range** of results, including values of x close to the largest and smallest possible values of x.

x/cm	I/mA	$1/I$ / A^{-1}
5	109	9.17
20.3	117	8.55
35.8	126	7.94
58.1	143	6.99
76.2	159	6.28
95.1	195	5.13

✔ All the column headings have the quantity and an appropriate unit, separated by a /.

✗ This value of $1/I$ has been rounded incorrectly – it should be 6.29.

✗ All the values of x must be recorded to the nearest mm (the precision of the metre rule). No mark for consistency – the first value should be 5.0.

✔ All the values of $1/I$ have been recorded to the same number of sig. figs. as I.

(e) (i) Plot a graph of $\dfrac{1}{I}$ on the y-axis against x on the x-axis.

(ii) Draw the straight line of best fit.

(iii) Determine the gradient and y-intercept of this line. [8]

Poor choice of scale for the y-axis (the scale is 'compressed' so the points only occupy 3 large squares vertically). ✗

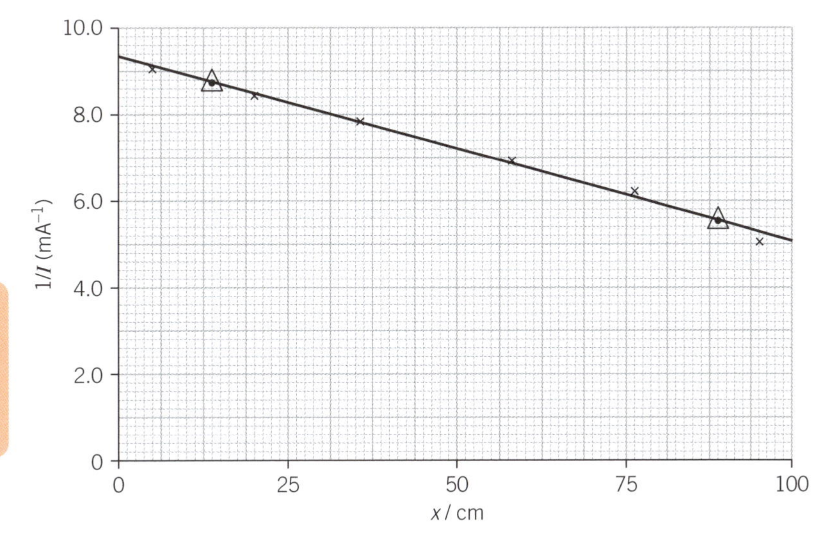

All 6 points plotted correctly with small crosses. ✔

A good 'best line', with some points above and some below the line. ✔

All the points are within ± 0.4 mA⁻¹ of the line. ✔

$$\text{gradient} = \frac{\Delta y}{\Delta x} = \frac{5.6 - 8.8}{88.75 - 13.75} = -0.043 \ \text{mA}^{-1}\text{cm}^{-1}$$ ✔

Using the point $(51.25, 7.2)$ in $y = mx + c$:

$y\text{-intercept} = c = 7.2 - (-0.043 \times 51.25) = 9.4 \ \text{mA}^{-1}$ ✔

(f) The quantities x and $\dfrac{1}{I}$ are related by the equation:

$$\frac{1}{I} = -Px + Q$$

Using your answers to (e)(iii), determine the values of P and Q.

$$P = -\text{gradient} = 0.043$$

$$Q = y\text{-intercept} = 9.4$$

Don't forget to include the units for P and Q – use the equation and the graph axes to work out what they should be.

As a check, see if your calculated value of the y-intercept is the same as the direct read-off from the graph.

One mark for matching P to -gradient and Q to the intercept. ✔

Does not score mark for units (should be mA⁻¹cm⁻¹ for P and mA⁻¹ for Q). ✗

1 Estimate the absolute uncertainty in measuring:

(a) the diameter of a copper wire of approximate diameter 0.5 mm using a micrometer [1]

(b) the period of a pendulum of approximate period 2.0 s using a stopwatch with a precision of 0.01 s. [1]

2 Determine the reading on the vernier scale. [1]

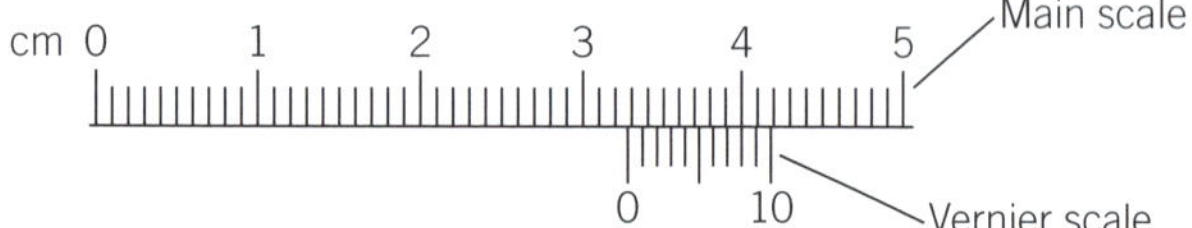

3 The resistance of a length of wire has a percentage uncertainty of 1.0%. The percentage uncertainty in measuring the diameter of the wire is 5% and the percentage uncertainty in measuring the length of the wire is 2%.

Calculate the percentage uncertainty in calculating the resistivity of the wire. [2]

4 Describe the difference between systematic errors and random errors, and give one example of each. [4]

5 A student carries out an experiment to measure the time t taken for a table-tennis ball to fall to the ground, bounce several times and finally come to rest, when released from a height h.

h / cm	t_1 / s	t_2 / s	t_3 / s	mean t / s	$\sqrt{h}$ / cm$^{1/2}$
96.5	5.2	5.5	5.4		
87.3	5.0	4.9	4.8		
69.5	4.5	4.3	4.3		
54.4	3.6	3.8	3.7		
40.0	3.3	3.3	3.1		
29.8	2.7	2.8	2.9		

Copy and complete the table. [3]

6 The table shows the results of an experiment in which the period of oscillation T of a compound pendulum is recorded for different values of a variable x.

x / cm	T / s
2.5	2.8
2.3	3.3
1.8	5.7
1.5	7.8
1.3	10.2
1.1	14.7

(a) Plot a graph of T^2 on the y-axis against x on the x-axis. [2]

(b) Draw the straight line of best fit. [1]

(c) Determine the gradient and y-intercept of this line. [2]

7 The table shows the results of an electrical experiment, recording the current I in a circuit for different values of a resistor of resistance R.

R / Ω	I / mA
100	35.4
220	19.0
330	14.4
470	13.7
560	10.7
820	9.0

(a) Plot a graph of I on the y-axis against $1/R$ on the x-axis. [2]

(b) Identify the anomalous result and draw a circle around it. Draw the straight line of best fit for the remaining five points. [2]

(c) Determine the gradient and y-intercept of this line. [2]

8 Measurements of the viscosity η of a gas are made at two different temperatures.

T/K	η/10^{-6} Pa s
323	18.9
373	21.2

It is suggested that the relationship between T and η is:

$$T - k\eta^2$$

where T is the absolute temperature and k is a constant.

(a) Using the data in the table, calculate two values of k. [2]

(b) The overall uncertainty in the measurements used to calculate k is estimated to be 10%. State and explain whether the measurements support the suggested relationship. [1]

Paper 5: Planning, Analysis and Evaluation

This paper, together with Paper 4, are A level papers, taken after completing the whole course. There are two questions on Paper 5 *Planning, Analysis and Evaluation*, which lasts 1 hour 15 minutes. There are 15 marks for each question. The first question is designed to test your ability to plan an investigation and is likely to be concerned with a topic that you have not seen before. The second question looks at analysing the results of an experiment, identifying relationships, and evaluating the reliability of experimental evidence.

Question 1

The question will ask you to design a laboratory experiment to test a mathematical relationship between two variables. It may also list some of the apparatus that is available to you. You need to describe the procedure to be followed and the measurements to be taken.

Defining the problem (2 marks)

Start by identifying the **independent variable**, the **dependent variable**, and any other relevant variables that could affect the results, and so need to be kept constant. Some simple examples are given in Table 27.8.

▼ **Table 27.8** Identifying variables

Investigation	Independent variable	Dependent variable	Variables to keep constant
How does the resistance of a metal wire vary with diameter?	diameter	resistance	temperature, length of wire
How does the period of oscillation of a mass on a spring vary with mass?	mass	period	spring stiffness, amplitude
How does the viscosity of oil vary with temperature?	temperature	viscosity	

Outlining the procedure (4 marks)

Draw a clear, labelled diagram of the apparatus you intend to use; if the question is an electrical one, draw a circuit diagram. Outline in detail the measurements you intend to make. Draw the column headings (with units) of the table that you will use to record your results

Some key points to consider:

- How will you vary the independent variable?

- How will you **measure** the dependent and independent variables?

- What is the **range** of measurements you intend to make? What are the largest and smallest values of the independent variable?

- What is the **precision** of your measurements?

- Is it necessary to repeat your measurements and calculate an average? If so, when?

> **Remember**
>
> The **independent variable** is the one **you change and control**; for example the length of a resistance wire or the number of masses hung on a spring.
>
> The **dependent variable** is the one that is **altered** by the change you make; for example the current in a circuit or the extension of a spring.

> ★ **Exam tip**
>
> It is a good idea to leave a couple of 'spare' columns in your table for any calculated quantities you might need to test the relationship.

> 🔗 **Link**
>
> For more on precision see Unit 2 *Measurement techniques*.

Analysing the data (3 marks)

A mathematical relationship between the dependent and independent variables will be suggested in the question. You should state the graph you would plot to test whether the relationship is valid. Normally you should plot the graph that would give you a straight line. See Figure 27.10 for some strategies.

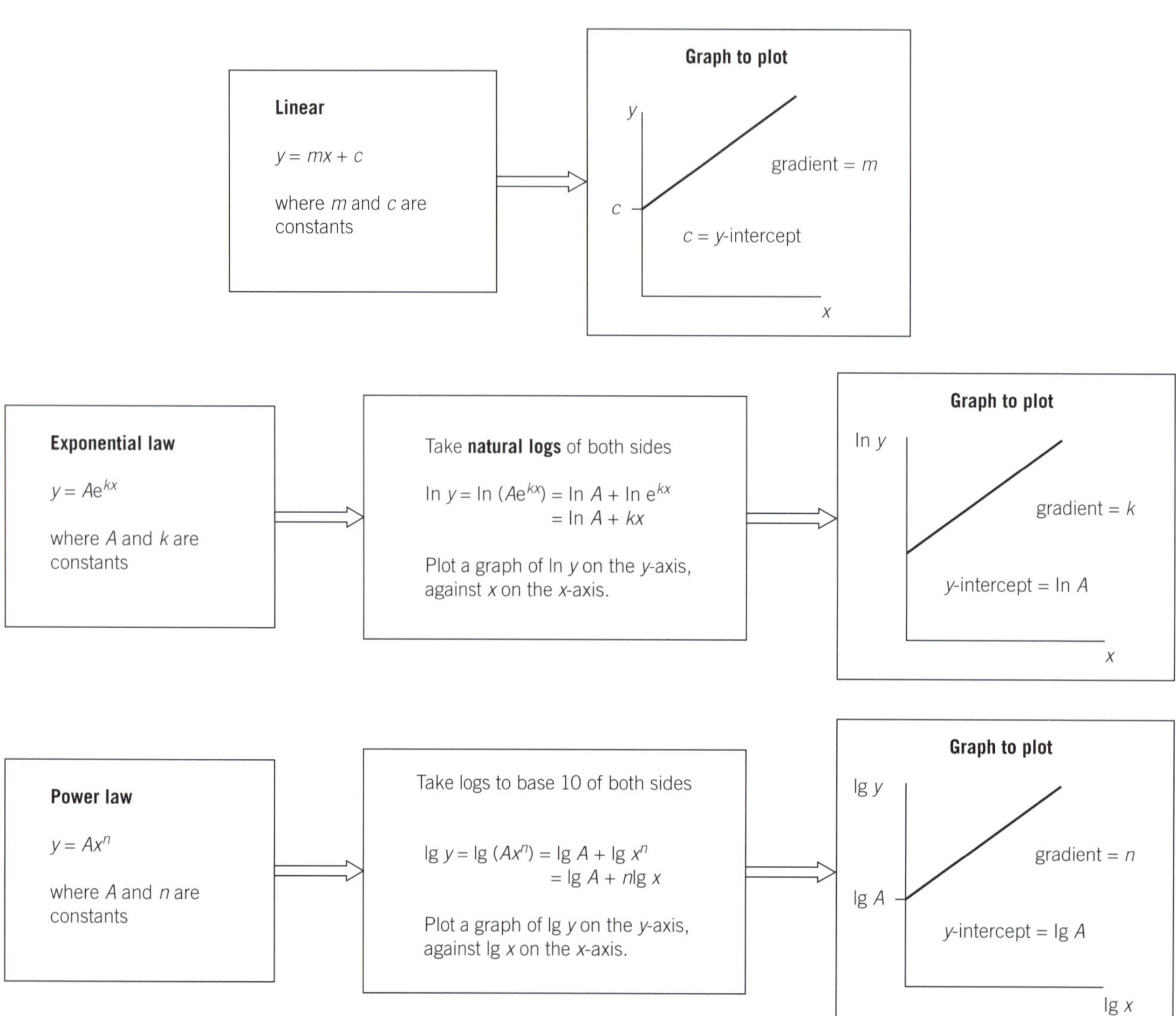

▲ **Figure 27.10** Which graph?

Additional details, including safety (6 marks)

Up to six marks are awarded for additional details. This might include a description of:

- how specific variables are to be kept constant
- initial experiments to establish a suitable range of values
- the use of an oscilloscope (or storage oscilloscope) to measure voltage, current, time, and frequency
- how to use light gates connected to a data logger to determine time, velocity, and acceleration
- how other sensors, such as motion or pressure sensors, can be used with a data logger.

Additional marks are also awarded for identifying any potential hazards in the experiment and the safety procedures that should be followed. Table 27.9 gives some examples.

▼ **Table 27.9** Examples of safety precautions

Activity	Hazard	Safety procedure
Stretching metal wires	Wires break and hit eyes	Wear safety spectacles/goggles
Measuring radioactive decay	Exposure to harmful radiation	Handle the radioactive source using tongs; replace the source in a lead-lined box when not in use
Using heavy weights	Weights fall onto feet/floor	Ensure the weights are as near to the ground as possible; place a sand tray or foam rubber block underneath the weights
Investigating sound waves	Damage to ears	Use ear defenders/switch off loudspeaker(s) when not in use
Investigating diffraction using a laser	Damage to eyes	Do not look directly at the laser

Worked example

A student is investigating how the resistance of a particular type of thermistor varies with temperature. It is suggested that:

$$R = Ae^{\frac{k}{T}}$$

where R is the resistance of the thermistor at temperature T, and A and k are constants.

Design a laboratory experiment to test the relationship between R and T. Explain how your results could be used to determine values for A and k. You should draw a diagram showing the arrangement of your equipment. In your account pay particular attention to the:

- procedure to be followed,
- measurements to be taken,
- control of variables,
- analysis of the data,
- safety precautions to be taken.

Answer

Defining the problem

The independent variable is the temperature of the thermistor; the dependent variable is the resistance of the thermistor.

Method

- Fill a 250 ml beaker with approximately $200\,cm^3$ of cooking oil. Place a thermometer in the oil and gently heat the oil over a Bunsen burner (or using an electrical immersion heater) until the temperature reaches approximately 120 °C.

- Place the thermistor (supported by a wooden board resting on the top of the beaker) into the oil.

- Use a stirrer to ensure all the oil is at the same temperature.

- After a few minutes (allowing the oil, the thermistor, and the thermometer to reach the same temperature) record the temperature on the thermometer, and the resistance as measured by the digital ohmmeter.

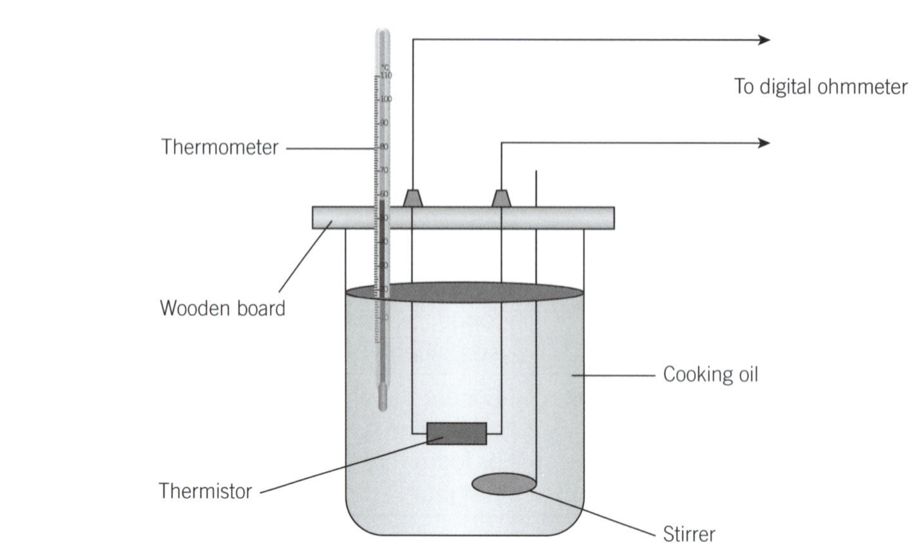

▲ **Figure 27.11**

Results

Record the results as shown in Table 27.10.

▼ **Table 27.10**

T/K	R/Ω	$\ln(R/\Omega)$	$\dfrac{1}{T}/s^{-1}$

Include columns for $\ln R$ and $\dfrac{1}{T}$.

Analysis

Plot a graph of $\ln R$ on the *y*-axis against $1/T$ on the *x*-axis. The gradient of the graph is k and the *y*-intercept is $\ln A$.

Safety

Wear safety goggles (in case the hot oil splashes). Remove the heat source before placing the thermistor inside the beaker of oil (to avoid connecting wires being melted).

Question 2

This question provides you with a set of results for analysis and evaluation. A graph is plotted (with error bars) and a line of best fit is drawn together with the worst acceptable straight line so that a calculation of the absolute uncertainty in the gradient or the *y*-intercept of the graph can be calculated.

Data analysis (1 mark)

Table 27.11 shows the different relationships you will need to be familiar with.

▼ **Table 27.11** Relationships between two variables

Relationship	Graph to plot	Gradient	*y*-intercept
Linear ($y = mx + c$)	y against x	m	c
Exponential ($y = a\,e^{kx}$)	$\ln y$ against x	k	$\ln a$
Power law ($y = ax^n$)	$\lg y$ against $\lg x$	n	$\lg a$

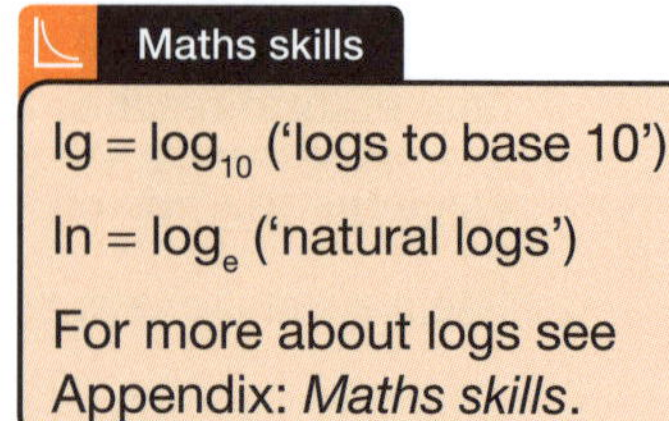

$\lg = \log_{10}$ ('logs to base 10')

$\ln = \log_e$ ('natural logs')

For more about logs see Appendix: *Maths skills*.

Worked example

When a load is suspended from a metal wire such as copper, the wire stretches. If the load is left on the wire it stretches further over time, a process known as **creep**.

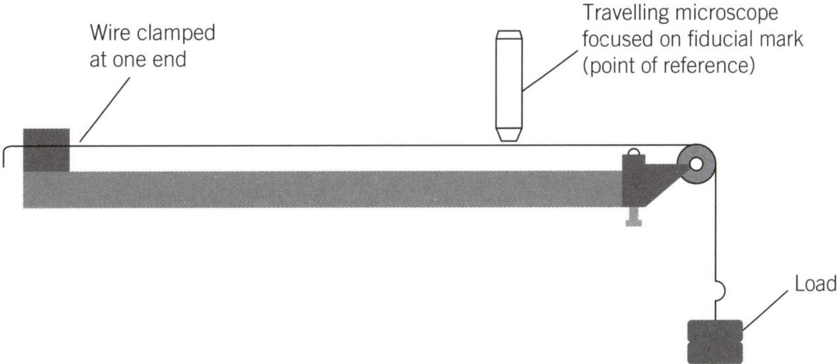

▲ **Figure 27.12** Creep of copper wire

It is suggested that the relationship between the extension x of a wire (excluding the initial extension) and the time t measured from when the load is first placed on the wire, is given by the expression:

$$x = At^n$$

where A and n are constants.

A graph is plotted of $\lg t$ on the x-axis against $\lg x$ on the y-axis. Determine expressions for the gradient and y-intercept.

Answer

Taking logs of both sides of the equation:

$$\lg x = \lg(At^n) = \lg A + \lg(t^n) = \lg A + n\lg t$$

$$\underset{y}{\lg x} = \underset{m}{n}\ \underset{x}{\lg t} + \underset{c}{\lg A}$$

The gradient $m = n$; the y-intercept $c = \lg A$.

Table of results (1 mark)

A table of results will need to be completed, following the same guidelines as for Paper 3 *Advanced practical skills*:

- every column should have a heading, and each heading should have a quantity and a unit, separated by / or with the units in brackets; for example, l / cm or l (cm)

- calculated values should be recorded to the same number of significant figures (s.f.) as (or one more than) the s.f. of the raw value(s) recorded to the least number of s.f.

- where appropriate, uncertainty estimates, in absolute terms, should be recorded beside every value in the table of results

- calculated values should be rounded correctly.

Worked example

The results of the experiment on creep, described earlier are given in **bold** Table 27.12.

Calculate and record values of lg (t /s) and lg (x /mm) in Table 27.12.

Include the absolute uncertainties in lg x.

Answer

Units: When the logarithm of a quantity is calculated, the units should be shown with the quantity; for example, lg (t /s) and lg (x /mm). The logarithm itself does not have a unit).

Uncertainty: To calculate the uncertainty (lg x_{max} − lg x_{min})/2

▼ Table 27.12

Time t / s	Extension x / mm	lg (t / s)	lg (x / mm)
10	2.60 ± 0.05	1.00	0.415 ± 0.008
20	3.35 ± 0.05	1.30	0.525 ± 0.006
30	3.90 ± 0.05	1.48	0.591 ± 0.006
40	4.30 ± 0.05	1.60	0.633 ± 0.005
50	4.65 ± 0.05	1.70	0.667 ± 0.005

Decimal places: When calculating the logarithmic value of a quantity, the number of decimal places should be the same as (or one more than) the number of s.f. of the quantity itself. (e.g., if x /mm is 4.65 (3 s.f.) then lg (x /mm) should be 0.667 (3 s.f.) or 0.6675 (4 s.f.).)

Graph (2 marks)

The axes of the graph will already be labelled with the quantities to be plotted, and the scales marked on the axes. You will need to:

- plot the points on the graph correctly, including **error bars**

- draw a straight **line of best fit** and a straight **worst acceptable line** through the points on the graph when the trend on the graph is linear

- draw a curved trend line and a tangent to the curve where appropriate.

The **worst acceptable** line can be either the steepest possible line or the shallowest possible line that still passes through the error bars of all the data points. Draw this as a broken line and label it.

Error bars

When plotting measurements on a graph, the uncertainty in the measurements can be shown by including **error bars** ('uncertainty bars' is a better name for them).

Suppose, for example, you are plotting a value of $y = 6.2 \pm 0.1$ s (see Figure 27.13). The bar should extend 0.1 s (using the scale on the y-axis) either side of the nominal value of 6.2 s (the bar should always be the same height above and below the plotted point). If the uncertainty in both the x values and y values are known, error bars can be drawn both vertically and horizontally.

This question may ask you to plot the log of a quantity. To calculate the uncertainty, calculate the log of the largest value and the log of the measured value – the difference is the uncertainty. Alternatively, calculate:

$$\frac{(\log(\text{max value}) - \log(\text{min value}))}{2}$$

A **best fit line** should pass through all the error bars, with an even and balanced distribution of points above and below the line. To estimate the uncertainty in the gradient, a **worst acceptable line** (e.g., the steepest line that *just* passes through all the error bars, should also be drawn). The uncertainty in the gradient is then:

$$\text{gradient of steepest line} - \text{gradient of best fit line}$$

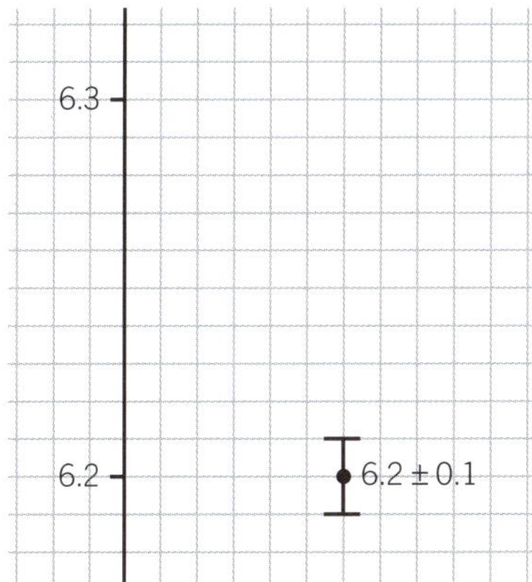

▲ **Figure 27.13** Plotting error bars

Conclusion (3 marks)

You will need to:

- find the gradient and y-intercept of a straight-line graph or a tangent to a curve,

- derive quantities that equate to the gradient and the y-intercept

- draw conclusions from these quantities.

> **Link**
>
> Use the same rules as in Paper 3 *Advanced practical skills*.

Treatment of uncertainties (3 marks)

You need to be able to convert absolute uncertainty estimates of a quantity into fractional or percentage uncertainty estimates and vice versa. A quantity should be expressed as a value, with an uncertainty estimate and a unit (e.g., $6.4 \pm 0.2\,\text{V}$).

You may also need to calculate uncertainty estimates in derived quantities.

> **Remember**

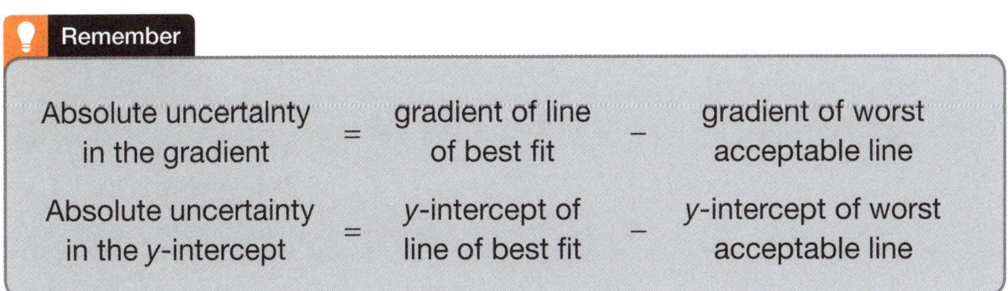

$$\frac{\text{Absolute uncertainty}}{\text{in the gradient}} = \frac{\text{gradient of line}}{\text{of best fit}} - \frac{\text{gradient of worst}}{\text{acceptable line}}$$

$$\frac{\text{Absolute uncertainty}}{\text{in the } y\text{-intercept}} = \frac{y\text{-intercept of}}{\text{line of best fit}} - \frac{y\text{-intercept of worst}}{\text{acceptable line}}$$

Remaining marks (5 marks)

The remaining five marks for this question are allocated across the different skill areas, and their allocation will vary.

Worked example

For the data in Table 27.12 (p. 234) plot a graph of lg (x/mm) against lg (t/s). Include error bars for x.

Draw the straight line of best fit and a worst acceptable straight line on your graph. Both lines should be clearly labelled.

Determine the gradient and the y-intercept of the line of best fit.

Use your answers to determine the values of A and n.

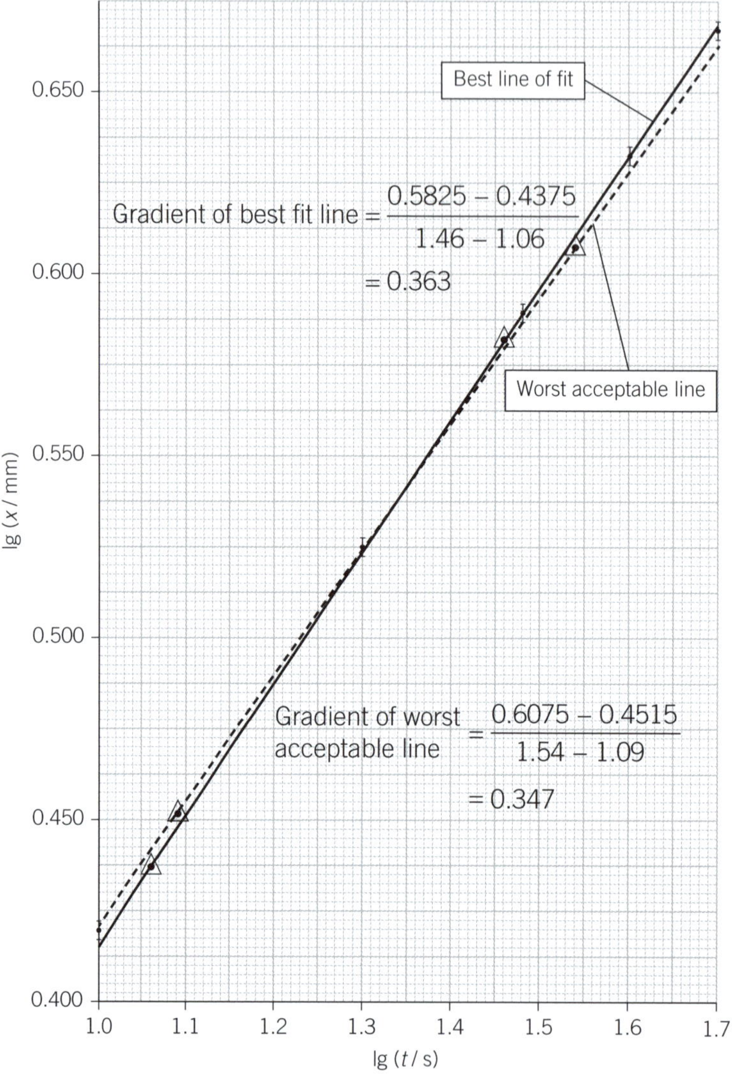

▲ **Figure 27.14**

Answer

The best fit line is the solid line and the worst acceptable line is the dotted line.

Using the point (1.35, 0.5425) in $y = mx + c$: y-intercept $= c = y - mx$

$$= 0.5425 - 0.363 \times 1.35$$

$$= 5.25 \times 10^{-2}$$

$n = m = 0.363$ lg $A = c = 0.0525$ so $A = 10^c = 1.13$ $x = 1.13\,t^{0.363}$

Determine the percentage uncertainty in the value of n.

Since $n = m$, the uncertainty in n is the same as the uncertainty in the gradient.

Absolute uncertainty in the gradient $= 0.363 - 0.347 = 0.016$

% uncertainty in $n = \dfrac{\text{absolute uncertainty in } n}{\text{nominal value of } n} \times 100 = \dfrac{0.016}{0.363} \times 100 = 4.4\,\%$

❓ Exam-style questions

In the style of Question 1

1 A student is investigating the deflection of a cantilever beam when a load is suspended from one end of the beam.

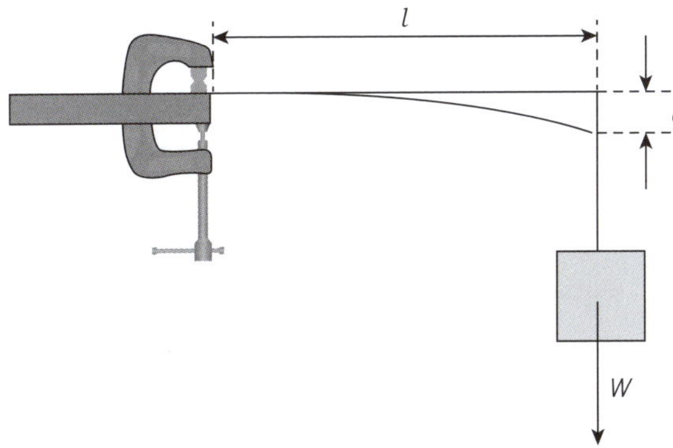

He varies the length l of the beam and measures the deflection d in each case.

(a) State the dependent and independent variables in this experiment. [1]

(b) State any other variables that should be controlled. [1]

(c) It is suggested that:

$$d = kl^3$$

State the graph you would plot to test this relationship. Explain how the value of k could be found from your graph. [3]

(d) State any safety precautions that should be taken when carrying out the experiment. [2]

2 An experiment is carried out to investigate the factors affecting the time taken for a capacitor to discharge through a resistor, using the circuit shown.

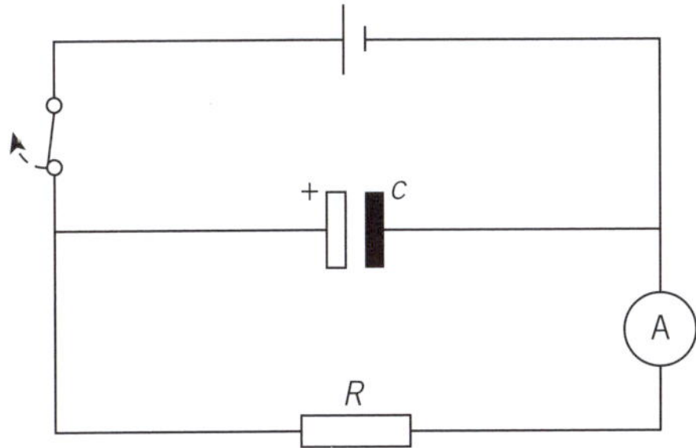

When the switch is closed the reading on the ammeter is I_0. When the switch is opened the capacitor gradually discharges through the resistor of resistance R.

The time t taken for the initial current to fall to $I_0/2$ (half its original value) is measured. The experiment is repeated for different values of R.

Theory suggests that:

$$t = kCR$$

where k is a constant.

(a) State which graph you would plot to test this relationship. [1]

(b) Explain how the value of k could be determined from your graph. [2]

3 The variables x and y are believed to be related by an equation of the form:

$$y = Ax^n \qquad \text{(a power law)}$$

where A and n are constants.

x	1.0	1.7	2.1	2.8	3.5	3.9
y	3.0	6.6	9.0	14.1	19.5	23.1

(a) Plot a suitable graph using the results shown in the table.

(b) Determine the value of n.

4 A student is investigating how the count rate C from a radioactive source emitting γ-rays varies with distance d from the source.

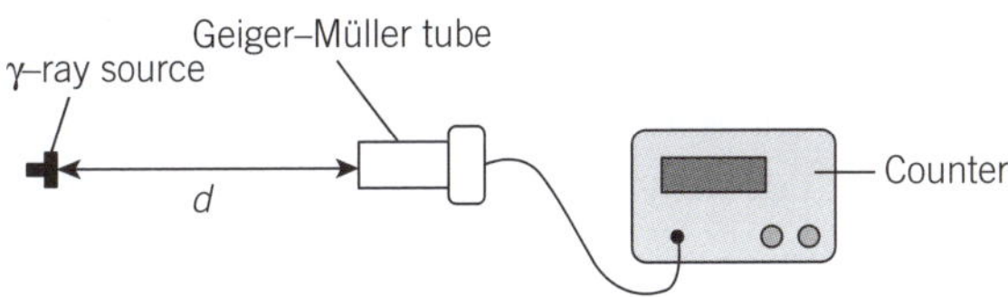

It is suggested that the count rate is related to the distance d between the radioactive source and the Geiger–Müller tube by the equation:

$$C = \frac{A}{d^2}$$

where A is a constant.

Design a laboratory experiment to test the relationship between C and d. Explain how your results could be used to find a value for A.

In your account, you should pay particular attention to:

- the procedure to be followed
- the measurements to be taken
- the control of variables
- the analysis of the data
- any safety precautions to be taken. [15]

5 In an investigation to measure the absorption of β-rays by aluminium, different thicknesses of aluminium were placed in front of a β-emitting radioactive source and the count rate measured, as shown. The measurements were adjusted for background radiation.

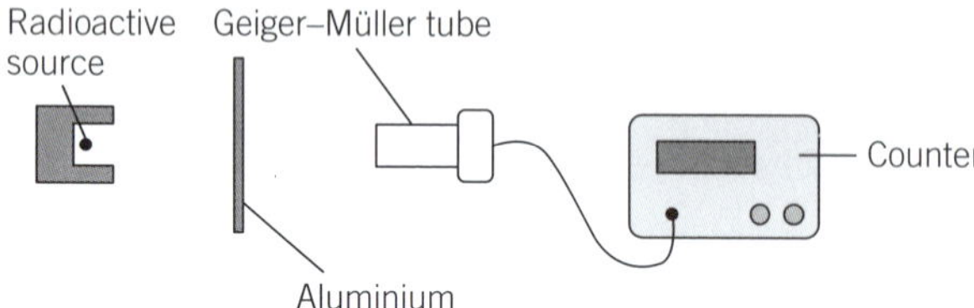

The count rate C and the thickness of aluminium x are related by the equation:

$$C = C_0\, e^{-\mu x}$$

where C_0 and μ are constants.

(a) A graph is plotted of $\ln C$ on the y-axis against x on the x-axis. Determine expressions for the gradient and intercept. [1]

(b) Values of C and x are given in the table.

C / counts min^{-1}	x / mm	$\ln(C$ / counts min$^{-1})$
2284	1.0 ± 0.2	
1987	1.9 ± 0.2	
1682	2.8 ± 0.2	
1326	4.1 ± 0.2	
1145	4.9 ± 0.2	
982	6.0 ± 0.2	

Calculate values of $\ln(C$ / counts min$^{-1})$. Record your values on a copy of the table. [1]

(c) (i) Plot a graph of $\ln C$ on the y-axis against x on the x-axis.

(ii) Draw the straight line of best fit and a worst acceptable straight line on your graph. Both lines should be clearly labelled.

(iii) Determine the gradient of the line of best fit. Include the absolute uncertainty in your answer.

(iv) Determine the y-intercept of the line of best fit. Include the absolute uncertainty in your answer. [8]

(d) (i) Using your answers to (c)(iii) and (c)(iv), determine the values of C_0 and μ. Include appropriate units.

(ii) Calculate the percentage uncertainty in your value of μ. [4]

6 A student investigates how the stiffness of a spring varies with the diameter d of the wire.

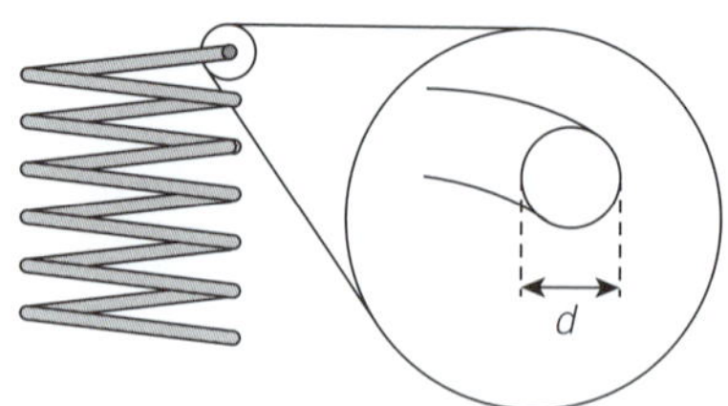

A fixed weight is suspended from a spring and the extension x of the spring is measured. The experiment is repeated using springs with different diameters.

Theory suggests that x and d are related by the equation:

$$x = kd^n$$

where k and n are constants.

(a) A graph is plotted of $\lg x$ on the y-axis against $\lg d$ on the x-axis. Determine expressions for the gradient and intercept. [1]

(b) Values of d and x are given in the table.

d / mm	x / cm	$\lg(d$ / mm$)$	$\lg(x$ / cm$)$
6.4	5.8 ± 0.2		
7.0	4.0 ± 0.2		
7.6	2.9 ± 0.2		
8.2	2.1 ± 0.2		
8.8	1.6 ± 0.2		
9.5	1.2 ± 0.2		

Calculate values of $\lg(x$ / cm$)$ and $\lg(d$ / mm$)$. Record your values on a copy of the table. Include the absolute uncertainties in $\lg(x$ / cm$)$. [2]

(c) (i) Plot a graph of $\lg x$ on the y-axis against $\lg d$ on the x-axis. Include error bars for $\lg(x$ / cm$)$.

(ii) Draw the straight line of best fit and a worst acceptable straight line on your graph.

(iii) Determine the gradient of the line of best fit. Include the absolute uncertainty.

(iv) Determine the y-intercept of the line of best fit. Include the absolute uncertainty. [8]

(d) (i) Using your answers to (c)(iii) and (c)(iv), determine the values of n and k.

(ii) Calculate the percentage uncertainty in your value of n. [4]

28 Exam-style questions

Paper 1 style questions: Multiple choice

The actual paper will have 40 multiple choice questions and you will have 1 hour 15 minutes to answer them. You should aim to answer the questions in this sample paper in about 40 minutes. The exam paper will include the standard list of data and formulae, and you will be provided with a separate answer sheet.

1 The Reynolds number Re is a dimensionless constant used in studying the flow of liquids in pipes. It is given by the equation:

$$\mathrm{Re} = \frac{\rho v D}{\mu}$$

where ρ is the density of the liquid, v its velocity, and D the diameter of the pipe. What are the SI base units of μ, the viscosity of the liquid? [1]

A $\mathrm{kg\,m\,s}$ **B** $\mathrm{kg\,m^{-1}\,s}$
C $\mathrm{kg\,m\,s^{-1}}$ **D** $\mathrm{kg\,m^{-1}\,s^{-1}}$

2 Forces **p** and **q** are represented by two vectors.

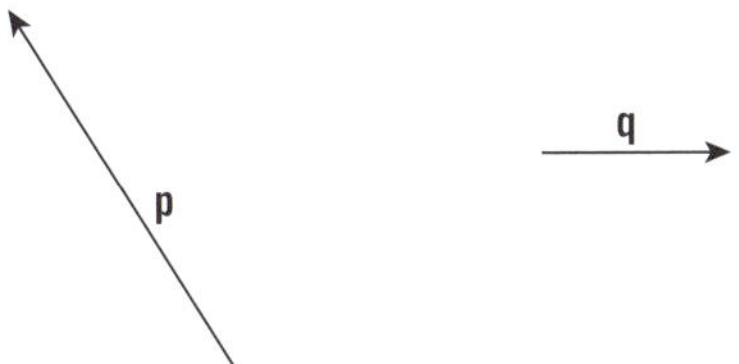

Which diagram shows **p−q**? [1]

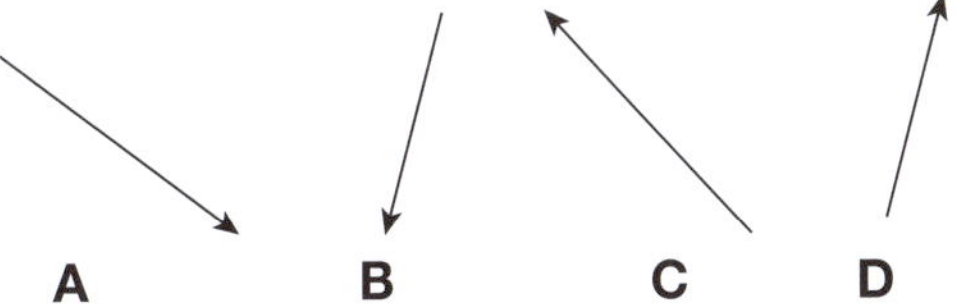

3 In an experiment to measure the resistivity of nichrome using a long thin wire, the following measurements were made:

length of wire l	$99.5 \pm 0.4\,\mathrm{cm}$
diameter of wire d	$0.38 \pm 0.01\,\mathrm{mm}$
resistance of wire R	$49 \pm 1\,\Omega$

The resistivity ρ was calculated using the equation $\rho = \dfrac{RA}{l}$, where A is the cross-sectional area of the wire.

What is the percentage uncertainty in the value of ρ? [1]

A 4.2%   **B** 5.0%   **C** 6.9%   **D** 7.7%

4 A stone is thrown vertically down from the top of a tall building. It passes one window travelling at a speed of $7.0\,\mathrm{m\,s^{-1}}$. It then passes a lower window travelling at a speed of $11.0\,\mathrm{m\,s^{-1}}$.

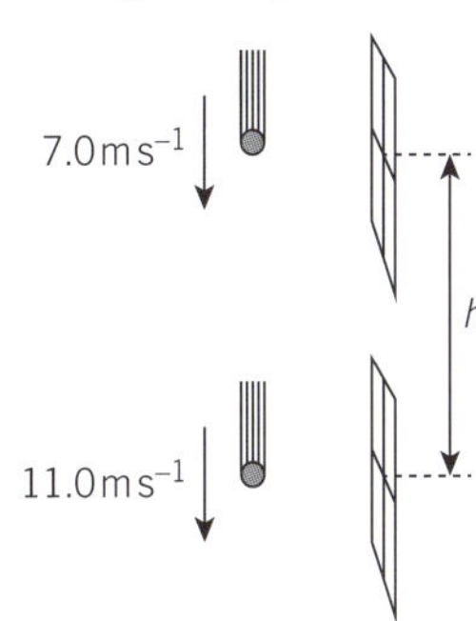

What is the height h between the two windows? [1]

A 0.82 m   **B** 1.6 m
C 3.7 m   **D** 7.3 m

5 A force of 120 N is needed to operate a foot pedal brake, as shown.

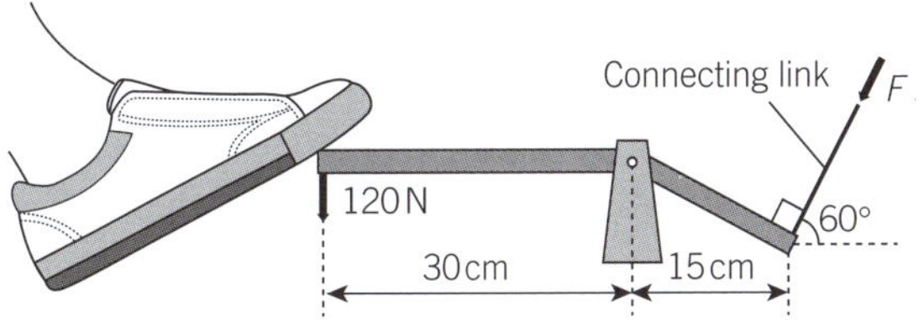

What is the force F in the connecting link? [1]

A 120 N   **B** 210 N
C 280 N   **D** 480 N

6 A lorry of mass $3.0 \times 10^{3}\,\mathrm{kg}$ is moving at a constant speed of $18\,\mathrm{m\,s^{-1}}$ up a slope. The slope is inclined at an angle of 10° to the horizontal. The frictional forces exert a force of 6.0 kN down the slope.

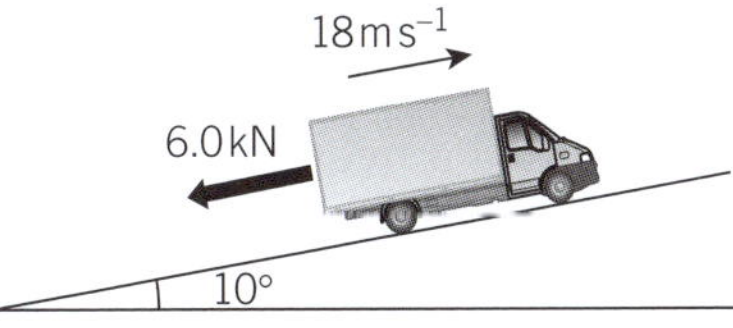

What is the output power of the lorry's engine? [1]

A 92.0 kW   **B** 108 kW
C 101 kW   **D** 200 kW

7 A deep-sea submersible used for exploring the bottom of the deepest oceans can withstand pressures up to 150 MPa. What is the maximum depth the submersible can descend? [1]

A 10 km   **B** 15 km
C 20 km   **D** 30 km

8 A loudspeaker connected to a signal generator emits a single frequency of sound waves. It is placed above a closed pipe, as shown. The pipe is filled with water. When a valve is opened the water level gradually falls.

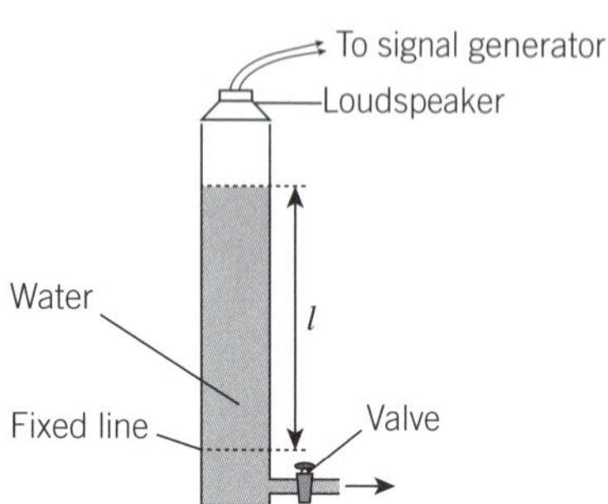

A louder sound is first heard when the length l above a fixed line, drawn on the pipe, is 67.0 cm. A second louder sound is heard when $l = 42.0$ cm. The speed of sound is 330 m s^{-1}.

What is the frequency of the sound emitted from the loudspeaker? [1]

A 330 Hz **B** 660 Hz
C 990 Hz **D** 1300 Hz

9 An oil drop with charge $-q$ is held stationary in the uniform electric field between two parallel plates. The potential difference between the plates is V and the distance between the two plates is d.

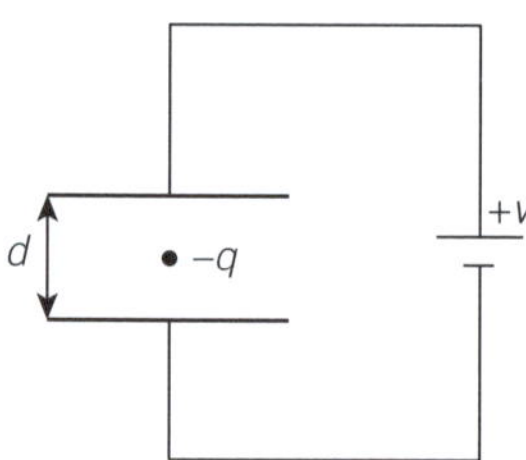

What is the mass m of the drop? [1]

A $\dfrac{dg}{qv}$ **B** $\dfrac{dq}{gv}$ **C** $\dfrac{gV}{qd}$ **D** $\dfrac{qV}{dg}$

10 The siren of an ambulance emits two notes, the higher note having a frequency of 960 Hz. What is the frequency of the higher note heard by a stationary observer when the ambulance is moving directly away from him at a speed of 30 m s^{-1}? [1]

[The speed of sound in air is 330 m s^{-1}.]

A 870 Hz **B** 880 Hz
C 1050 Hz **D** 1060 Hz

11 Two cells, with e.m.f.s of 14 V and 10 V and negligible internal resistance, are connected to three 1 Ω resistors, as shown.

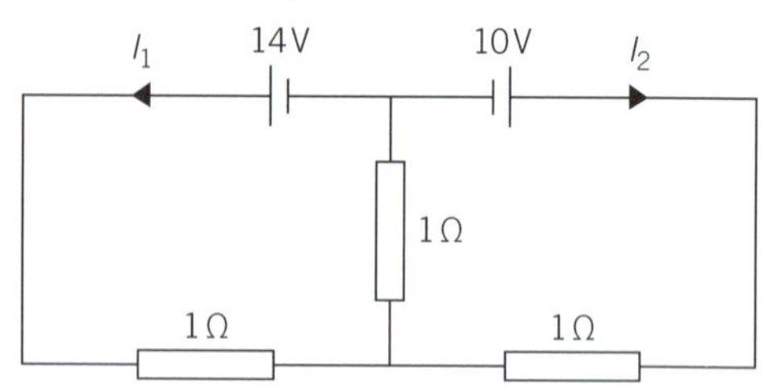

What are the values of I_1 and I_2? [1]

	I_1/A	I_2/A
A	2	10
B	3	8
C	5	4
D	6	2

12 A dry cell, with internal resistance r, is connected to a 5 Ω resistor and an ammeter, as shown.

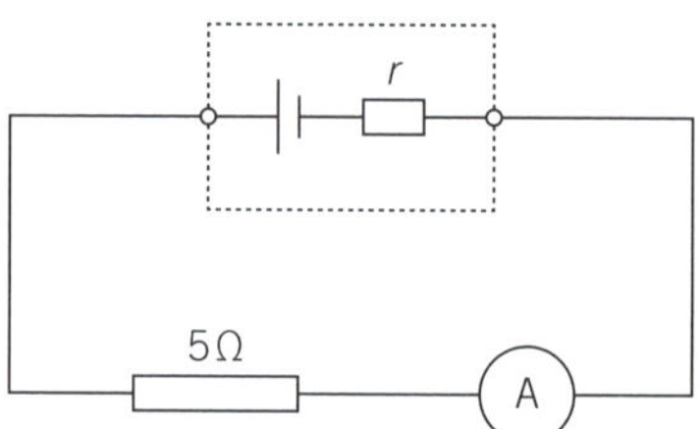

The reading on the ammeter is 2.0 A. When the 5 Ω resistor is replaced by a 7 Ω resistor, the reading on the ammeter falls to 1.5 A.

What is the value of r? [1]

A 0.5 Ω **B** 1.0 Ω **C** 1.5 Ω **D** 2.0 Ω

13 $^{214}_{83}$Bi is a radioactive isotope of bismuth which decays by β^--decay into an isotope of polonium. The polonium decays by α-decay into an isotope of lead.

What are the proton number and the nucleon number of the isotope of lead? [1]

	proton number	nucleon number
A	81	210
B	81	214
C	82	210
D	82	214

14 Which elementary particle is **not** a lepton? [1]

A an electron **B** a neutrino
C a positron **D** a quark

Paper 2 style questions: AS structured questions

You should aim to complete this sample paper in 1 hour 15 minutes. In the actual exam, the paper will include the standard list of data and formulae, and there will be spaces in the paper for you to write your answers.

1 **(a)** Explain what is meant by *work done.* [1]

(b) A lorry of total mass 2000 kg is travelling along a road with a constant uphill gradient, as shown. The angle of the road to the horizontal is 6°.

Calculate the component of the weight of the lorry down the slope. [2]

(c) The lorry is travelling at a speed of $20\,\mathrm{m\,s^{-1}}$ when the driver applies the brakes. The braking force resisting the motion of the lorry is 7200 N.

(i) Show that the deceleration of the lorry when the brakes are applied is $4.6\,\mathrm{m\,s^{-2}}$. [2]

(ii) Calculate the distance travelled by the lorry from the moment the brakes are applied until it comes to rest. [2]

(d) **(i)** Calculate:

1. the kinetic energy lost by the lorry [1]

2. the work done by the braking force. [1]

(ii) Explain why your answers to (d)(i) 1. and (d)(i) 2. are not the same. [1]

2 **(a)** **(i)** State the principle of conservation of momentum. [2]

(ii) Explain what is meant by an *elastic* collision. [1]

(b) A snooker ball of mass m, travelling at speed u, collides elastically with a second, stationary ball of equal mass, as shown. The speed of the first ball after the collision is $\dfrac{u}{2}$. The speed of the second ball after the collision is v at an angle θ to the original direction of the first ball.

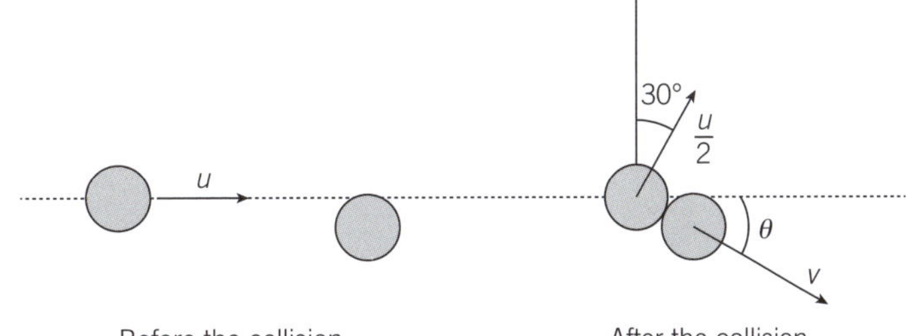

(i) State expressions for:

1. the kinetic energy of the first ball before the collision [1]

2. the kinetic energy of the first ball after the collision [1]

3. the kinetic energy of the second ball after the collision. [1]

(ii) Hence show that $v = \dfrac{u\sqrt{3}}{2}$. [1]

(c) Using the principle of conservation of momentum, determine the value of θ. [3]

3 **(a)** Define, for a metal wire:

    (i) *stress* [1]

    (ii) *strain* [1]

    (iii) *Young modulus.* [1]

(b) An aluminium wire, of diameter 0.28 mm, is fixed at one end and passes over a pulley at the other end. A mark is made on the wire 1.50 m from the fixed end and a travelling microscope is placed above the mark, as shown.

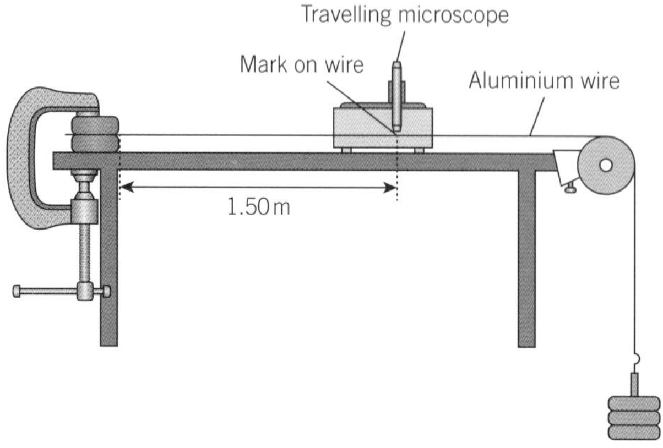

When a load of 3.5 N is hung on the free end of the wire, the mark moves 1.24 mm.

    (i) Show that the stress on the wire is 5.7×10^7 Pa. [2]

    (ii) Calculate the Young modulus of aluminium. [3]

(c) A student suggests that a wire of aluminium of the same length, but with diameter 0.14 mm, will stretch four times as far with the same load.

State whether, or not, the student is correct. Explain your answer. [2]

4 **(a)** Explain what is meant by the terms:

    (i) *diffraction* [2]

    (ii) *interference.* [1]

(b) A student attempts to measure the wavelength of blue light using Young's double-slits experiment, as shown. The two slits are 0.60 mm apart.

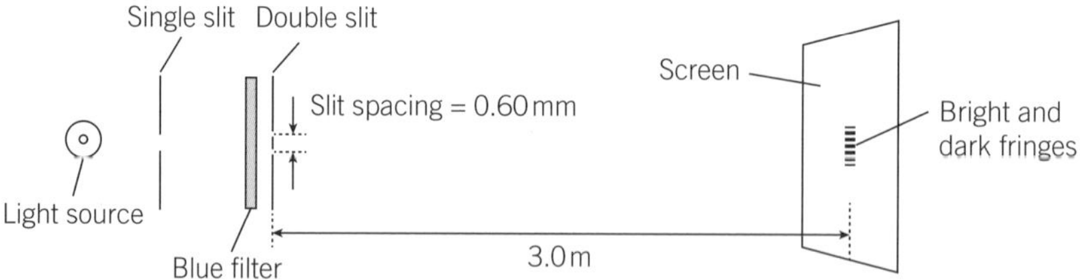

    (i) The light emerging from the two slits is coherent. Explain what is meant by *coherent*. [1]

    (ii) Bright and dark fringes are observed on a screen placed 3.0 m away from the double slits. Explain why bright and dark fringes are observed. [2]

(c) The distance between adjacent bright fringes is 2.4 mm. Calculate the wavelength of the blue light. [2]

(d) State what would happen to the fringes if the blue filter is replaced by a red filter. Explain your answer. [1]

5 **(a)** **(i)** State Kirchhoff's first law. [1]

(ii) Kirchhoff's first law is linked with the conservation of a physical quantity. State this quantity. [1]

(b) A 9.0 V dry cell, with an internal resistance of 0.5 Ω, is connected to a network of resistors, as shown.

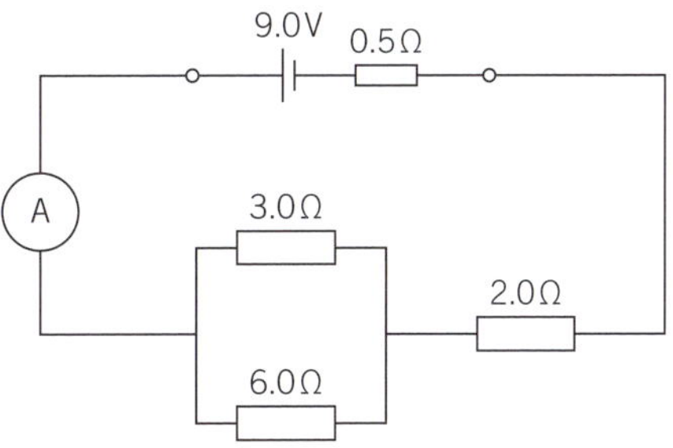

Show that the reading on the ammeter is 2.0 A. [2]

(c) Calculate:

(i) the current in the 6 Ω resistor [1]

(ii) the power dissipated in the 2 Ω resistor [2]

(iii) the potential difference across the terminals of the cell. [2]

6 **(a)** The radioactive decay of an isotope of cobalt is shown by the nuclear equation:

$$^{60}_{27}\text{Co} \rightarrow {}^{60}_{28}\text{Ni} + \text{X}$$

(i) Explain the meaning of the term *isotope*. [2]

(ii) State

1. the number of protons in the cobalt nucleus, [1]

2. the number of neutrons in the nickel nucleus. [1]

(b) **(i)** X is a type of radiation. State the name of this radiation. [2]

(ii) Describe **two** properties of this radiation. [2]

(c) The mass of the cobalt (Co) nucleus is greater than the combined mass of the nucleus of nickel (Ni) and X. Use a conservation law to explain how this is possible. [3]

Paper 3 style questions: Advanced practical skills

The actual exam will last for 2 hours, but in this sample paper you are not expected to complete the practical work; you are provided with specimen results. You should aim to complete this sample paper in 1 hour.

1 *Part (a) of this question asks candidates to set up the electrical circuit shown. The movable lead is first placed approximately half-way along the resistance wire. The potential difference V across the fixed resistor and the value of x are recorded. The experiment is then repeated for different values of x – **a set of specimen results is provided for you**.*

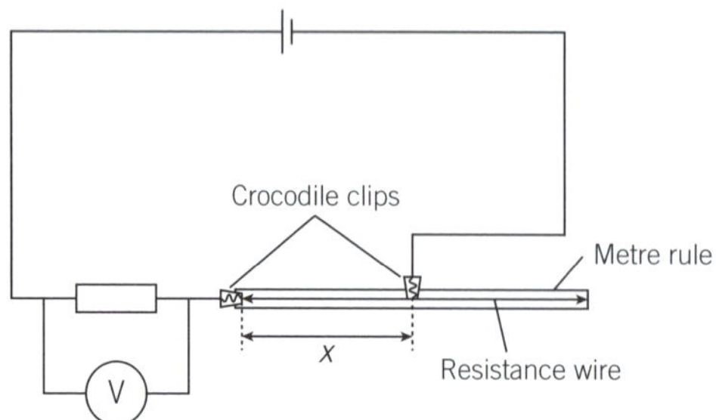

(b) Change x and repeat the experiment until you have six sets of values of x and V.

Record your results in a table. Include values of $\dfrac{1}{V}$ in your table.

$x\,/\,cm$	$V\,/\,V$	$\dfrac{1}{V}\,/\,V^{-1}$
5.1	1.43	0.699
16.5	1.24	0.806
36.1	1.01	0.990
59.8	0.87	1.15
72.3	0.80	1.25
96.3	0.72	1.39

(c) (i) Using graph paper, plot a graph of $\dfrac{1}{V}$ on the y-axis against x on the x-axis. [3]

   (ii) Draw the straight line of best fit. [1]

   (iii) Determine the gradient and y-intercept of this line. [2]

(d) It is suggested that the quantities $\dfrac{1}{V}$ and x are related by the equation:

$$\frac{1}{V} = Px + Q$$

where P and Q are constants.

Using your answers in **(c)(iii)**, determine the values of P and Q.

Give appropriate units. [2]

2 In this experiment, you will investigate the deflection of a thin wooden beam when a load is suspended from it. *Specimen results are provided when needed.*

(a) (i) Set up the apparatus as shown.

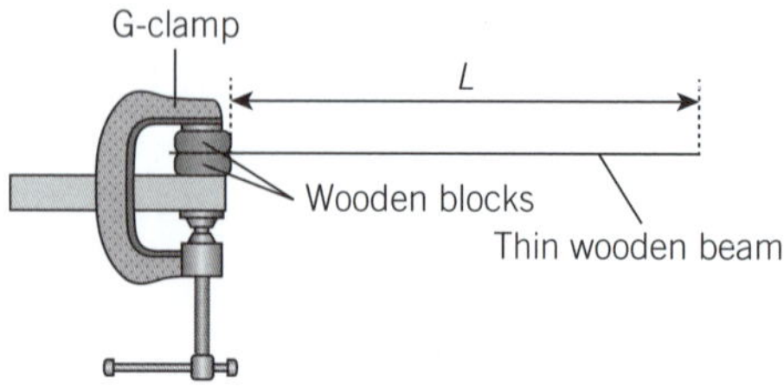

Adjust the beam so that the length L is approximately 90 cm.

   (ii) Measure and record L. $L = $ 89.7 cm

(b) (i) Suspend the mass M from the beam, as shown below.

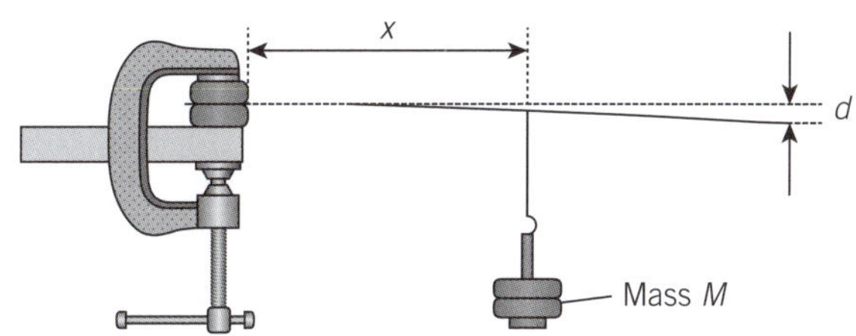

Adjust the position of M so that x is approximately 50 cm.

(ii) Record x.

$x =$ ......50.2 cm......

(c) (i) Calculate the value of P, where:

$$P = x^2(3L - x)$$ [1]

(ii) Justify the number of significant figures that you have given for your value of P. [1]

(d) (i) Measure and record the deflection d of the end of the beam.

$d =$ ......8 mm......

(ii) Estimate the percentage uncertainty in your value of d. [1]

(e) Without changing L, change the position of the mass M so that x is approximately 80 cm.

Repeat (b)(ii), (c)(i) and (d)(i).

$x =$ ......79.6 cm......

$d =$ ......17 mm...... [1]

(f) It is suggested that the relationship between d and P is

$$d = kP$$

where k is a constant.

(i) Using your data, calculate two values of k. [1]

(ii) Explain whether your results in (f)(i) support the suggested relationship. [1]

(iii) Using your first value of k, estimate the deflection of the beam when $x = 65$ cm. [1]

(g) (i) Describe four sources of uncertainty or limitations of the procedure for this experiment. [4]

(ii) Describe four improvements that could be made to this experiment. You may suggest the use of other apparatus or different procedures. [4]

Paper 4 style questions: A Level structured questions

You should aim to complete this sample paper in 2 hours. In the actual exam, the paper will include the standard list of data and formulae, and there will be spaces in the paper for you to write your answers.

1 (a) State Newton's law of gravitation. [2]

(b) A satellite orbits the Earth at a constant speed, at a height of 220 km above the Earth's surface.

A student states that '*the satellite is travelling at constant speed, so the net force on the satellite must be zero*'. Explain why the student is incorrect. [2]

(c) The mass of the satellite is 2.5×10^3 kg. The mass of the Earth is 6.0×10^{24} kg. The radius of the Earth is 6570 km.

Show that:

(i) the gravitational force acting on the satellite is 2.2×10^4 N [2]

(ii) the centripetal acceleration of the satellite is $8.7\,\mathrm{m\,s^{-2}}$. [1]

(d) Determine:

(i) the speed of the satellite [2]

(ii) the time taken for one complete orbit of the Earth. [1]

(e) Suggest one use of this type of satellite. [1]

2 (a) Define *simple harmonic motion*. [2]

(b) A torsion pendulum consists of a long, thin metal wire supporting a horizontal bar of length 6.0 cm, as shown. The bar is twisted in a horizontal plane through an angle of 20° and released. The bar rotates back and forth with simple harmonic motion. The period T of the oscillation is 1.2 s.

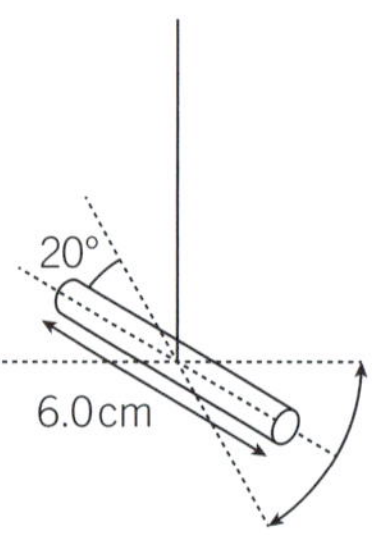

(i) Calculate the frequency of the oscillation. [2]

(ii) Show that the angular frequency of the oscillation is $5.2\,\mathrm{rad\,s^{-1}}$. [1]

(iii) Determine the position of the bar 5.7 s after it is released. [1]

(c) Calculate the maximum speed of one end of the bar. Assume the oscillation is undamped. [3]

(d) The graph shows the variation of the kinetic energy of the bar and wire over half a cycle.

On a copy of the axes, sketch:

(i) the variation of potential energy with time over ½ a cycle [1]

(ii) the variation of total energy of the bar and wire with time over ½ a cycle. [1]

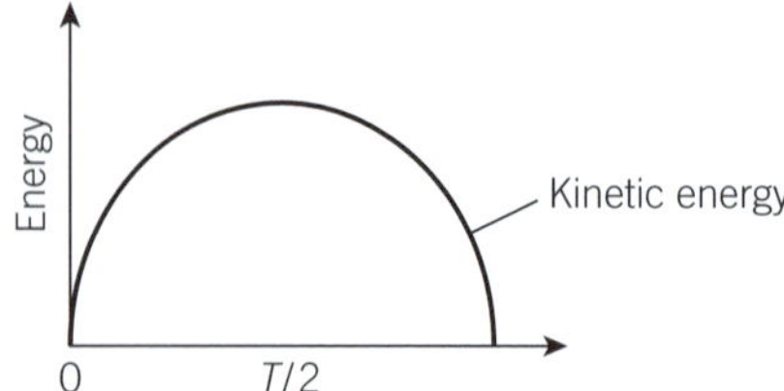

3 **(a)** Define *capacitance*. [1]

(b) A student says that '*capacitors are for storing charge*' Explain why the student is incorrect. [2]

(c) Three parallel-plate capacitors are connected to a 12 V d.c. supply, as shown.

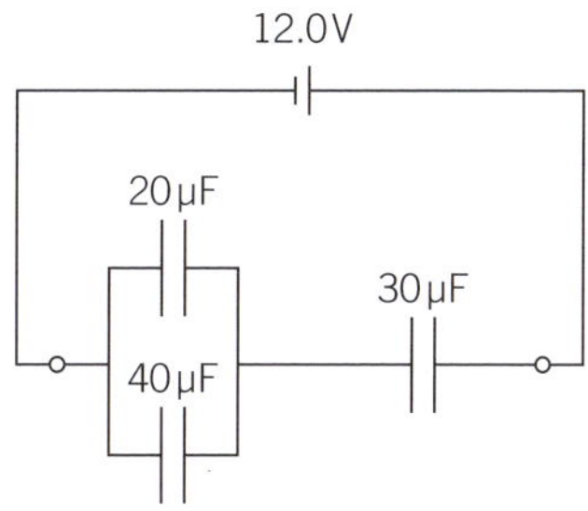

    (i) Calculate the combined capacitance of the three capacitors. [2]

    (ii) Show that the potential difference across the 30 µF capacitor is 8.0 V. [2]

    (iii) Determine the charge on one plate of the 20 µF capacitor. [1]

(d) **(i)** Calculate the energy stored in the 40 µF capacitor. [2]

    (ii) Describe how energy is stored in a capacitor. [1]

    (iii) Suggest one use of capacitors as energy storage devices. [1]

4 **(a)** An experiment is carried out to calibrate a negative temperature coefficient thermistor as a thermometer.

What is meant by the term *negative temperature coefficient*? [1]

(b) The circuit used is shown.

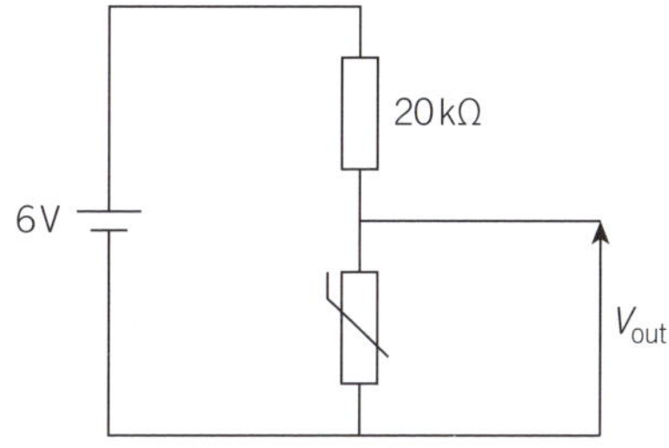

    (i) What is the name given to this type of circuit? [1]

    (ii) At 30°C the output p.d. V_{out} is 4 V. Show that the resistance of the thermistor at 30°C is 40 kΩ. [2]

(c) The calibration curve for the thermistor is shown.

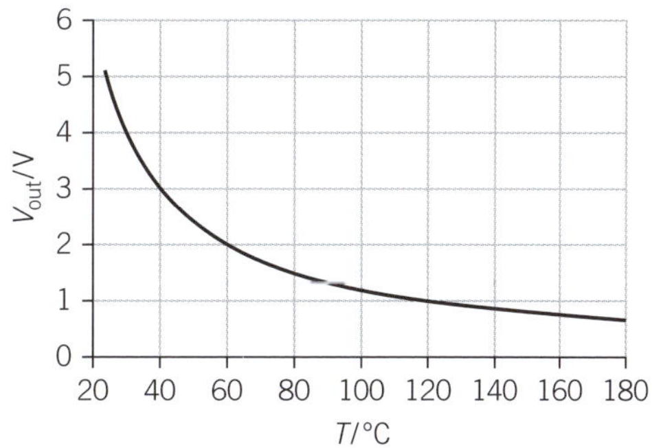

    (i) Explain what is meant by *calibration curve*. [1]

    (ii) Determine the output p.d. when the temperature is 70°C. [2]

    (iii) Determine the temperature of the thermistor when V_{out} is 0.8 V. [1]

(d) Suggest a reason why the thermistor is a less reliable thermometer when used at high temperatures. [1]

5 **(a)** **(i)** State Faraday's law. [1]

    (ii) Define *magnetic flux*. [2]

(b) A metal bar of length l is moving at a constant speed v along a frictionless wire loop. The bar is pulled by a weight W connected to the bar by a string passing over a frictionless pulley, as shown. A uniform magnetic field B acts vertically downwards.

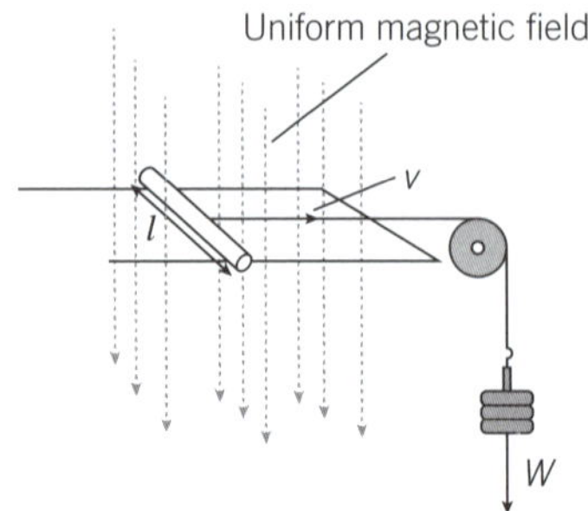

    (i) State the area 'swept out' by the bar in one second. [1]

    (ii) Hence show that the e.m.f. E induced across the ends of the bar is given by the equation $E = Blv$. [1]

(c) The electrical resistance of the bar is R ohms. Determine:

    (i) the current I in the bar [1]

    (ii) the direction of the current [1]

    (iii) the magnitude and direction of the magnetic force F on the bar. [2]

(d) State what would happen to the direction of the magnetic force if the direction of the magnetic field was reversed. [1]

(e) Show that the speed v of the bar is given by the equation: $v = \dfrac{RW}{B^2 l^2}$ [3]

Paper 5 style questions: Planning, analysis, and evaluation

The actual exam will last for 1 hour 15 minutes; you should aim to complete this sample paper in the same time. There will be spaces in the exam paper for you to write your answers.

1 A student is investigating the stiffness of metal springs.

It is suggested that the spring stiffness k of the spring is related to the diameter d of the coils of the spring by the equation:

$$k = \frac{C}{d^3}$$

where C is a constant.

Design a laboratory experiment to test the relationship between k and d. Explain how your results could be used to determine the value of C. You should draw a diagram showing the arrangement of your equipment. In your account you should pay particular attention to:

- the procedure to be followed
- the measurements to be taken
- the control of variables
- the analysis of the data
- any safety precautions to be taken. [15]

2   A student is investigating how the time taken for a water tank to empty is related to the initial height h of water in the tank, as shown.

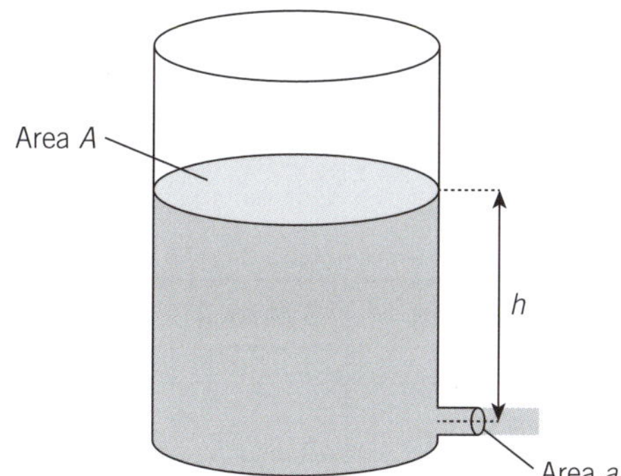

The student records the time t taken for the tank to empty from different heights.

Theory suggests that h and t are related by the equation:

$$t = \frac{A}{a}\sqrt{\frac{2}{g}}h^{n}$$

where A is the cross-sectional area of the tank and a is the area of the water outlet. g is the acceleration of free fall and n is a constant.

(a) A graph is plotted of lg t on the y-axis against lg h on the x-axis.

Determine expressions for the gradient and y-intercept. [2]

(b) Values of h and t are given in the table.

h/cm	t/s	lg (h/cm)	lg(t/s)
75.4	98 ± 2		
63.6	86 ± 2		
51.5	74 ± 2		
46.9	70 ± 2		
37.1	59 ± 2		
25.7	45 ± 2		

Calculate and record values of lg (h/cm) and lg (t/s).

Include the absolute uncertainties in lg (t/s). [3]

(c) (i) On graph paper, plot a graph of lg (t/s) against lg (h/cm). Include error bars for lg (t/s). [2]

(ii) Draw the straight line of best fit and a worst acceptable straight line on your graph.

Both lines should be clearly labelled. [2]

(iii) Determine the gradient of the line of best fit. Include the absolute uncertainty in your answer. [2]

(iv) Determine the y-intercept of the line of best fit. Include the absolute uncertainty in your answer. [2]

(d) Using your answers to (a), (c)(iii) and (c)(iv), determine the values of n and $\dfrac{A}{a}$. [2]

Take $g = 981\,\text{cm s}^{-2}$

Appendix: Maths skills for AS and A Level Physics

Signs and symbols

You will need to be familiar with the meanings of the symbols in Table A.1.

▼ **Table A.1** Symbols

Symbol	Meaning	Symbol	Meaning	Symbol	Meaning
$<$	is less than	$>>$	is much greater than	Σ	the sum of
$>$	is more than	$\approx$	is approximately equal to	Δx	a small amount of **or** a change in
$\leq$	is less than or equal to	$/$	is divided by	δx	an infinitesimal change in
$\geq$	is greater than or equal to	$\propto$	is proportional to	$\sqrt{}$	square root of
$<<$	is much less than	$<x>$ or $\bar{x}$	mean value of		

Standard form (scientific) notation

Physics is often concerned with very large and very small numbers (e.g., the mass of the Earth or the diameter of a nucleus). For convenience, these numbers are usually written in **standard form** (also called **scientific notation**).

A number written in standard form consists of a number with one digit only before the decimal point, multiplied by a power of 10.

An alternative way of recording very large or very small values is to use prefixes such as m (milli), p (pico), M (mega) and G (giga).

Perimeters, areas, and volumes

Make sure you know how to calculate the quantities shown in Figure A.1.

Examples:

charge on an electron = 1.6×10^{-19} C

diameter of Sun = 1.39×10^{9} m

Link

A full list of prefixes and their meanings is given in Table 1.4 in Unit 1 *Physical quantities and units*, p.4.

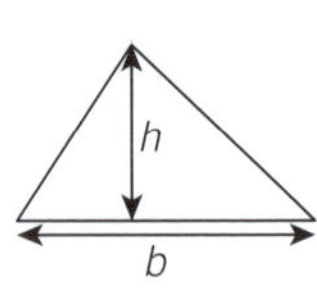

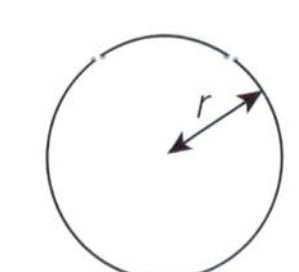

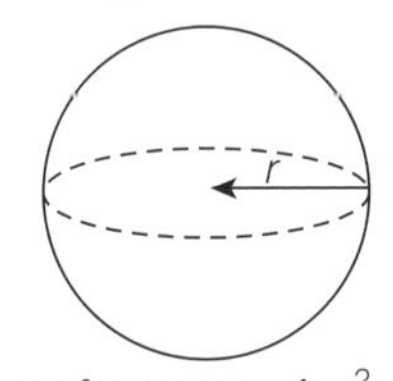

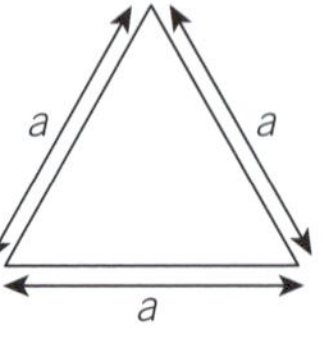

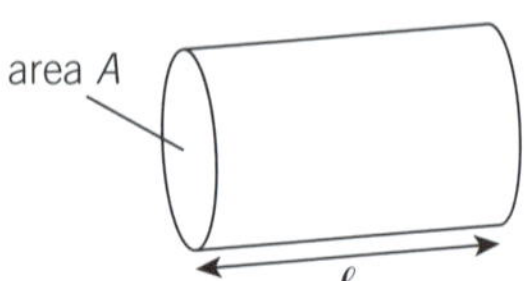

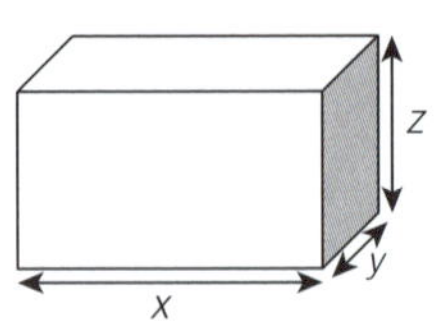

▲ **Figure A.1** Calculating perimeters, areas, and volumes

Trigonometry

Measuring angles

Angles can be measured in degrees (°) or radians (rad).

360° is equivalent to 2π rad (see Figure A.2), and so to convert an angle in degrees to radians:

$$\theta\,(\text{in rad}) = \theta\,(\text{in degrees}) \times \frac{2\pi}{360}$$

$$\theta\,(\text{in rad}) = \frac{l}{r}$$

Right-angled triangles only

Sin, cos, and tan

For all right-angled triangles (see Figure A.3):

$$\sin\theta = \frac{\text{opposite}}{\text{hypotenuse}} = \frac{b}{a}$$

$$\cos\theta = \frac{\text{adjacent}}{\text{hypotenuse}} = \frac{c}{a}$$

$$\tan\theta = \frac{\text{opposite}}{\text{adjacent}} = \frac{b}{c} = \frac{\sin\theta}{\cos\theta}$$

Pythagoras' theorem

For all right-angled triangles (see Figure A.3):

$$a^2 = b^2 + c^2 \rightarrow a = \sqrt{b^2 + c^2}$$

All triangles

For any triangle (see Figure A.4):

cos rule: $\qquad a^2 = b^2 + c^2 - 2bc\cos A$

sin rule: $\qquad \dfrac{a}{\sin A} = \dfrac{b}{\sin B} = \dfrac{c}{\sin C}$

$A + B + C = 180°$

Trigonometric identities

These statements are true for all values of θ:

$\sin(90° - \theta) = \cos\theta \qquad\qquad \sin(90° + \theta) = \cos\theta$

$\cos(90° - \theta) = \sin\theta \qquad\qquad \cos(90° + \theta) = -\sin\theta$

$\sin(180° - \theta) = \sin\theta \qquad\qquad \sin(180° + \theta) = -\sin\theta$

$\cos(180° - \theta) = -\cos\theta \qquad\qquad \cos(180° + \theta) = -\cos\theta$

$\sin 2\theta = 2\sin\theta\cos\theta \qquad\qquad \cos 2\theta = \cos^2\theta - \sin^2\theta$

$\qquad\qquad\qquad\qquad\qquad\qquad\quad = 2\cos^2\theta - 1$

$\cos^2\theta + \sin^2\theta = 1 \qquad\qquad\qquad = 1 - 2\sin^2\theta$

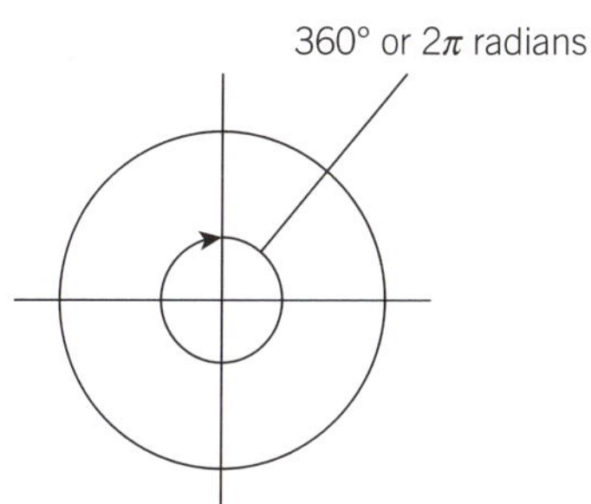

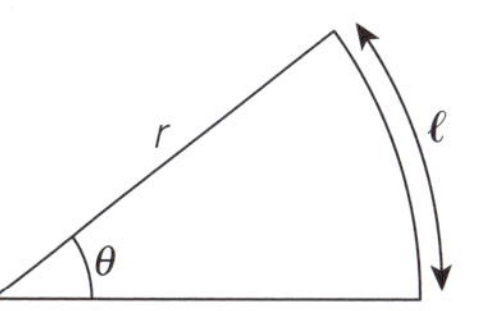

▲ **Figure A.2** Converting degrees to radians

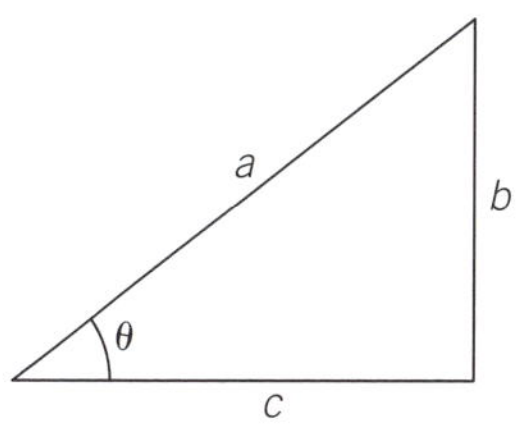

▲ **Figure A.3** Right-angled triangle

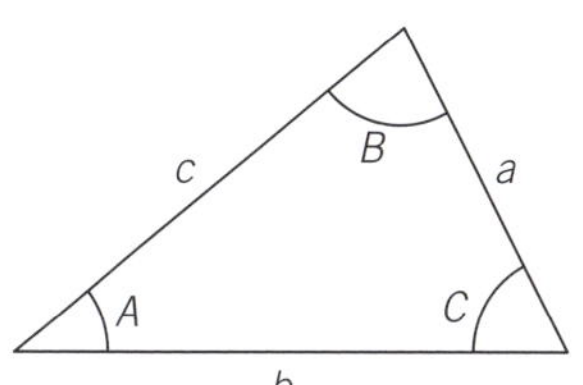

▲ **Figure A.4** Any triangle

Small angles

For **small angles**, with θ measured in **radians**:

$$\theta \approx \sin\theta \approx \tan\theta \qquad\qquad \cos\theta = 1.$$

Similar triangles

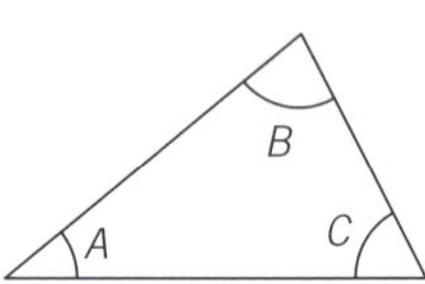
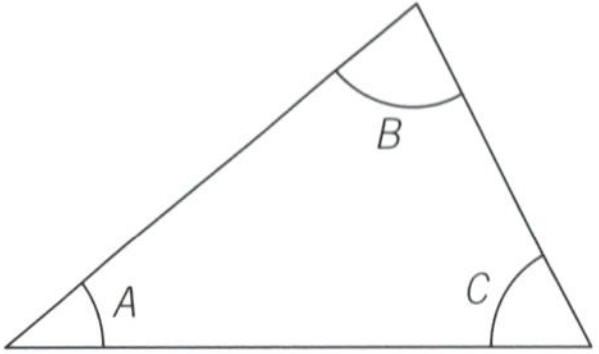

▲ **Figure A.5** Similar triangles

Two triangles are **similar** if one is a simple magnification of the other (see Figure A.5). The three angles, A, B and C will be the same in both triangles.

Vectors

Vector quantities, such as force and velocity, have both magnitude and direction, so can be represented by arrows, the length of the arrow representing the magnitude of the quantity and the direction of the arrow indicating the direction of the quantity.

Adding vectors

Two vectors can be added by sliding one vector so that the arrowhead of one of the vectors coincides with the start of the other. The resultant vector is represented by a line from the start of one vector to the arrowhead of the other (see Figure A.6a). Figure A.6b shows how two vectors can be subtracted.

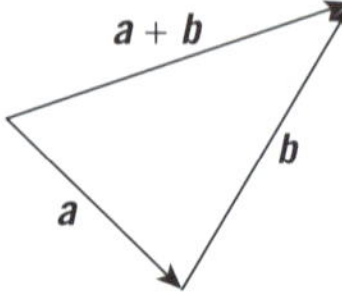

a Adding

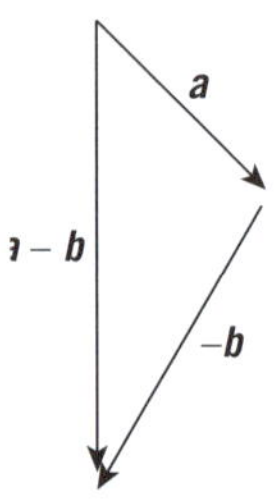

b Subtracting

▲ **Figure A.6** Adding and subtracting vectors

Resolving vectors

It is often useful to be able to **resolve** vectors – to split them into two perpendicular components: $v\sin\theta$ and $v\cos\theta$ (see Figure A.7).

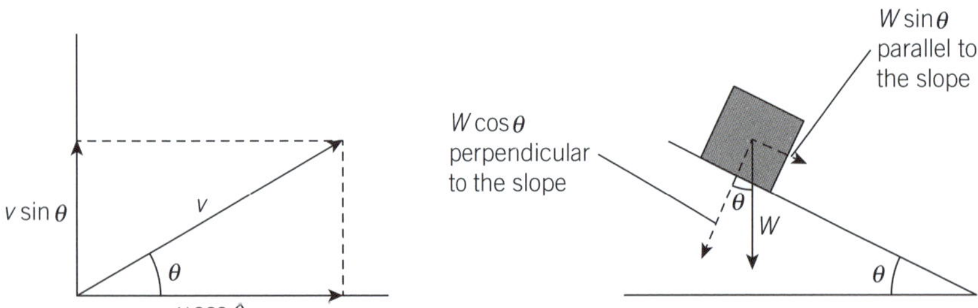

▲ **Figure A.7** Resolving vectors into two perpendicular components

The resultant of the two perpendicular components can be found using Pythagoras' theorem:

$$v = \sqrt{[(v\cos\theta)^2 + (v\sin\theta)^2]} = \sqrt{[v^2(\cos^2\theta + \sin^2\theta)]} = \sqrt{v^2} = v$$

Graphs

Direct proportionality

Some important relationships in physics, such as Hooke's law and Ohm's law, involve **direct proportionality**. If one quantity is directly proportional to the other, then doubling one will double the other. A graph of one quantity plotted against the other will be a straight line through the origin (see Figure A.8).

The symbol $\propto$ means 'proportional to'.

Linear relationships

Linear relationships take the form:

$$y = mx + c$$

where m is the gradient of the line and c is the intercept of the line on the y axis (see Figure A.9).

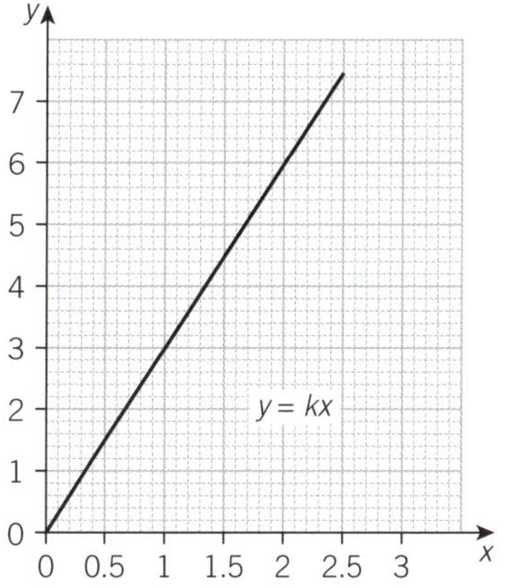

▲ **Figure A.8** Direct proportionality

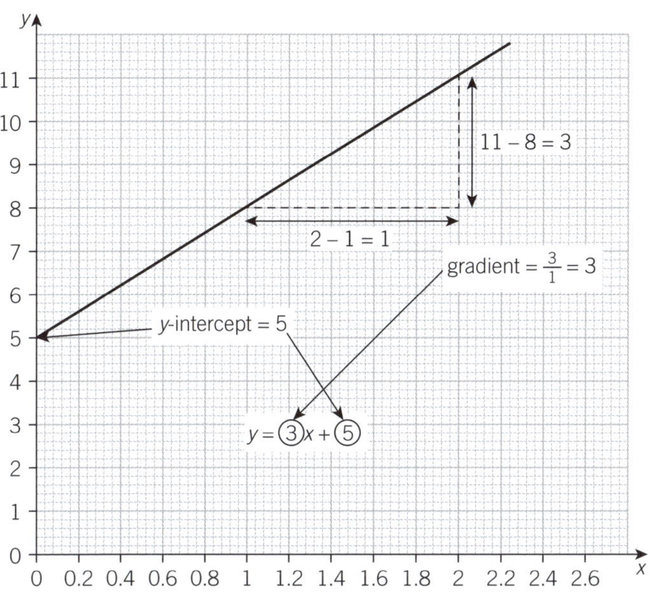

▲ **Figure A.9** Linear relationship

Linear relationships

Figure A.10 gives examples of some non-linear graphs that you may encounter.

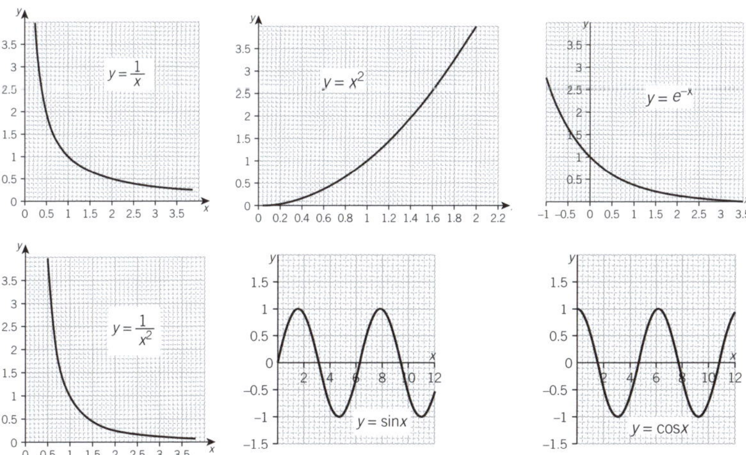

▲ **Figure A.10** Important non-linear graphs

Tangents, gradients, and areas

Tangents and gradient

It is often informative to find the gradient of a graph at a particular point (e.g., if a graph of the velocity of an object against time is plotted, its gradient at any time is the acceleration at that moment).

When drawing a tangent, the tangent line should just touch but not 'cut' the curved line at one point and the 'angle' between the tangent and the curved line should be about the same on either side (see Figure A.11).

To find the gradient draw a large triangle and then calculate $\frac{\Delta y}{\Delta x}$.

Area under a graph

It's also sometimes useful to find the area underneath a graph between two points (e.g., for a graph of the velocity of an object against time, the area under the graph between two times is the displacement of the object in that time) (see Figure A.12).

To find the area under a graph between two points, either:

- count the number of squares underneath the graph, counting part squares that are more than half a square as '1' and squares which are less than half a square as '0', or

- divide the area up into a series of rectangles and triangles – the smaller the squares and triangles, the more accurate the value.

Logarithms

If $y = a^x$ then $x = \log_a y$; that's to say 'x is the logarithm, to base a, of y'.

Common logarithms (logs to base 10)

The most commonly used logarithms are **common logarithms** to base 10; written as $\log_{10}$ or **lg**. For example:

$1000 = 10^3$	so	$\log_{10} 1000 = 3$
$0.01 = 10^{-2}$	so	$\lg 0.01 = -2$
$28.7 = 10^{1.458}$	so	$\lg 28.7 = 1.458$

If $p = 10^x$, then $x = \lg p$; similarly if $q = 10^y$, then $y = \lg q$ so

$$pq = 10^x \times 10^y = 10^{x+y}$$

so $\qquad \lg pq = x + y = \lg p + \lg q$

It can similarly be shown that $\lg\left(\dfrac{p}{q}\right) = \lg p - \lg q$

Logarithms are particularly useful when testing if the relationship between two variables, say x and y, is in the form of a power law $y = Ax^n$ where A and n are constants. 'Taking logs' of both sides of this equation:

$$\log y = \log (Ax^n) = \log A + \log x^n$$

$$\log y = n \log x + \log A$$

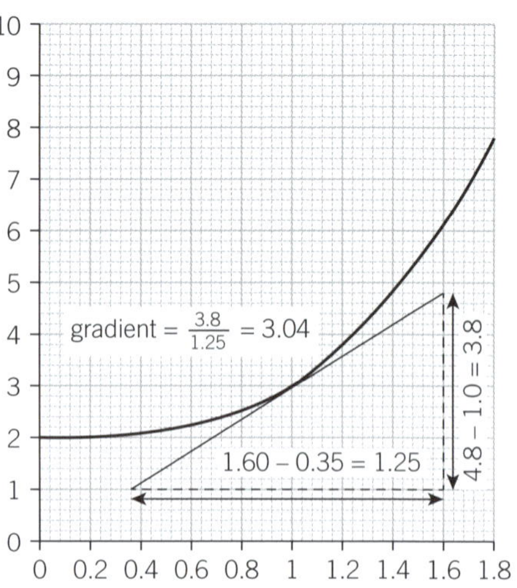

▲ **Figure A.11** Calculating the tangent at a point

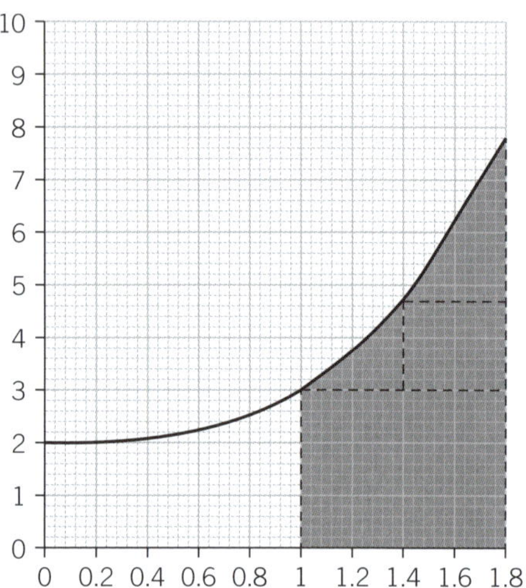

▲ **Figure A.12** Calculating the area under a graph

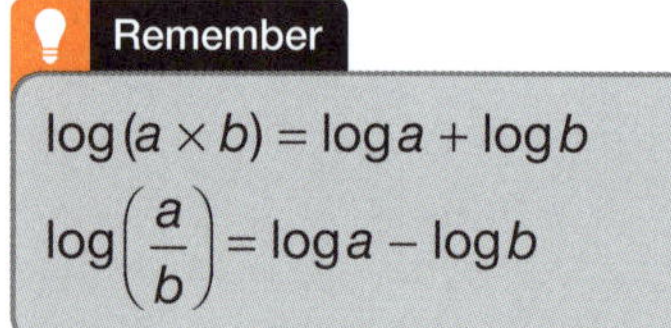

Compare this equation with $y = mx + c$, the equation for a straight line.

If the relationship is correct, the graph of $\log y$ on the y-axis against $\log x$ on the x-axis (a log–log graph) should be a straight line, with the gradient equal to n and the y-intercept equal to $\log A$. Note that the base of the logarithms has not been specified – the method would work using any base of logarithms – though it is usual to use base 10 logs.

Natural logarithms (logarithms to base e)

The **natural logarithm** of a number is its logarithm to base e, where e is the irrational constant 2.71828… Many physics phenomena, such as radioactive decay, the absorption of γ-rays by lead, and the discharge of a capacitor through a resistor, are described by the mathematical function e^{kx}, where k is a constant which can be negative (exponential decay) or positive (exponential growth). Logarithms to base e are usually written **ln**.

Natural logarithms are particularly useful when testing for an exponential relationship between two variables e.g., $y = A\,e^{kx}$, where x and y are the two variables, and A and k are constants. 'Taking natural logs' of this equation:

$$\ln y = \ln (A\,e^{kx}) = \ln A + \ln e^{kx}$$

$$\ln y = kx + \ln A$$

If the relationship between x and y is exponential, then the graph of $\ln y$ on the y-axis against x on the x-axis (a log-linear graph) should be a straight line, with gradient k and y-intercept $\ln A$.

> Compare this equation with $y = mx + c$, the equation for a straight line.

Summary

Table A.2 summarises which graph to use for which relationship.

▼ **Table A.2** Which graph?

Relationship	Graph to plot	Gradient and y-intercept
$p = mq + c$	p on y-axis, q on x-axis	m is the gradient, c is the y-intercept
$p = Aq^n$	$\log p$ on y-axis, $\log q$ on x-axis	n is the gradient, $\log A$ is the y-intercept
$p = Ae^{Bq}$	$\ln p$ on y-axis, q on x-axis	B is the gradient, $\ln A$ is the y-intercept

> Logarithms are also useful when describing quantities with values which range over several orders of magnitude, such as noise levels, electronic amplification and earthquakes (the Richter scale is an example of a logarithmic scale).

Uncertainties and significant figures

Measurements taken during an experiment may vary when the experiment is repeated. Suppose, for example, five measurements of the time t taken for an object to fall a fixed distance through a viscous liquid are measured as 7.1 s, 7.4 s, 6.8 s, 6.7 s, and 7.5 s. The average time is 7.1 s with a range (largest value – smallest value) of 0.8 s. The value of t can be expressed as 7.1 ± 0.4 s.

As well as random variations, uncertainties arise both from the limitations of the instrument being used (both its accuracy and its precision) and the difficulty in judging the measurement (e.g., deciding when an oscillation of a pendulum is completed). If a metre rule is calibrated in millimetres, it is tempting to judge the uncertainty in a measurement made with the rule as $\pm$ 1 mm, but if it is used to measure the length of a resistance wire, for example, there may be kinks in the wire. The actual overall uncertainty in the measurement may be 2–5 mm.

> ★ **Exam tip**
>
> If a calculation involves a number of stages, or uses several different measured values, do not 'round' any intermediate values in the calculation. Instead, only round the final answer.

When calculating a quantity using measured values, the calculated quantity also has an uncertainty. For example, the resistance of a resistance wire is found by measuring the current I through the wire and the p.d. V across it.

$$I = 0.15 \pm 0.01\,\text{A} \qquad V = 1.54 \pm 0.02\,\text{V}$$

The nominal resistance of the wire, $R = \dfrac{V}{I} = \dfrac{1.54}{0.15} = 10.2667\,\Omega$

$$\text{largest value of } R = \frac{V_{max}}{I_{min}} = \frac{1.56}{0.14} = 11.1\,\Omega$$

$$\text{smallest value of } R = \frac{V_{min}}{I_{max}} = \frac{1.52}{0.16} = 9.5\,\Omega$$

The large range of possible values of R (between $9.5\,\Omega$ and $11.1\,\Omega$) means the resistance cannot be known as precisely as $10.2667\,\Omega$! Instead, the value of R should be given as:

$$R = 10.3 \pm 0.8\,\Omega$$

Calculating percentage uncertainties

If the absolute uncertainty in a measurement of x is Δx, then the percentage uncertainty is:

$$\text{percentage uncertainty} = \frac{\Delta x}{x} \times 100\%$$

For example, the inside diameter d of a glass tube is measured using Vernier calipers. In this case the absolute uncertainty is just the precision of the calipers ($\pm\, 0.1\,\text{mm}$), as the tube is likely to have the same diameter all the way along its length.

$$d = 14.3\,\text{mm}$$

The percentage uncertainty $= \dfrac{\Delta d}{d} \times 100\% = \dfrac{0.1}{14.3} \times 100 = 0.7\%$

d lies between 14.25 mm and 14.34 mm.

Index

electronics 154, 158–9
 electronic sensors 151
 ideal operational amplifiers 154
 operational amplifier circuits 155–6
 output devices 157
electrons 206
 electron energy levels 192–3
 photoelectrons 189
emission spectra 193
 emission line spectrum 193
energy 43, 49–50
 change in potential energy in a
  gravitational field 47
 change in potential energy in an electric
  field 47
 conservation of energy 86
 efficiency 46–7
 electron energy levels 192–3
 energy gap 194
 energy released in nuclear fission 211
 energy released in nuclear fusion 211–12
 energy stored in capacitors 134
 gravitational potential energy 44
 internal energy of a gas 71
 internal energy of a system 86
 kinetic energy of a gas molecule 72
 principle of conservation of energy 43
 simple harmonic motion (SHM) and
  energy 93–4
 springs 63
 types of energy 44–5
equations of motion 18
 using the equations of motion 19
equilibrium 37, 37–8
 equilibrium position 89
 equilibrium under three forces 38–9
errors 11
escape velocity 59
excited state 192
external potential difference 145, 145–6

F

Faraday's law 175, 176
farads 131
filament lamps 141
first harmonic 107
first law of thermodynamics 86
Fleming's left-hand rule 162
forbidden band 194
forces 34, 41–2
 centre of gravity 36
 deformation of solids 62
 electrical forces 34
 equilibrium of forces 37–9
 frictional and viscous forces 35–6
 gravitational forces 34
 turning effects of forces 36–7
 upthrust (buoyancy) forces 34–5
forward-bias 141
free fall and air resistance 31
frequency 90, 95, 99, 181, 189, 189–90
frequency modulation (FM) **116**, 116–17
 comparing FM and AM transmission 117
friction **35**, 35–6
full-wave rectification 185
fundamental mode of vibration 107
fundamental particles 208, 208–9
 quarks and beta decay 209

G

gamma radiation 207–8
gases 40, 46, 71, 72, 81
 see ideal gases
geostationary satellites 58, **120**
gradient 15
graphs 4, 253–4
 displacement–time graphs 15–16

drawing graphs 221–2
 using graphs to analyse simple harmonic
  motion (SHM) 93
 velocity–time graphs 17
gravitational fields 56, 60–1
 comparing electrical and gravitational
  fields 128
 geostationary satellites 58
 gravitational field and gravitational
  potential difference 59
 gravitational field strength 34, 57
 gravitational potential 58, 58–9
 gravitational potential energy 44
 Newton's law of gravitation 56
 orbital motion 57–8
gravity 19–20
 centre of gravity 36
ground state 192

H

hadrons 209
half-life 213
half-thickness 197
half-wave rectification 185
Hall effect 167, 167–8
 Hall voltage 167
 using a Hall probe 168
hardness 196
heat **75**
 see specific heat capacity; specific
  latent heat
homogeneity 3
Hooke's law 63
 Hooke's law limit 64

I

I–V characteristics 140–1
ideal gases 69, 69–70, 73–4
 Brownian motion 70
 equation of state for an ideal gas 69
 internal energy of a gas 71
 kinetic energy of a gas molecule 72
 kinetic theory of gases 71–2
ideal operational amplifiers 154
in phase 106
independent variables 216, 229
inelastic collisions 29
inertia 24
insulators 194
intensity 100
 intensity reflection coefficient 103
 X-rays 196
interference 106
 two-source interference 110–12
internal energy of a system 86
internal resistance 145, 145–6
inverting amplifiers 155–6
inverting inputs 154
ionisation 193
isotopes 206

K

key points
 alternating currents 181
 capacitance 131
 circular motion 51
 communications 115
 current of electricity 138
 direct current circuits (d.c.) 145
 deformation of solids 62
 dynamics 24
 electric fields 124
 electromagnetic induction 173
 electronics 154
 forces, density and pressure 34
 gravitational fields 56

ideal gases 69
kinematics 15
magnetic fields 160
measurements 8
oscillations 89
particle and nuclear physics 204
physical quantities and units 1
quantum physics 189
superposition 106
temperature 75
thermal properties of materials 81
waves 98
work, energy and power 43
kinematics 15, 22–3
 describing motion 15
 equations of motion 18–19
 motion under gravity 19–20
 projectiles 20–1
 uniform circular motion 51
 using graphs 15–17
kinetic energy 45
kinetic theory of gases 70–2
Kirchhoff's laws 146–9
 first law (conservation of charge) 146–7
 **second law (conservation of
  energy) 147**

L

length 8–9
Lenz's law 176
leptons 208
light 191
 matter waves 191–2
light sensors 151
light-dependent resistors (LDRs) **142**
limiting friction force 35
liquid-in-glass thermometers 76, 77
liquids 40, 81, 84, 85
logarithms 254
 common logarithms 254, 254–5
 natural logarithms 255
longitudinal (compression) waves 99

M

magnetic fields 160, 160–1, 171–2
 comparing the forces in different types
  of field 169
 deflecting charged particles in a
  magnetic field 165
 force on a charged particle moving in a
  magnetic field 164
 force on a current-carrying conductor in
  a magnetic field 161–3
 Hall effect 167–8
 magnetic field strength 162, 173
 measuring e/m$_e$ 166
 measuring magnetic field strength using
  a current balance 163
 nuclear magnetic resonance imaging
  (NMRI) 170
 separating particles with different
  velocities 167
magnetic flux 175
 flux cutting, flux linking and induced
  e.m.f. 177
 **magnetic flux density 161, 162,
  173, 175**
 magnetic flux linkage 175
magnitude 34
mass 9, 169
 mass and inertia 24
 mass and weight 26
 molar mass 5
mass defect 206, 210
mass excess 211
materials 64